Pervasive Computing in IoT

Pervasive Computing in IoT

Editors

Spyros Panagiotakis
Evangelos K. Markakis

MDPI

Basel • Beijing • Wuhan • Barcelona • Belgrade • Novi Sad • Cluj • Manchester

Editors

Spyros Panagiotakis
Hellenic Mediterranean University
Heraklion
Greece

Evangelos K. Markakis
Hellenic Mediterranean University
Heraklion
Greece

Editorial Office
MDPI AG
Grosspeteranlage 5
4052 Basel, Switzerland

This is a reprint of articles from the Special Issue published online in the open access journal *Information* (ISSN 2078-2489) (available at: https://www.mdpi.com/journal/information/special_issues/pervasive_computing_in_iot).

For citation purposes, cite each article independently as indicated on the article page online and as indicated below:

Lastname, A.A.; Lastname, B.B. Article Title. *Journal Name* **Year**, *Volume Number*, Page Range.

ISBN 9978-3-7258-1487-9 (Hbk)
ISBN 978-3-7258-1488-6 (PDF)
doi.org/10.3390/books978-3-7258-1488-6

Contents

information

MDPI

Editorial

An Editorial for the Special Issue "Pervasive Computing in IoT"

Spyros Panagiotakis * and Evangelos K. Markakis

Department of Electrical & Computer Engineering, Hellenic Mediterranean University, 71410 Heraklion, Greece; emarkakis@hmu.gr
* Correspondence: spanag@hmu.gr

In the era of Internet of Things (IoT) we have entered, the "Monitoring–Decision–Execution" cycle of typical autonomic and automation systems is extended, so it includes distributed developments that might scale from a smart home or greenhouse to a smart city and from autonomous driving to emergency management. In such highly distributed and scalable architectures, each of the three processes can take place in isolation from the others, situated at any physical or virtual computing system and located, ideally, at any place. Hence, communication and interoperation between the subsystems that comprise the total system, namely extreme edge, edge, fog, and cloud deployments, is of critical importance. To this end, new machine-to-machine protocols, as well as emerging serverless and decentralized architectures, enable the formation of ad hoc user groups for personalized communication and interaction.

Context awareness is of equal importance, since it enables situation awareness, event recognition, and pervasiveness across the system. The latter can dynamically provide customized service provision to end users via content adaptation to the user's situation and needs. Toward this direction, modern sensor technology extends typical ambient sensing to social and cyber sensing, triggering various interactions among connected devices or human beings. The same happens with modern human–computer interfaces that bring input and output capabilities to a plethora of everyday items, transforming them to enchanting and intelligent ones. In parallel, the crowdsensing paradigm vividly emerges on top of various networking topologies as a means for rapidly enabling social sensing. Very recently, researchers have promised the implementation of a full Internet of Senses (IoS) by 2030, where not only typical data will be transferred over the network but also data that will trigger senses like taste, smell, touch, etc.

Despite the richness of the available data, the key problem for application designers remains the same: How to fuse and mine reliable information from the data collected from largely unknown and possibly unreliable sources or how to dynamically extract user preferences, behaviors, and needs from the received events beyond the maintenance of static user profiles. Furthermore, recent advances in IoT management platforms, microcontrollers, and data science bring machine learning and computational intelligence closer to the source of data generation (end users, fog layers, edge, and extreme edge), enabling broader context awareness. However, despite the progress made to date, we are still far from providing low-power autonomous IoT devices, which could deal with a large amount of data processing or a frequent need for communication, or both.

This Special Issue presents a collection of research papers, each providing insights into the multifaceted landscape of this wide and transformative research area. These high-quality, state-of-the-art papers deal with challenging issues in pervasive computing across the different parts of the IoT ecosystem. A short introduction to the contributions of these collected works follows.

A notable theme of the articles in this Special Issue is the focus on customized IoT frameworks. Four papers falling in this thematic area can be found in this collection. Agapi Tsironi Lamari et al., in (Contribution 1), propose a methodology for the low-cost crafting of an interactive layered dashboard using domestic materials that are easily available in

Citation: Panagiotakis, S.; Markakis, E.K. An Editorial for the Special Issue "Pervasive Computing in IoT". *Information* **2024**, *15*, 320. https://doi.org/10.3390/info15060320

Received: 23 May 2024
Accepted: 27 May 2024
Published: 30 May 2024

every household. For demonstration purposes, they developed projection mapping for the pervasive and interactive projection of multimedia content to the users of this tangible interface. Manos Garefalakis et al., in (Contribution 2), at first summarize the common architectural characteristics found in most modern remote laboratories specializing in teaching microcontroller programming. Then, they propose the extension of this architecture with features for monitoring and assessing users' activities over remote labs in the context of pervasive and supervised learning. For the latter, the experience API (xAPI) standard is exploited to store users' learning analytics. Panduman Yohanes Yohanie Fridelin et al., in (Contribution 3), propose an edge framework for remotely optimizing and configuring edge devices in three phases. With this framework, they extend the functionality and usability of their IoT application server platform for smart environmental monitoring and analytics in real time. Koball Carson et al., in (Contribution 4), propose an unsupervised machine learning approach for correctly identifying each unique device in an IoT network. Machine learning-assisted approaches are promising for device identification since they can capture dynamic device behaviors and traffic patterns to this end.

In this collection, we can also find three papers dealing with issues of ambient intelligence and assisted living of the aging population. Thakur Nirmalya et al., in (Contribution 5), present innovative machine learning-driven methodologies that analyze the data from BLE beacons and scanners or from accelerometers and gyroscopes to detect users' indoor locations in a specific 'activity-based zone' during their daily activities. Also, in (Contribution 6), they present an intelligent decision-making algorithm that can analyze behavioral patterns and their relationships with the contextual and spatial features of the environment to detect any anomalies in user behavior that could constitute an emergency. Chen Lei et al., in (Contribution 7), investigate activity recognition with postural transition awareness. Three feature selection algorithms are considered to select the optimal feature subset from inertial sensor data for posture classification.

The next three papers in this Special Issue consider blockchain technology in various IoT applications. Calo James et al., in (Contribution 8), propose a method leveraging blockchain and federated learning to train neural networks at the edge, effectively bypassing limited computational resources of edge devices and privacy concerns. The decentralized nature of blockchain enables the authors to replace the centralized server in typical federated learning scenarios with a P2P network, providing distributed training across multiple devices. Samia Masood Awan et al., in (Contribution 9), discuss the cyberthreats and vulnerabilities in IoT environments and propose a novel secure framework that monitors and facilitates device-to-device communications with different levels of access/control based on environmental parameters and device behaviors. A zero-trust system provides dynamic behavioral analysis of IoT devices by calculating devices' trust levels and enforcing variable policies specifically generated for each instance. Blockchain is used to ensure that anonymous devices and users are registered, as well as confirming that immutable activity logs are recorded. Kristin Cornelius, in (Contribution 10), analyzes records produced by non-fungible token (NFT) blockchain applications and compares them to 'document standards' to see if they act to the extent that has been set by a body of literature concerned with authentic documents. Through a close reading of the current policies on transparency, compliance, and recordkeeping, as well as the consideration of blockchain records (such as user-facing interfaces), this study concludes that without an effort to design these records with the outlined concerns in mind and from the perspectives of all three stakeholders (Users, Firms, and Regulators), any transparency will only be illusory and could serve the opposite purpose for bad actors if not resolved.

The collection closes with two survey papers related to sensors and their applications. Paul D. Rosero-Montalvo et al., in (Contribution 11), survey various sensor and filtering technologies and propose a new sensor taxonomy, which deploys data pre-processing on an IoT device by using a specific filter for each sensor type. Statistical and functional performance metrics are defined to support filter selection. Kim Anh Phung et al., in (Contribution 12), conduct a comprehensive survey of pervasive computing in various healthcare IoT applications and provide a broad view of the key components, their roles, and connections in such use cases. In total, 118 research works are surveyed and summarized into categories concerning sensors, communication technologies, artificial intelligence, infrastructure, and security methods.

As this Special Issue demonstrates, the intersection of pervasive computing with the Internet of Things continues to be a thriving hub of innovation and discovery. This Special Issue provides a snapshot of the progress in this research domain, which is aimed at inspiring future work. Collectively, the curated papers contribute to the expanding knowledge in this realm and offer insights in the evolving landscape. By sharing diverse views, we hope to highlight the potential that can be found at their intersection.

Author Contributions: Conceptualization, S.P. and E.K.M.; methodology, S.P. and E.K.M.; writing, S.P. and E.K.M. All authors have read and agreed to the published version of the manuscript.

Funding: This research received no external funding.

Conflicts of Interest: The authors declare no conflict of interest.

List of Contributions:

1. Tsironi Lamari, A.; Panagiotakis, S.; Kamarianakis, Z.; Loukas, G.; Malamos, A.; Markakis, E. Construction of a Low-Cost Layered Interactive Dashboard with Capacitive Sensing. *Information* **2022**, *13*, 304. https://doi.org/10.3390/info13060304.
2. Garefalakis, M.; Kamarianakis, Z.; Panagiotakis, S. Towards a Supervised Remote Laboratory Platform for Teaching Microcontroller Programming. *Information* **2024**, *15*, 209. https://doi.org/10.3390/info15040209.
3. Panduman, Y.Y.F.; Funabiki, N.; Ito, S.; Husna, R.; Kuribayashi, M.; Okayasu, M.; Shimazu, J.; Sukaridhoto, S. An Edge Device Framework in SEMAR IoT Application Server Platform. *Information* **2023**, *14*, 312. https://doi.org/10.3390/info14060312.
4. Koball, C.; Rimal, B.P.; Wang, Y.; Salmen, T.; Ford, C. IoT Device Identification Using Unsupervised Machine Learning. *Information* **2023**, *14*, 320. https://doi.org/10.3390/info14060320.
5. Thakur, N.; Han, C.Y. Multimodal Approaches for Indoor Localization for Ambient Assisted Living in Smart Homes. *Information* **2021**, *12*, 114. https://doi.org/10.3390/info12030114.
6. Thakur, N.; Han, C.Y. An Ambient Intelligence-Based Human Behavior Monitoring Framework for Ubiquitous Environments. *Information* **2021**, *12*, 81. https://doi.org/10.3390/info12020081.
7. Chen, L.; Fan, S.; Kumar, V.; Jia, Y. A Method of Human Activity Recognition in Transitional Period. *Information* **2020**, *11*, 416. https://doi.org/10.3390/info11090416.
8. Calo, J.; Lo, B. Federated Blockchain Learning at the Edge. *Information* **2023**, *14*, 318. https://doi.org/10.3390/info14060318.
9. Awan, S.M.; Azad, M.A.; Arshad, J.; Waheed, U.; Sharif, T. A Blockchain-Inspired Attribute-Based Zero-Trust Access Control Model for IoT. *Information* **2023**, *14*, 129. https://doi.org/10.3390/info14020129.
10. Cornelius, K. Betraying Blockchain: Accountability, Transparency and Document Standards for Non-Fungible Tokens (NFTs). *Information* **2021**, *12*, 358. https://doi.org/10.3390/info12090358.
11. Rosero-Montalvo, P.D.; López-Batista, V.F.; Peluffo-Ordóñez, D.H. A New Data-Preprocessing-Related Taxonomy of Sensors for IoT Applications. *Information* **2022**, *13*, 241. https://doi.org/10.3390/info13050241.
12. Phung, K.A.; Kirbas, C.; Dereci, L.; Nguyen, T.V. Pervasive Healthcare Internet of Things: A Survey. *Information* **2022**, *13*, 360. https://doi.org/10.3390/info13080360.

information

MDPI

Article

Construction of a Low-Cost Layered Interactive Dashboard with Capacitive Sensing

Agapi Tsironi Lamari [1], Spyros Panagiotakis [1,*], Zacharias Kamarianakis [1,2], George Loukas [1], Athanasios Malamos [1] and Evangelos Markakis [1]

[1] Department of Electrical & Computer Engineering, Hellenic Mediterranean University, 71410 Heraklion, Greece; agapitslm@gmail.com (A.T.L.); zkamar@hmu.gr (Z.K.); giorgioloukas@gmail.com (G.L.); amalamos@hmu.gr (A.M.); emarkakis@hmu.gr (E.M.)

[2] Institute of Agri-Food and Life Sciences, University Research Centre, Hellenic Mediterranean University, 71410 Heraklion, Greece

* Correspondence: spanag@hmu.gr

Abstract: In the present work, a methodology for the low-cost crafting of an interactive layered dashboard is proposed. Our aim is that the tangible surface be constructed using domestic materials that are easily available in every household. Several tests were performed on different capacitive materials before the selection of the most suitable one for use as a capacitive touch sensor. Various calibration methods were evaluated so that the behavior of the constructed capacitive touch sensors is smooth and reliable. The layered approach is achieved by a menu of few touch buttons on the left side of the dashboard. Thus, various different layers of content can be projected over the same construction, offering extendibility and ease of use to the users. For demonstration purposes, we developed an entertaining plus an educational application of projection mapping for the pervasive and interactive projection of multimedia content to the users of the presented tangible interface. The whole design and implementation approach are thoroughly analyzed in the paper and are presented through the illustration and application of various multimedia layers over the dashboard. An evaluation of the final construction proves the feasibility of the proposed work.

Keywords: capacitive sensing; low cost; tangible sensing; interactive surface; tactile sensing; human-computer interaction (HCI); IoT; pervasive computing

Citation: Tsironi Lamari, A.; Panagiotakis, S.; Kamarianakis, Z.; Loukas, G.; Malamos, A.; Markakis, E. Construction of a Low-Cost Layered Interactive Dashboard with Capacitive Sensing. *Information* **2022**, *13*, 304. https://doi.org/10.3390/info13060304

Academic Editor: Gholamreza Anbarjafari (Shahab)

Received: 18 April 2022
Accepted: 13 June 2022
Published: 17 June 2022

Publisher's Note: MDPI stays neutral with regard to jurisdictional claims in published maps and institutional affiliations.

1. Introduction

With the continuous development in the technological sector and especially in the Internet of Things industry in recent years, the way we think is changing drastically. The Internet of Things (IoT) is one of the top three technological developments of the next decade and is becoming an increasingly debated topic, especially as an enabler for the implementation of pervasive cyber-physical applications [1–4]. The interaction between user and computer was initially limited to the simple use of input devices such as a mouse and keyboard. In recent years, however, there has been rapid progress and new ways of communicating and interacting with computers have emerged [5]. In particular, the use of touch is an important sensory organ that provides multiple possibilities to the interaction with machines. The operation of an interactive surface with capacitive sense is based on the use of touch as a means for human-computer interaction and is a more integrated way of communication. Interactive tangible surfaces often found in public spaces having informational, advertising, educational, and entertainment purposes [6,7]. They are a very smart means of advertising and learning as they arouse intense interest and reach a large number of people who aim to know their functions and capabilities.

Although capacitive sensing is a very popular technology for electronically implementing the sense of touch [8], existing products are far too expensive for considering them as easily replaceable consuming goods. On the other hand, there are several daily uses for such dashboards that would benefit from low-cost implementations of this type. For example, schools of all levels are always asking for accessible interactive panels for use in their curricula or announcing news for their students but they cannot afford buying industrial grade solutions with the risk to be damaged by such daily use. The same states also for hospitals, public transportation stations, airports, etc. Hence, building cost-effective interactive dashboards that can, however, be reliable and robust for use in a public space can be a challenging task [9].

The main purpose of the present work is to propose a methodology for crafting an interactive tangible and mutlifunctional dashboard integrating capacitive sensing that can be easily and inexpensively implemented from electronics novices in a 'do-it-yourself' (DIY) manner. Such a dashboard, combined with a typical projector to display graphics on it, can find several uses from information kiosks and advertising to education and amusement. The key target of our application is the accommodation of different layers of functionality over the same construction. To this end, a menu is provided that leads to different layers of content. This means that for each layer, the touch sensors arranged in the dashboard are assigned different roles, providing us with more touch events, thus eliminating the necessity for the use of more physical sensors. In our pilot construction, the physical touch sensors in place are sixteen, but with the use of the menu thirty-nine different actions can be supported. Our intention is not to present a new way for creating capacitive sensors, but to use the existing science and methods in order to find the optimal case for our implementation using materials that can be found in every home. Different materials with capacitive behavior were tested in order to select the most suitable one for the creation of the touch sensors that control the interaction events. Along with the above, which mainly concern the hardware implementation of the approach, a filtering method were applied in a software basis, so that the signal produced from the touch sensor is stable and reliable. In order for our work to be attributed, an indicative educational use case that deploys projection mapping for the pervasive and interactive projection of multimedia content to the users of our tangible surface was selected.

The main motivation of the present work can be summarized in the following research questions:

- Is it feasible to construct an interacive dashboard by using everyday materials?
- Is it possible for this interactive construction to be made multifunctional so different layers of information are projected over the dashboard?
- Is it possible for this development to operate reliably and be robust for use in a public space?
- Is it possible this contruction to be content agnostic so several use cases can be accommodated over the same dashboard?

The present paper is divided into five sections covering the study, design, and construction of an interactive surface, as well as of the information system for the support of our use case. The first section refers to the tangible technology and the interaction between user and computer. In the second section, we discuss related work, how other researchers approach crafting and capacitive sensing in terms of technology. In section three the architecture of the proposed implementation is presented. Subsequently, section four introduces the construction and operation of our information system and finally, in section five, the conclusions that emerged as well as suggestions for future improvement of the system are discussed.

2. Related Work

Many of the implementations in the literature that we studied were a source of inspiration and a motivation for our work. Researchers at Carnegie Mellon University in collaboration with Disney Research, Disney, Pittsburgh, presented Wall ++ [10], a wall with a capacitive feel, a large-scale project with low installation costs. As they mention, walls are everywhere and often occupy more than half the area in buildings, offices, homes, museums, hospitals, and almost any interior. Nevertheless, to this day they remain static, with the sole function of separating spaces and hiding the infrastructure of buildings. Their goal was to install a surface on walls, thus giving them multiple possibilities such as body position detection, touch, and even electromagnetic waves. The basic principle of Wall ++ was based on drawing large electrodes on a wall using conductive paint. Thus, as a first step, it was necessary to develop a reliable and economically feasible way to place large electrodes on the walls. To identify the appropriate materials and procedures, the team performed a series of tests with different conductive paints, different application methods, and a number of coats. They then researched the different electrode standards suitable for the applications they wanted and optimized them for detection range and analysis.

The Dalziel and Pow [11] studio as part of the London Retail Design Expo, in February 2015, aroused special interest having implemented an interactive surface created from conductive ink. Large sheets of plywood were used for the application as canvases. Dalziel and Pow then collaborated with the K2 printing lab to print the conductive ink on the canvases, which formed the interaction surfaces. The custom design allowed the team to have multiple points of contact and create interactions around them. Starting with the content, the team compiled a list of stories and possible interactions based on "The Future of Retail." Having the stories, they laid the foundation of the screen and were used to depict a series of 48 cartoons, the number of which then rose to 250. After printing the canvases with the base layer of conductive ink, the team applied a layer of non-conductive white ink on top so they could project the animations there. The conductive ink was then connected to a capacitive touch board called Ototo, designed specifically to convert touch to sound. With the installation of Ototo, the plywood walls became a living circuit of entrances, which would cause various sounds and visual elements with each contact. To project the various animations on each canvas, multiple projectors were used, which were mounted on the ceiling and were controlled through an existing Projection Mapping software.

An earlier implementation of Dalziel and Pow's, which inspired the above installation, was the new Zippy [12] children's store in Setúbal, Portugal. They designed two interactive installations and built both inside D&P for testing, before heading to Portugal to present and install the project. 'Sound Poster' is a panel with printable characters made of conductive ink and is used to make sounds. 'Fun Receipt' is a children's receipt, which you print from a giant mouth on the store's counter and includes characters for painting, mazes, and other toys.

Sam Jacoby and Leah Buenchley looked at conductive ink as a means of expressing storytelling and interaction design with children and presented StoryClip [13], a toolbox that incorporates functional everyday materials, calculations, and drawings. It consists of conductive ink, ordinary painting inks, and a hardware-software tool, allowing a child's drawing to function as an interface for recording and playing audio. Taking advantage of the artistic nature of children to motivate them in technological exploration, turning a conventional display into a multimedia interface that promotes multilevel interaction with children.

The Living Wall [14] project explores the construction and implementation of interactive wallpaper. Using conductive, durable, and magnetic colors, they created a wallpaper that allows the creation of dynamic, remodelable, and programmable spaces. The wallpaper consists of circuits that are painted on a sheet of paper and a set of electrodes are attached to it with the help of magnets. Wallpaper can be used for a variety of functional and stunning applications that can include lighting, environment detection, device control, and environmental information display. Additionally, they contain a set of detachable electronic modules for processing, detection, and wireless communication.

Jie Qi and Leah Buechley developed an interactive pop-up book called Electronic Popables [15] to explore paper-based computing. Their book incorporates traditional emerging mechanisms with thin, flexible paper-based electronics and the result looks and works such as a regular emerging book except that interaction elements have been added. They first made individual pop-up interactive cards and then assembled them into a book. They used three basic materials, self-adhesive copper tape, conductive fabric, and conductive paint, to create the circuits on the paper.

Researchers from the MIT Media Lab presented the implementation of Sticking Together [16]. They built sticky sensors and actuators that children can use to create handmade personalized remote communication interfaces. By attaching I/O stickers to special wireless cards, children can invent ways to communicate with their loved ones over long distances. A special interactive way of communication for children while learning new technologies in a fun and creative way.

Pen-on-Paper Flexible Electronics [17] offers a unique approach to making flexible devices using a configuration instrument that is as ubiquitous and portable as paper. Rollerball pens are commercially available and are specially designed for precision writing on paper. Using a rollerball pen filled with conductive silver ink, it is possible to write and draw conductive text, diode interfaces, electronic circuits, LED arrays, and 3D antennas on paper.

3. Materials and Methods

3.1. Low-Cost DIY Capacitive Sensors

3.1.1. Introduction to Capacitive Sensing

In electrical engineering, capacitive sensing is a technology based on the capacitive coupling that can detect and measure anything that is conductive or has a dielectric different from that of air, such as the human body or hand. This is achieved by the effect of each object on the electric field created around the active face of a capacitive sensor. A capacitive sensor works like an open capacitor. An electric field is formed between the measuring electrode and the ground electrode. If a material with a dielectric constant greater than air enters the electric field, the field capacitance increases according to the dielectric constant of that material. The electrodes measure the increase in capacitance and generate an output signal that corresponds to the trigger. Figure 1 illustrates the operating principle behind capacitive sensing. Such metering is based on the RC circuits' time constant. The time constant of an RC circuit is defined as the time required for the capacitor's voltage to reach 63.2% of its maximum value when the capacitor is fully charged [18,19].

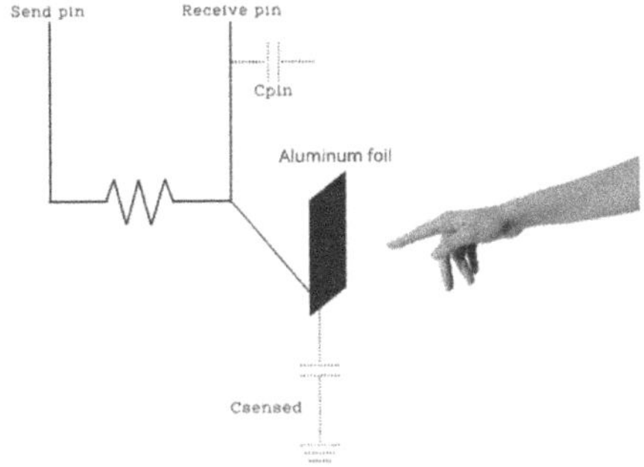

Figure 1. Operating principle of capacitive sensing.

3.1.2. Selection of Conductive Materials

To carry out the present work, three conductive materials, easily accessible and economically affordable, were evaluated. These are: conductive paint, self-adhesive aluminum tape, and pencil graphite. In order to highlight the material with the best properties, some experiments were performed. The experiments were conducted over a piece of paper on top of which the materials under test were applied. Each material was placed on the paper with two different widths, 0.5 cm and 1 cm. The aim was to find the material with the lowest ohmic sheet resistance, making it the best conductor. Measurements were made for each material to find the ohmic resistance using a multimeter. In the first test, the terminals of the multimeter were 3.5 cm apart, while in the second they were twice as far, in 7 cm.

Starting with the graphite (Figure 2), the ohmic sheet resistance of the Thin Line (0.5 cm), consisting of a set of two hundred pencil strokes, was first measured. The result was quite high as in the first distance (3.5 cm) the ohmic resistance was measured at 178 kOhms, while at twice the distance (7 cm) the result was 321 kOhms. In the Wide Line (1 cm), sheet ohmic resistances of 150 kOhms and 306 kOhms were obtained for the short and the long distance correspondingly. Finally, in the third and last line, the strokes were less, and the resistance measured was 5.32 MOhms (3.5 cm) and 12.3 MOhms (7 cm).

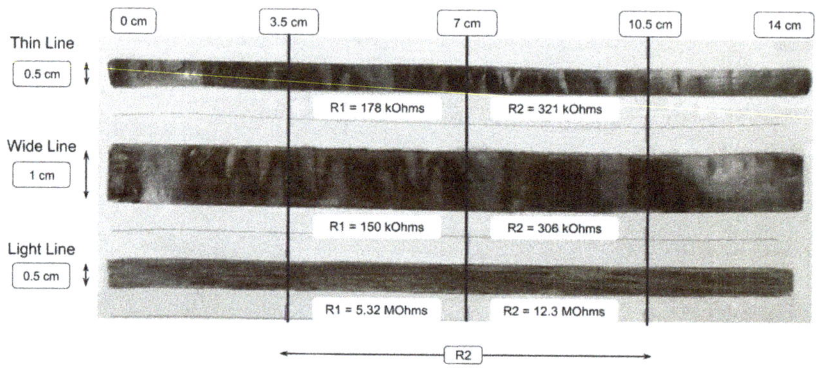

Figure 2. Ohmic sheet resistances for Graphite.

A noticeable difference was observed during the ohmic sheet resistance measurement of the aluminum tape's strips (Figure 3), in which the result for the Thin Line (0.5 cm) was 3.2 Ohms (3.5 cm) and 3.3 Ohms (7 cm), respectively. In the Wide Line (1 cm), the measurements did not show large variations from those of the Thin Line and as a result, we obtained 3 Ohms and 3.1 Ohms (7 cm), respectively. As a result, we can notice that the amount of material in the given case has minimal effect on the change of its conductivity.

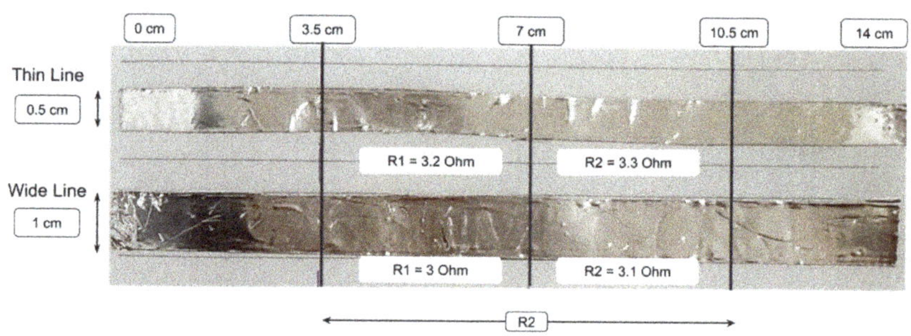

Figure 3. Ohmic sheet resistances for aluminum Tape.

To measure the sheet ohmic resistance of conductive paint (Figure 4), two coats of paint were applied. The measurement results correspond to 0.65 kOhms (3.5 cm) and 1.39 kOhms (7 cm) for the Thin Line. For the Wide Line, 0.5 kOhms (3.5 cm) and 1.1 KOhms (7 cm) were measured respectively.

Examining the above measurements, we can notice that our results confirm what is known theoretically: (a) The longer a material, the larger its resistance, and hence lower its conductivity. (b) The wider a material, the lower its resistance, and hence higher its conductivity. After the successful completion of the experiments for the materials, it was obvious that the aluminum tape had the lowest sheet ohmic resistance and consequently the best conductivity (Table 1). Therefore, the aluminum tape was selected for the construction of the DIY capacitive sensors in this work.

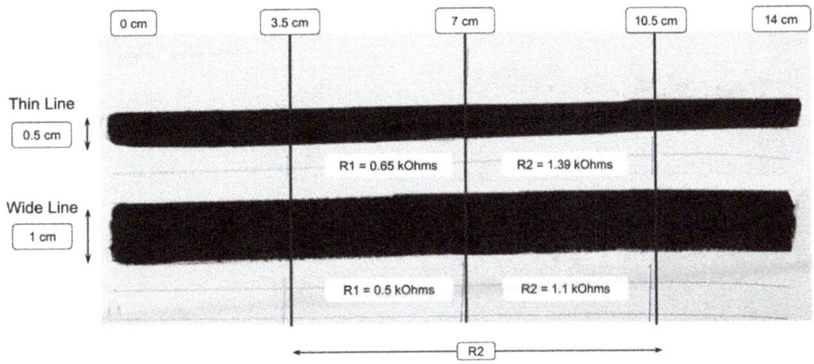

Figure 4. Ohmic sheet resistances for conductive paint.

Table 1. Conductivity tests in the materials used in this study.

		Graphite	Aluminum Tape	Conductive Paint
Thin Line	R1	178 kOhms	3.2 Ohms	0.65 kOhms
	R2	321 kOhms	3.3 Ohms	1.39 kOhms
Wide Line	R1	150 kOhms	3 Ohms	0.5 kOhms
	R2	306 kOhms	3.1 Ohms	1.1 kOhms
Light Line	R1	5.32 kOhms	-	-
	R2	12.3 kOhms	-	-

3.1.3. Calibration of the Sensors

Figure 5 illustrates the connection of our DIY capacitive sensors with the touch pins of an Espressif ESP32 development board [20]. The integrated touch pins of the EPS32 microcontroller were found to be a great advantage as it was not necessary to use two different pins as shown in Figure 1 in order to create an equivalent RC circuit for reading the capacitance change, nor a capacitive sensing library. This process (e.g., reading) is automatically completed inside the ESP32′s firmware thus the touch pins are easily and reliably processed just using the touchRead() function, as is depicted in the following quote from our code. This also saves pins on the microcontroller for future use.

Figure 5. Connection of our capacitive sensors with an ESP32 microcontroller.

```
// reading input values
touch_sensor_value = touchRead(touch_pin);
```

```
// process value with Median Filter Library
test.in(touch_sensor_value);
touch_sensor_value = test.out();
```

Quote 1. Reading the Capacitive Sensors

As it is depicted in Quote 1, the Median Filter Library [21] is used as a means for smoothing sensor readings and outliers' cancellation. In Figure 6a,c,e the noise that occurs during the continuous reading of the sensor is noticeable counter to Figure 6b,d,f, that show the application of the Median Filter over the input signals. The values read by the touch pins are displayed on the vertical axis and the time is displayed on the horizontal axis. To achieve the best possible filtering, sample windows of different sizes were tested (10, 20, 30 number of samples). Regarding the smallest sample window (10 samples), a faster response of all three was observed but with a slight instability. Then, for the sample window of 20 samples, it was observed that its responsiveness was slightly slower but with improved stability and finally, for the value window of 30 samples the responsiveness was slower but with better stability in random disturbances compared to the other three. Taking into consideration that the application domains of our sensors require relatively fast response from the sensors, we concluded that the best choice for our case was a value window of 20 samples. Hence, this is the size of the Median filter window used hereafter. Figure 6 illustrates the behavior of the proposed capacitive sensor without filtering and when applying filtering with the Median filter window of 20 samples. It is obvious that the behavior of the sensor after such filtering is quite stable, and the readings acquired very reliable.

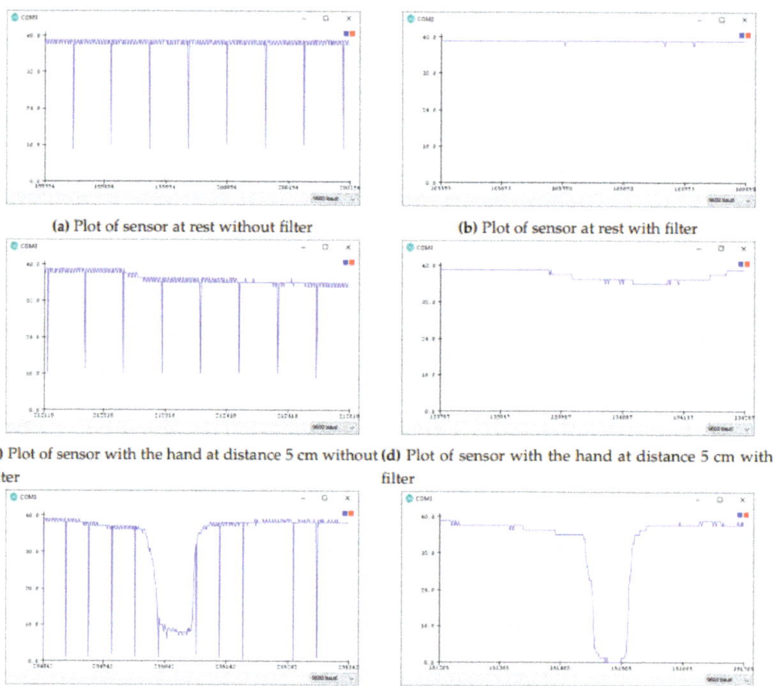

(a) Plot of sensor at rest without filter

(b) Plot of sensor at rest with filter

(c) Plot of sensor with the hand at distance 5 cm without filter

(d) Plot of sensor with the hand at distance 5 cm with filter

(e) Plot of sensor when touching without filter

(f) Plot of sensor when touching with filter

Figure 6. Plots of our sensors with and without median filtering. In (**a**,**b**) a plot of sensor at rest without/with filter, is showing. In (**c**,**d**) similar plots but with the hand at 5cm distance and without/with filter, respectively. Finally, in (**e**,**f**) the plots of the sensor are when touching, again without/with median filter applied, respectively.

3.2. Development of the Interactive Dashboard with Capabilities for Projection Mapping

3.2.1. High-Level Architecture and Requirement Analysis

As it was mentioned in the Introduction, the present work focuses on the craft of an interactive tangible and layered dashboard integrating capacitive sensing that can be easily and inexpensively implemented from electronics novices, in a DIY way. Our aim is this dashboard, along with a typical projector displaying graphics on it, to be used as an interactive surface oriented mainly to educational, advertising, or entertaining purposes. The projection of graphics over the surface will take place by applying the Projection Mapping technique, so an impressive result is achieved for the users [22,23]. The purpose is capacitive sensing in combination with the Projection Mapping technique to compose a fully functional and interactive system. Figure 7 illustrates the high-level architecture of our system. For demonstration purposes, an educational scenario was selected, that is oriented mainly to children.

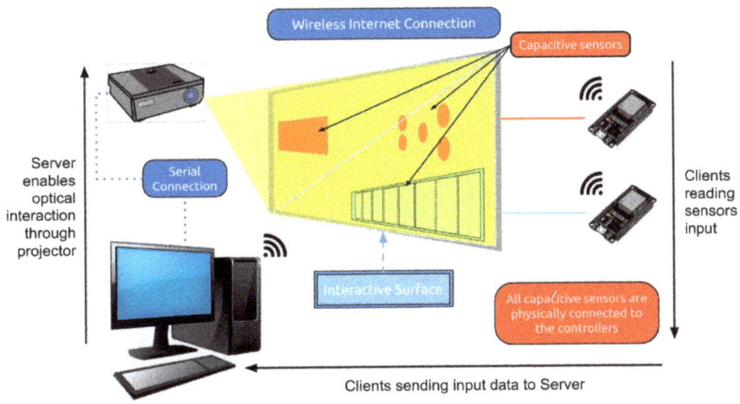

Figure 7. High-level architecture of our system.

For the proper development of the project, it was necessary to design it having in mind the requirements towards the final application. A step-by-step analysis was logically followed. The first step involved the creation and evaluation of our tangible surface. To this end, as was shown in Figure 5, a smaller-scale simulation was created at first, where the performance of sensors to each press by the users were tested. As it was mentioned previously, the ESP32 was selected to be the "heart" of our implementation. A basic requirement for the final system was the microcontroller to have built-in Wi-Fi support, which the ESP32 microcontroller meets. Also, it is powerful enough to meet soft real time requirements, since the final system is expected to operate with several users in parallel, hence, simultaneous touches are expected to take place. Finally, an open-source software implementation was considered that would be able to undertake the Projection Mapping functionality and the rendering of graphics. Processing [24] is a very powerful programmable software, with many enriched libraries and features, that could help us complete the project.

Definitely, the communication between the individual parts of the system determines the proper operation of the final application. According to the adopted architecture, the communication of the microcontrollers behind the tangible surface with our server running Processing is based on TCP/IP socket communication following the Client-Server paradigm and takes place on top of a WLAN network. The role of the Server is undertaken by a Processing-based Internet application and the role of the Clients by two ESP32 microcontrollers situated behind the surface. Each microcontroller running the client application is responsible for: (a) detecting the change in capacity of various aluminum touch sensors spread over the surface; (b) receiving the status of the sensors, and then (c) sending the status to the Processing application only when the capacity change meets the desired conditions (less than a predefined threshold). Once the data are received, the Processing application manages each interaction independently and orchestrates the projection of the respective audio and multimedia content over the surface. The projector used to play back the media content over the interactive surface is serially connected to the computer hosting the server application.

3.2.2. Layout and Layered Design of the Tangible Surface

Having decided on the high-level architecture of our system, the next step was to assure that the dashboard could be multi-functional and able to accommodate several different scenarios. Layers are the appropriate solution to this end. To enable them, two different levels of operation on our surface were developed. In the first level, which is called the Menu Area, the user is given the opportunity via buttons to choose the scenario/layer she/he wants to interact with on the second level. The Menu is permanently exposed to the user so the navigation between scenarios/layers can be performed at will. When the user selects the scenario/layer she/he wishes to (e.g., by pressing one of the available touch sensors in the Menu Area), this leads her/him to the second level where the basic interactions of the selected scenario/layer has been enabled. Hence, the second level is the Working Area of the surface. With this visual separation of the surface at two levels, emphasis is given on the sensors of each scenario/layer that implement its functionality. This enhances the usability of the dashboard and the convenience for the users. Figure 8 illustrates the leveled architecture of the surface.

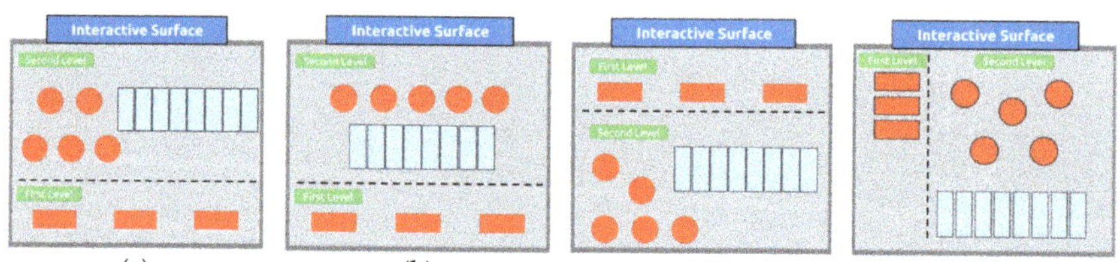

Figure 8. The different layouts tested for the surface; each with a discrete Menu (first level) and Working Area (second level). In (**a**,**b**) the second level of the surface is exposed higher than the first level. In (**c**) the first level is situated higher than the second and in (**d**) both levels of operation are exposed at the same height.

Then, the optimal layout for the placement of the sensors on the surface had to be found. To this end, four different layouts were considered, which are shown in Figure 8. Among them, the first two (Figure 8a,b) expose the second level of the surface, higher than the menu level. These seem to be ideal for users with medium height, since they prohibit shorter persons from reaching the sensors on the second level, while forcing taller persons to stoop. Exactly the opposite takes place with the third layout (Figure 8c), where the first level is situated higher than the second one. The fourth layout (Figure 8d) exposes both levels of operation at the same height, so it seems to be more convenient for domestic use. Taking into consideration that the target group of the proposed construction is mostly children, we decided to go with the fourth layout (Figure 8d).

3.2.3. Crafting of the Tangible Surface

As a basis for the construction of our dashboard, a piece of thick brown cardboard about 70 cm × 150 cm in size, was selected. This is a very affordable solution, trivial to find, and at the same time easy to manage. Moreover, its easy portability is a plus, considering the final implementation. The cardboard offers this as it can be wrapped even in a roll without affecting the sensors. For the latter, pieces of adhesive aluminum tape were placed on the front side of the cardboard, which were thereafter the touch sensors (Figure 9). To keep the front side as simple and flat as possible, slits were made in the cardboard, so the self-adhesive aluminum tape passes through to the backside of the surface. With this technique continuity was achieved between the front and the backside of the construction. All the wiring between sensors and microcontrollers, as well as the microcontrollers themselves, were placed at the backside, thus keeping the front-side of the surface tidy. Figure 10 illustrates the front and the back side of the surface. The front-side was finally covered with white paper for hiding the sensors from users and demonstrating a uniform layout ready for projection.

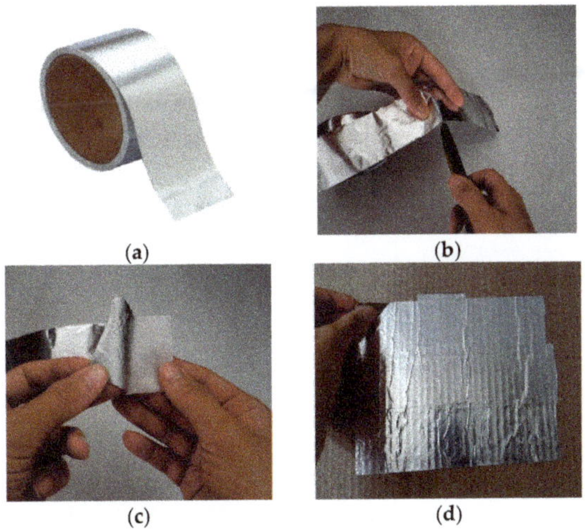

Figure 9. (**a**) Aluminum adhesive tape (**b**) Cutting the tape (**c**) Preparing for installation (**d**) Placement on the surface.

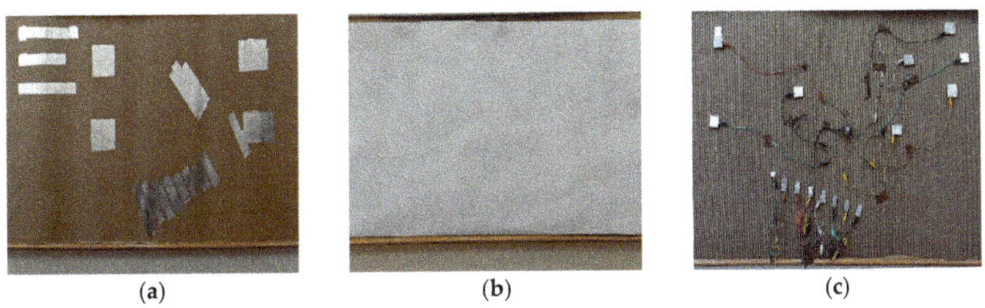

Figure 10. (**a**) Final layout of sensors on the front-side of the surface (**b**) Front-side of the surface covered with white paper (**c**) Wiring at the backside of the surface.

3.3. *Programming of the System*

Having decided on the high-level architecture of the system, the next step was to define the size, shape, and position on the surface of each sensor to cover with aluminum tape the necessary area. This required the design of each sensor individually based on the scenarios that had to be implemented. Knowing the scenarios that make up the final application, it is possible to find how many, and on the same time of what type, interactions each sensor should recognize. The scenarios considered for construction are as follows:

- Scenario 1: Music Wall with six different music instruments
- Scenario 2: English Alphabet Wall
- Scenario 3: Non-Interactive animation based on projection mapping

In our case, the total number of sensors on the tangible surface are sixteen and are grouped according to the performed scenario, each time. The scenarios are interchanged via a three-button menu at the first level of the surface; one button for each scenario. For example, when Scenario 1 is activated by pressing the associated button, thirteen different interactions are enabled at the second level of the surface. The same sensors are also available for Scenarios 2 and 3, so the total interactions that can be potentially supported over our surface are thirty-nine. However, not all of them have been enabled, since it was the authors' choice for each scenario to demonstrate a different number of interactions to the users. So, thirteen interactions are enabled for the first scenario, four for the second and none for the third one. This layered architecture of the second level of the surface, also spares the inputs of the microcontrollers since each touch sensor does not need to correspond to just one action. The distinction between layers is achieved via software. Hence, in the current version of the proposed surface three different layers of operations are supported at present, one layer for each menu button. For each layer, as it was mentioned, up to thirteen different actions can be supported. With one more menu button, one more layer of operation with the same number of actions could be supported, etc.

As it was mentioned in Section 3.2.1, the implementation is based on the Client-Server paradigm: as Clients act two Esp32 microcontrollers and as a Server a PC running the Processing software. Initially, when the association of microcontrollers with the WLAN network is successful, a message is printed on the serial monitor with the IP addresses assigned to them by the access point. The microcontrollers use the IP address and port number assigned to the Server for communicating with it. Running the Clients and the Server in the same network, definitely simplifies the communication among them. The microcontrollers are responsible for receiving the status of each touch sensor (sensor touched, sensor released) and sending it to the server. On the other hand, when the server receives the touch events, it manages each interaction and orchestrates the projection of the respective audio and multimedia content over the second level of the surface.

Figure 11 illustrates the interactions considered for Scenario 1. The Scenario starts when the user touches the corresponding button in the Menu Area. Then, a Music Wall with various available instruments is projected in the Working Area and the thirteen designated touch interactions are enabled. Whenever each of the thirteen sensors is touched, a sound is played-back, and some colors appear on the sensor area with projection mapping. Figures 12 and 13 illustrate the interactions that have developed for Scenarios 2 and 3, respectively. Scenario 2 activates four touch sensors and Scenario 3 none.

As it was discussed in Section 3.1.3, due to the 'prone to noise' nature of our touch sensors, all readings from the sensors are filtered via a median filter with window size of 20 samples. This eliminates the outliers in a certain degree and makes the readings very reliable. The output of this filter is stored into a variable and when its value is less than the threshold already set, the microcontroller understands that a sensor has been activated. As it is shown in Figure 14, the sensors implemented in the final construction return values close to 40 when they are not touched (Figure 14a). On the contrary, when they are touched, the values they return decrease close to 13 (Figure 14b). After several tests it was decided that the appropriate threshold for safely differentiating a sensor touch event from a sensor release, was the value of 30. A separate variable is assigned to each touch sensor, so the server can identify the sensor that its status has changed.

As the sensor values are read continuously, it was necessary in the implementation to find a way to avoid debouncing and understand the prolonged touch events on the sensors. In our implementation, a prolonged touch of a sensor is treated as a single touch and not as multiple, so every interaction with our Server remains active as long as a sensor is touched. In order to properly recognize touch events, release events and prolonged touch events, over our sensors, the two following checks are performed: (a) threshold check and (b) counter check. As it was previously discussed, the threshold differentiates a touch event from a release event. In addition, a counter is used as a flag to keep only the first reading from the sensor and ignore the rest, until the status of the sensor changes again. This way a prolonged touch of a sensor is treated as an individual one. Figure 15 depicts the algorithm that determines when a touch event or a release event is sent to the server.

When the reading from the sensor is below the threshold and the counter equals zero, a touch event is sent to the Server. Then the counter increases to 1 so all next sensor readings with values below the threshold are not communicated to the Server. A release event is sent to the Server only when the reading from the sensor is above the threshold and the counter equals 1. Then the counter decreases again to zero so only one release event is sent to the Server.

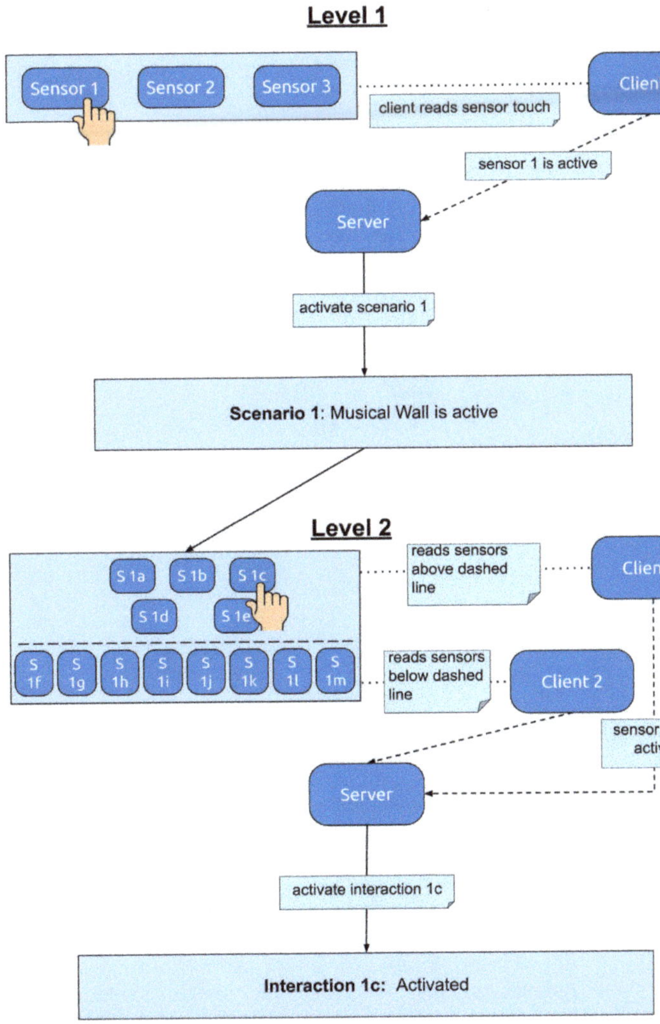

Figure 11. Interactions for Scenario 1 (Music Wall).

Figure 16 illustrates the behavior of the system across distinct and prolonged sensor touches. In the left side plot, two distinct touches are made over a sensor and hence two curves with sensor readings less than 30 are captured, respectively. We observe the response time of the sensor to reach its minimum value, which depends on the pressure that is exercised over it and how it is exercised over time. In the right plot it can be seen that the touch is one but longer in duration. This represents a prolonged touch of the sensor.

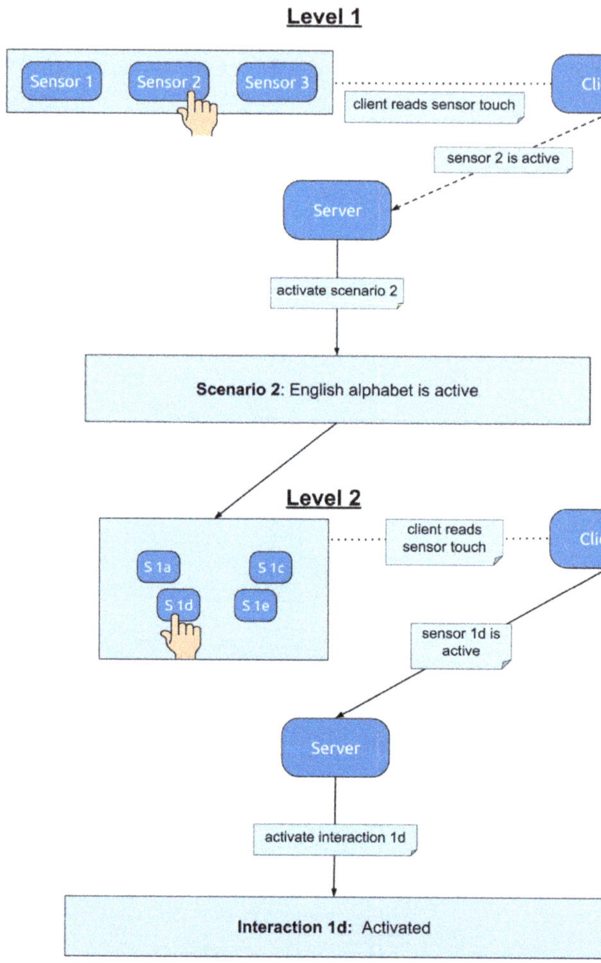

Figure 12. Interactions for Scenario 2 (English Alphabet).

Figure 17 illustrates the result of the counter flag in our implementation. In the left plot of Figure 17, the counter flag is not used, so a large amount of touch and release events are recognized by our system. On the right plot of Figure 17, the use of the counter flag helps our system to differentiate distinct from prolonged touches and releases, so the amount of touch and release events that are recognized are far less. As we can also see here, the first values obtained in each touch event differ from touch to touch. This is again due to the level of pressure that is exercised on the sensor with each touch. The use of this counter flag also enables our surface to accommodate several users simultaneously since it enhances the availability of the server.

Figures 18 and 19 illustrate the Finite State Machines (FSM) for our Client and Server, respectively. The value sent by the Client to the Server after each touch event is unique for each sensor, so the Server is able to activate the corresponding multimedia content. In Figure 19, with x1 the sensor touch is represented and with x2 the sensor, the release events, respectively.

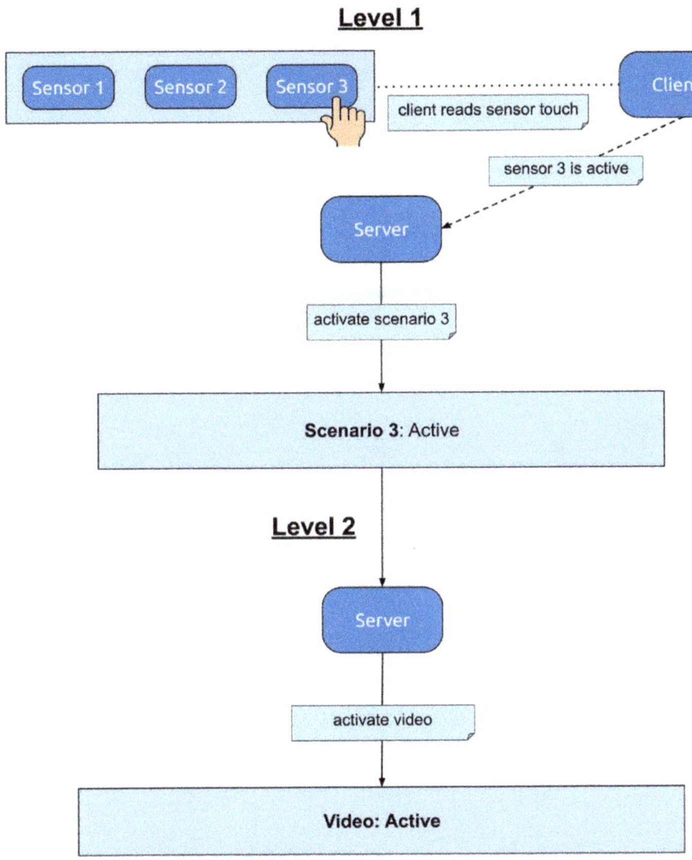

Figure 13. Interactions for Scenario 3 (Non-Interactive animation).

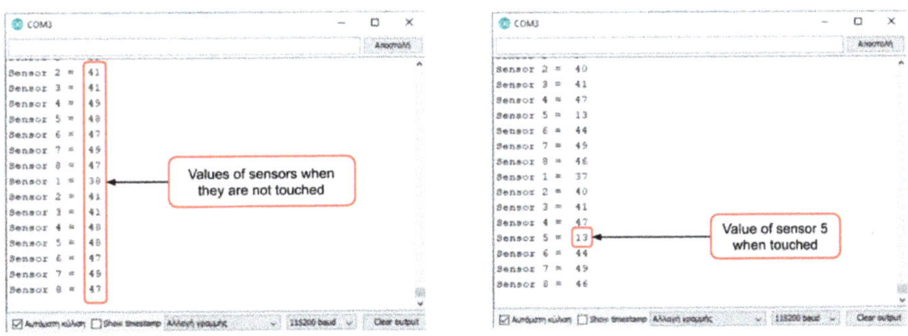

(**a**) (**b**)

Figure 14. Sensor readings when not touched (**a**) and when touched (**b**), shown side by side.

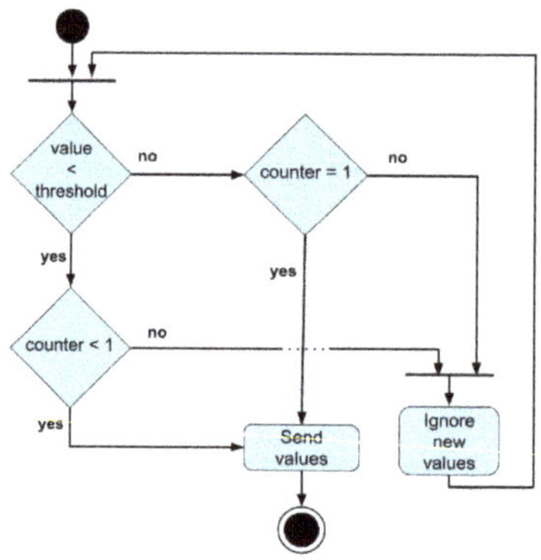

Figure 15. Touch and release events sent to the server.

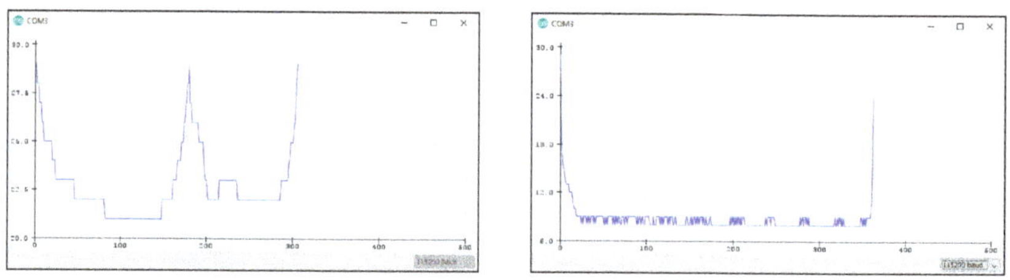

Figure 16. Behavior of the system across distinct and prolonged sensor touches. On the left, a plot with filter and threshold when touched is shown. On the right, plot of the sensor when touched with longer duration.

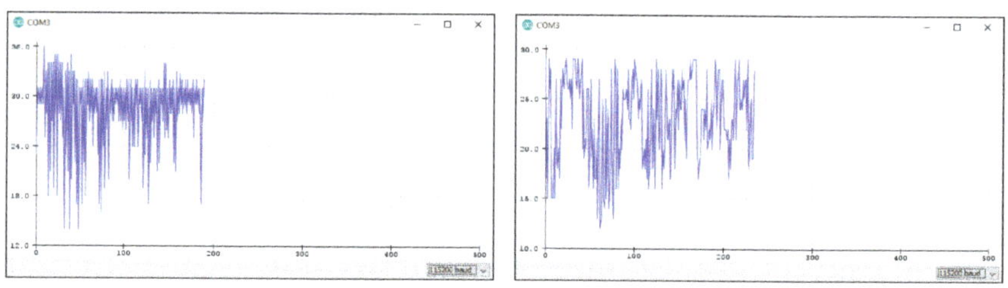

Figure 17. Result of the counter flag in our implementation.

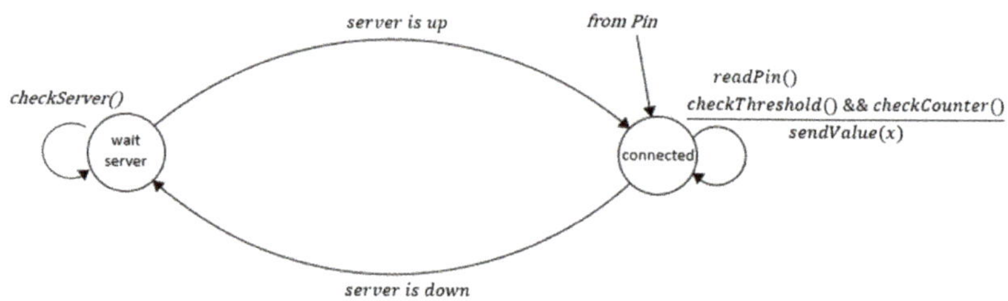

Figure 18. Finite state machine of our client.

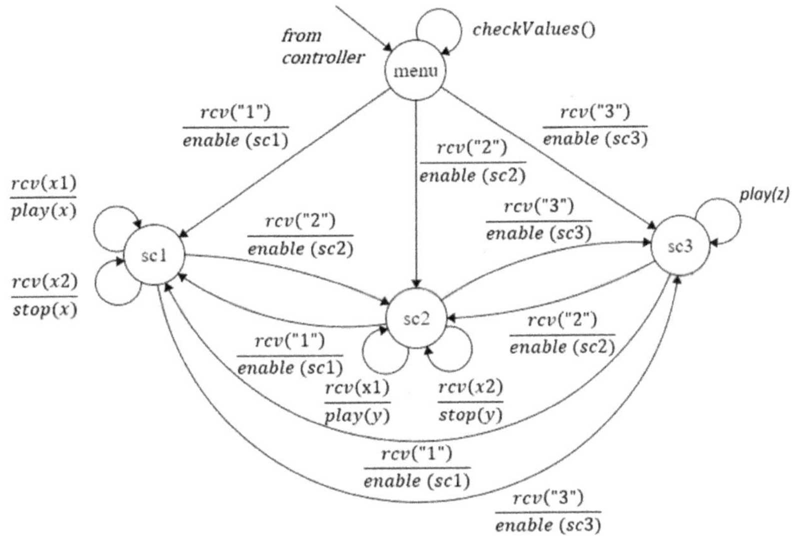

Figure 19. Finite state machine of our server.

4. Results

4.1. Use Cases

As happens with applications that involve projection mapping, the alignment of projectors with the projection area is critical since this type of application is supported by the illusion it provides to the user. So, several tests were performed before the optimal position and orientation of both projector and tangible surface, were found. Also, the displayed content was tested in many different layouts to find the best one. Then, the content was mapped using Processing with the touch sensors of the surface, so each sensor matches to the corresponding interactive content. The tests were completed, and the position, height, and orientation of the projector were noted on the floor of the room for easy repositioning for future use. The audio playback was achieved through the speakers of the computer unit, which was quite convenient. Initially, the computer's USB ports were used in the test application to power the microcontrollers. However, this was not possible in the final deployment as the personal computer was placed at least two meters away from the dashboard and the microcontrollers. For this reason, power was supplied to the microcontrollers with the help of power banks, which were placed at the back of the construction along with the wiring and the microcontrollers. Figure 20 illustrates the welcome page of our application which calls the users to start interacting with it using the left-sided menu.

The first scenario that is demonstrated is called Music Wall and from Figure 21 one can easily understand what she/he is going to encounter. Our Music Wall consists of six interactive musical instruments (tambourine, mandolin, trumpet, maracas, accordion, and metallophone) projected by the projector over the tangible surface. The first five correspond to one touch sensor each, as opposed to the last one which uses eight, one sensor for each note of the metallophone. Figure 21a illustrates the mapping of each instrument with the corresponding touch sensor behind the white paper. Each touch sensor activates an acoustic and a visual interaction. The musical interaction corresponds to the sound of each musical instrument, and the playback is carried out with the help of the speakers of the computer unit. Visual interaction is achieved with the help of the projector and makes each object change color with each touch (Figure 21b–d). Figure 22 reveals the simultaneous use of our interactive surface by several users as it was discussed in Section 3.3. The music wall is a very interesting interaction game for children and adults, it is fun and at the same time easy to understand.

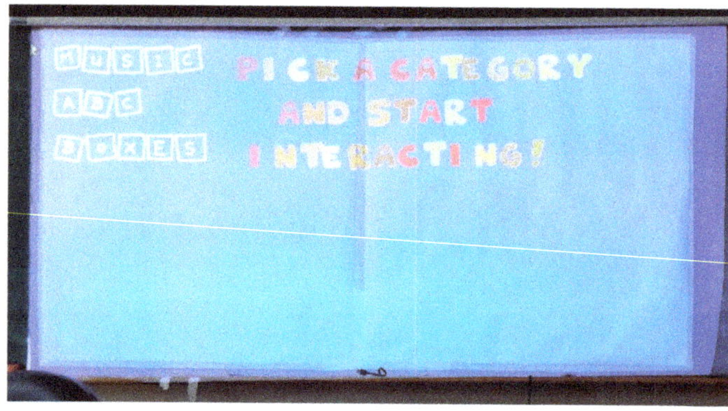

Figure 20. Welcome page of our application.

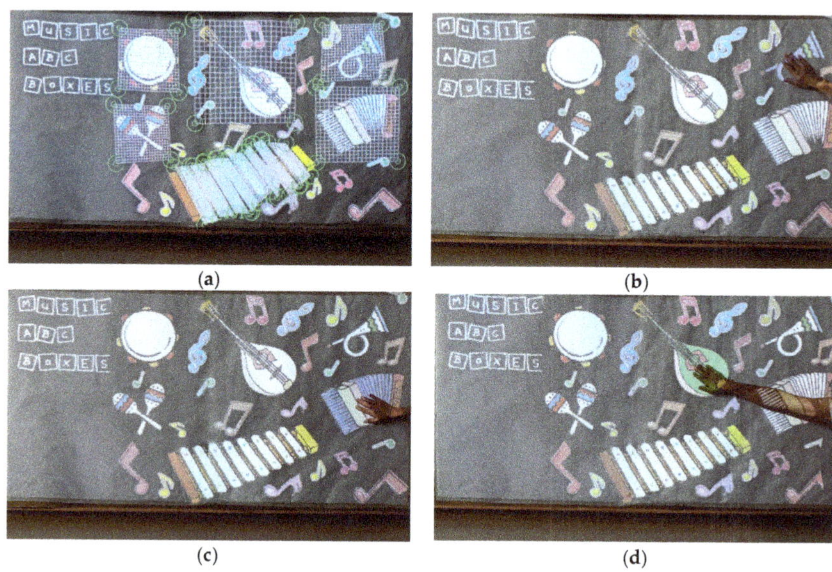

Figure 21. Musical wall scenario. In (**a**), the mapping of instruments with the corresponding touch sensor behind the white paper is shown. Visual interaction achieved with the projector making each object to change color. Touching the trumpet (**b**), the accordion (**c**) and the mandolin (**d**), respectively.

The second scenario (Figure 23) is based on the English alphabet and aims to let the children interact with the English alphabet, so they learn English words in an interactive way. This layer consists of four touch sensors each associated with a letter of the English alphabet. The letters are projected by the projector over specific areas of the surface and each time they are randomly selected. By selecting one of the four letters, the user reveals the corresponding word. Then, the word appears on the screen along with a related image.

Figure 22. Simultaneous use of our interactive surface by several users.

Figure 23. English alphabet scenario. The layer consists of four touch sensors, each associated with a letter of the English alphabet. Here is an example run of the scenario with the letters "L" in (**a**), "U" in (**b**), "N" in (**c**) and "F" in (**d**), projecting the corresponding English words associated with each letter.

The third and final scenario is not interactive and is dedicated to the projection mapping technique. Its aim is to demonstrate to the users what can be carried out with projection mapping on surfaces with different positions, angles, and inclinations in relation to the projector. To this end, a simple 2D animation downloaded from the internet was projected over a 3d construction made by cardboard and placed over our surface (Figure 24). The positions of the sensors used in the previous scenarios had also to be taken into account, so the construction did not cover the sensors and did not obstruct the use of the surface when another scenario is active. A part of the animation is projected on each board, and this gives the final result.

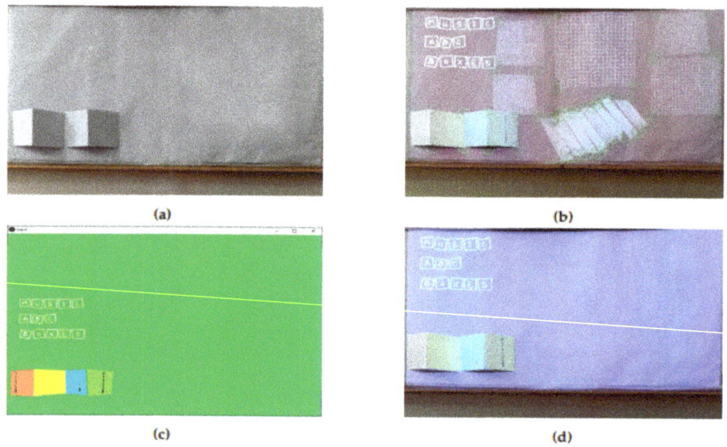

Figure 24. Non-interactive Scenario. Demonstration of projection mapping technique on surfaces with different positions, angles, and inclinations, that appear in (**b–d**) in relation to the projector. In (**a**) the 3d construction, made by cardboard and placed in the surface, is shown.

4.2. Evaluation

To evaluate the performance of the whole construction under different circumstances of use, we set up a small benchmark in our premises, emulating realistic user behaviors over the touch sensors. Figure 25 illustrates the touch sensors that were evaluated during this benchmark. Sensor 4 is the sensor behind the tambourine, Sensor 5 is the sensor behind the maracas, Sensor 6 is the sensor behind the mandolin, Sensor 7 is the sensor behind the trumpet, and Sensor 8 is the sensor behind the accordion. Each experiment was repeated eleven times and the values that are presented here correspond to the median value that was recorded for each set of measurements.

Figure 25. Sensors under evaluation.

4.2.1. Touching the Sensors with One or More Fingers

In this experiment the sensors were touched at first with just one finger and then with four. Figure 26 illustrates this experiment. The results proved that our touch sensors are not sensitive to the number of fingers pressed on them.

Figure 26. Touching the sensors with one or more fingers.

4.2.2. Touching the Sensors at Different Frequencies

In this experiment the behavior of our sensors was tested when they are touched frequently. This is some kind of stress test for the sensors. To perform this experiment a common music metronome was activated and the success rate for each sensor at different beats-per-minute (bpm) of the metronome was recorded. Table 2 shows the results from the evaluation of Touch Sensor 4 and Figure 27 illustrates instances from this experiment. Similar results were recorded also for the other four touch sensors. We believe that the recorded mean success rate of 95% proves that our touch sensors behave sufficiently steadily for the use under consideration.

Table 2. Success rate of touches for touch Sensor 4 under different BPM.

BPM	50	80	100	120	150
Success Rate	19/20	20/20	18/20	19/20	19/20

Figure 27. Touching the sensors with different frequencies.

4.2.3. Touching the Sensors with Lighter Pressure

Considering that this implementation will be used by children among others, we wished to evaluate the behavior of the sensors when touched with lighter pressure. Table 3 summarizes the success rate for the five sensors under evaluation. The results show less sensitivity of the sensors when less pressure is applied over them. We still believe, however, that this performance is adequate for the intended use. The slightly different performance of Touch Sensor 8 is attributed to the fact that at this area of the construction the white paper has slightly unstuck and air has intruded between the sensor and the white paper.

Table 3. Success rate of touches for touch sensors under lighter pressure.

Touch Sensor	4	5	6	7	8
Success Rate	86%	83%	86%	84%	80%

4.2.4. Simultaneous Use of the Touch Sensors

The last experiment that was conducted was the evaluation of sensors when they were used simultaneously by several users. To this end, two users started touching the sensors together, so four sensors were activated in parallel. More specifically, three sensors were permanently touched when the same time the sensor under evaluation was repeatedly touched and released to measure its success rate.

Figure 28 illustrates instances from this experiment, in particular the evaluation of Sensor 6 under simultaneous use is presented. The results showed similar behavior of the sensors with their independent use (that is, a success rate of 95% for each sensor was recorded again).

Figure 28. Simultaneous use of the touch sensors.

4.2.5. Comparison of Our Touch Sensors with a Commercial Touch Sensor

To compare the robustness of our touch sensors with commercial ones, we used a commercial 3.5″ touch display, connected directly to our system, as touch sensor. We used a touch pen (see Figure 29) and proceeded 100 times by repeatedly pressing the touch display, while we repeated the same process for some of our sensors independently. The results recorded, showed a comparable behavior of the sensors developed and a mean success rate of 95% for each one in comparison to the commercial one touch display that had a 100% success rate. In this test, no information is used from the projected content, but it is aimed as a quick comparison of our touch sensors with a commercial one.

Figure 29. Comparing our touch sensors with a commercial 3.5″ touch screen.

5. Conclusions

The present work showed that using simple low-cost materials in combination with open-source software tools, one is able to build a functional interactive surface. This surface was used as a simplified example demonstration of the research that is carried out by our Sensor Networks laboratory during the visits of various schools in our university. Figure 30 illustrates the use of our surface by the students. The comments received from them were very positive not only for the attractiveness and the simplicity of this construction but also for its performance.

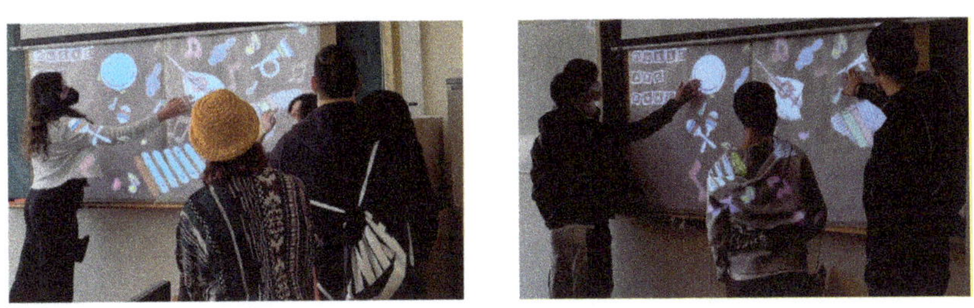

Figure 30. Use of our interactive surface by pupils during school visits in our laboratory.

Regarding the construction of the surface, the material chosen for this work was cardboard, mainly for the easy management and portability that it provides. The cardboard offered many advantages in the construction and made the connection of the front side and the backside easy. On the backside all the wiring was connected to the microcontrollers, this way the front side was not burdened with unnecessary materials and cables. However, the choice of cardboard also brings several disadvantages and some limitations to the final result. One such disadvantage is that since all the wiring is on the back of the surface, the cardboard was not always in a good contact with the wall behind the construction. As a result, in certain cases the touch sensors can be unstable. Another limitation seems to be that the way of sticking the white paper over the interactive surface, so we did not avoid air intruding in the middle. The addition of a frame where the cardboard can be placed would play an important role and would significantly improve the stability of the surface. Alternatively, a different rigid material can be used, for example plywood or plexiglass sheets. With these options, the surface will be more stable and efficient, but it will be difficult to be managed and easily moved.

As it concerns software, the design and programming of the multimedia interactions through Processing was also a challenge. For example, the playback of multiple videos at the same time caused significant delays in the system, resulting in avoiding the use of video content on the final implementation. Moreover, although processing is a very promising tool for the creative industries, its requirement for programming dexterity limits the audience it targets. Instead, a multimedia editing tool would offer more flexibility and would greatly speed up the development of such dashboards.

In our immediate future plans, now that our first proof-of-concept implementation has been positively evaluated, is to construct a revised version of this dashboard accommodating the above improvements. Such an interactive surface, made by more rigid materials than cardboard, and projecting guidance information, will be placed for use in a public place in our university campus to test its performance under non controllable conditions, as well as its acceptance by the audience. Apart from this alternative approach, our future plans also include the creation of an interactive touch grid, similar to the logic of the touch screens, using same low-cost philosophy. Such an implementation will eliminate the restriction of putting touch sensors in strict positions over the dashboard and will make our construction open to host dynamic scenarios for use in more demanding use cases.

Author Contributions: Conceptualization, S.P.; methodology, A.T.L. and S.P.; software, A.T.L.; validation, A.T.L., S.P. and Z.K.; formal analysis, A.T.L., S.P., Z.K. and G.L.; investigation, Z.K. and G.L.; resources, G.L., A.M. and E.M.; data curation, A.T.L., S.P.; writing—original draft preparation, A.T.L., S.P. and Z.K.; writing—review and editing, Z.K., G.L., A.M. and E.M.; visualization, A.T.L., S.P.; supervision, A.M. and E.M.; project administration, A.M. and E.M. All authors have read and agreed to the published version of the manuscript.

Funding: This research received no external funding.

Institutional Review Board Statement: Not applicable.

Informed Consent Statement: Not applicable.

Data Availability Statement: The data used to support this study's findings are available from the corresponding author upon request.

Conflicts of Interest: The authors declare no conflict of interest.

References

1. Atsali, G.; Panagiotakis, S.; Markakis, E.; Mastorakis, G.; Mavromoustakis, C.X.; Pallis, E.; Malamos, A. A mixed reality 3D system for the integration of X3DoM graphics with real-time data. *Multimed. Tools Appl.* **2018**, *77*, 4731–4752. [CrossRef]
2. Vakintis, I.; Panagiotakis, S.; Mastorakis, G.; Mavromoustakis, C.X. Evaluation of a Web Crowd-Sensing IoT Ecosystem Providing Big Data Analysis. In *Chapter Contribution in the "Resource Management for Big Data Platforms, Algorithms, Modelling, and High-Performance Computing Techniques"*; Florin, P., Joanna, K., Beniamino, D., Eds.; Springer: Berlin/Heidelberg, Germany, 2016; pp. 461–488.
3. Papadokostaki, K.; Panagiotakis, S.; Malamos, A.; Vassilakis, K. Mobile Learning in the Era of IoT: Is Ubiquitous Learning the Future of Learning? In *Early Childhood Education*; IGI Global: Hershey, PA, USA, 2020; pp. 252–280. [CrossRef]
4. Pinikas, N.; Panagiotakis, S.; Athanasaki, D.; Malamos, A. A Device Independent Platform for Synchronous Internet of Things collaboration and Mobile Devices Screen Casting. *Sci. Publ. Group Int. J. Inf. Commun. Sci.* **2017**, *2*, 59–67. [CrossRef]
5. Alexakis, G.; Panagiotakis, S.; Fragkakis, A.; Markakis, E.; Vassilakis, K. Control of Smart Home Operations Using Natural Language Processing, Voice Recognition and IoT Technologies in a Multi-Tier Architecture. *Designs* **2019**, *3*, 32. [CrossRef]
6. Qian, K.; Kakarala, R.; Akhtar, H. A review of sensing technologies for small and large-scale touch panels. In Proceedings of the Fifth International Conference on Optical and Photonics Engineering, Singapore, 4–7 April 2017; p. 1044918. [CrossRef]
7. Sathyan, A.; Manikandan, L.C. A Study and Analysis of Touch Screen Technologies. *Int. J. Sci. Res. Comput. Sci. Eng. Inf. Technol.* **2020**, *6*, 737–744. [CrossRef]
8. Grosse-Puppendahl, T.; Holz, C.; Cohn, G.; Wimmer, R.; Bechtold, O.; Hodges, S.; Reynolds, M.S.; Smith, J.R. Finding Common Ground: A Survey of Capacitive Sensing in Human-Computer Interaction. In Proceedings of the 2017 CHI Conference on Human Factors in Computing Systems, Montreal, QC, Canada, 21–26 April 2018.
9. Pourjafarian, N.; Withana, A.; Paradiso, J.A.; Steimle, J. Multi-Touch Kit: A Do-It-Yourself Technique for Capacitive Multi-Touch Sensing Using a Commodity Microcontroller. In Proceedings of the 32nd Annual ACM Symposium on User Interface Software and Technology (UIST'19), New Orleans, LA, USA, 20–23 October 2019; Association for Computing Machinery: New York, NY, USA, 2019; pp. 1071–1083. [CrossRef]

10. Zhang, Y.; Yang, C.; Hudson, S.E.; Harrison, C.; Sample, A. Wall++: Room-Scale Interactive and Context-Aware Sensing. In Proceedings of the 2018 CHI Conference on Human Factors in Computing Systems, Montreal, QC, Canada, 21–26 April 2018; Volume 273, pp. 1–15. [CrossRef]
11. Interactive Touch Wall. Available online: https://www.core77.com/posts/35697/How-Dalziel-and-Pow-Realized-This-Awesome-Interactive-Touch-Wall (accessed on 10 April 2022).
12. Zippy. Available online: https://www.dalziel-pow.com/news/zippy-digital-installations-making (accessed on 10 April 2022).
13. Jacoby, S.; Buechley, L. Drawing the Electric: Storytelling with Conductive Ink. In Proceedings of the 12th International Conference on Interaction Design and Children, New York, NY, USA, 24–27 June 2013; pp. 265–268. [CrossRef]
14. Buechley, L.; Mellis, D.; Perner-Wilson, H.; Lovell, E.; Kaufmann, B. Living Wall: Programmable Wallpaper for Interactive Spaces. In Proceedings of the 18th ACM International Conference on Multimedia, Firenze, Italy, 25–29 October 2010.
15. Qi, J.; Buechley, L. Electronic Popables: Exploring Paper-Based Computing through an Interactive Pop-Up Book. In Proceedings of the Fourth International Conference on Tangible, Embedded, and Embodied Interaction, Cambridge, MA, USA, 25–27 January 2010.
16. Freed, N.; Qi, J.; Setapen, A.; Breazeal, C.; Buechley, L.; Raffle, H. Sticking Together: Handcrafting Personalized Communication Interfaces. In Proceedings of the 10th International Conference on Interaction Design and Children, Ann Arbor, MI, USA, 20–23 June 2011.
17. Russo, A.; Ahn, B.Y.; Adams, J.J.; Duoss, E.B.; Bernhard, J.T.; Lewis, J.A. Pen-on-Paper Flexible Electronics. *Adv. Mater.* **2011**, *23*, 3426–3430. [CrossRef] [PubMed]
18. How to Make an Arduino Capacitance Meter. Available online: https://www.circuitbasics.com/how-to-make-an-arduino-capacitance-meter/ (accessed on 19 March 2022).
19. How Capacitive Sensors Work and How to Use Them Effectively. Available online: https://www.sensorland.com/HowPage070.html (accessed on 2 April 2022).
20. ESP32 Capacitive Touch Sensor Pins with Arduino IDE. Available online: https://randomnerdtutorials.com/esp32-touch-pins-arduino-ide/ (accessed on 19 March 2022).
21. Simple Median Filter Library Designed for the Arduino Platform. Available online: https://github.com/daPhoosa/MedianFilter (accessed on 10 April 2022).
22. Grundhöfer, A.; Iwai, D. Recent Advances in Projection Mapping Algorithms, Hardware and Applications. *Comput. Graph. Forum* **2018**, *37*, 653–675. [CrossRef]
23. The Illustrated History of Projection Mapping. Available online: http://projection-mapping.org/the-history-of-projection-mapping/ (accessed on 8 April 2022).
24. Welcome to Processing! Available online: https://processing.org/ (accessed on 10 April 2022).

information

MDPI

Article

Towards a Supervised Remote Laboratory Platform for Teaching Microcontroller Programming

Manos Garefalakis [1,*], Zacharias Kamarianakis [1,2] and Spyros Panagiotakis [1]

[1] Department of Electrical & Computer Engineering, Hellenic Mediterannean University, Estavromenos, GR71410 Heraklion, Greece; zkamar@hmu.gr (Z.K.); spanag@hmu.gr (S.P.)
[2] Institute of Agri-Food and Life Sciences, Research & Innovation Center, H.M.U.R.I.C., Hellenic Mediterranean University, GR71410 Heraklion, Greece
* Correspondence: mgarefal@gmail.com

Abstract: As it concerns remote laboratories (RLs) for teaching microcontroller programming, the related literature reveals several common characteristics and a common architecture. Our search of the literature was constrained to papers published in the period of 2020–2023 specifically on remote laboratories related to the subject of teaching microcontroller programming of the Arduino family. The objective of this search is to present, on the one hand, the extent to which the RL platform from the Hellenic Mediterranean University (HMU-RLP) for Arduino microcontroller programming conforms to this common architecture and, on the other hand, how it extends this architecture with new features for monitoring and assessing users' activities over remote labs in the context of pervasive and supervised learning. The HMU-RLP hosts a great number of experiments that can be practiced by RL users in the form of different scenarios provided by teachers as activities that users can perform in their self-learning process or assigned as exercises complementary to the theoretical part of a course. More importantly, it provides three types of assessments of the code users program during their experimentation with RLs. The first type monitors each action users perform over the web page offered by the RL. The second type monitors the activities of users at the hardware level. To this end, a shadow microcontroller is used that monitors the pins of the microcontroller programmed by the users. The third type automatically assesses the code uploaded by the users, checking its similarity with the prototype code uploaded by the instructors. A trained AI model is used to this end. For the assessments provided by the HMU-RLP, the experience API (xAPI) standard is exploited to store users' learning analytics (LAs). The LAs can be processed by the instructors for the students' evaluation and personalized learning. The xAPI reporting and visualization tools used in our prototype RLP implementation are also presented in the paper. We also discuss the planned development of such functionalities in the future for the use of the HMU-RLP as an adaptive tool for supervised distant learning.

Keywords: remote laboratory; Arduino microcontroller programming; xAPI; LRS; assessment types; pervasive learning; supervised learning

Citation: Garefalakis, M.; Kamarianakis, Z.; Panagiotakis, S. Towards a Supervised Remote Laboratory Platform for Teaching Microcontroller Programming. *Information* **2024**, *15*, 209. https://doi.org/10.3390/info15040209

Academic Editors: Luis Borges Gouveia and Aneta Poniszewska-Maranda

Received: 16 February 2024
Revised: 22 March 2024
Accepted: 1 April 2024
Published: 8 April 2024

1. Introduction

Remote laboratories (RLs) have been extensively studied in the literature, with a prominent highlighted issue being the lack of a common design pattern among online laboratory systems. This lack of standardization has constrained scalability, integration, interoperability, traceability, reliability, security, and privacy. Addressing this concern, the IEEE 1876-2019 standard [1] was developed to facilitate the design, implementation, and usage of online laboratories in education. The fundamental purpose of remote labs is to afford students the opportunity for practical experimentation (PEX) and laboratory experimentation (LEX), leveraging components and equipment typically found in university labs. Through remote labs, students can apply theoretical knowledge gained in the classroom, accommodating their learning styles, at their convenience. Whether utilizing PCs for E-learning or smartphones for M-learning, students can engage with learning objects ranging from simple tasks to more complex scenarios, within the parameters set by the remote lab. This educational approach fills a crucial gap between simulated environments (full virtual labs) and real-world lab experiences, enabling the provision of distant courses focused on hardware equipment usage and software programming. The demand for remote access to educational lab facilities was further underscored during the COVID-19 pandemic, where such labs became essential for mobile learning, enabling students to practice and experiment with specialized equipment and tools from any location to acquire practical skills and fulfill course requirements.

In the paper [2], the classification of laboratories is presented. They are divided into local and remote laboratories. The remote laboratories are classified into real remote laboratories (remote laboratories), hybrid remote laboratories, and virtual laboratories. In this paper, we review and propose a real remote laboratory, where the user works at a distance on real equipment, especially on programming microcontrollers like Arduino boards. Working on an RL is different from working on a remote simulation, like Tinkercad [3], which also provides learning skills (circuit design, coding, etc.), but only to some extent. An RL can implement more experiments, more complex and with real components, working in real situations and not based on a mathematical model [4], because they can connect to an existing infrastructure such as LoRa Gateway, MQTT servers, etc. It must be highlighted that there is no discrimination between the two types of laboratories, real or virtual, but on the contrary, they are used in a complementary way to each other and both provide skills to the user.

The HMU-RLP is a project designed and implemented by the Sensor Network LAB of the HMU. It started from an Erasmus+ KA2 project named SYS-STEM, which included five partners from European universities and is described in the paper [4]. During SYS-STEM, a qualitative evaluation from teachers and students that used the SYS-STEM RL was organized [4]. The evaluation highlighted some technical issues with the SYS-STEM infrastructure, such as the need for more inherent interaction of the user with the RL as well as the need for loading an initialization sketch in the experiment controller at the end of each user session. Also, teachers of secondary education declared their willingness to run pilot testing of the SYS-STEM methodology and ARDLAB with their students, although they welcomed a tool providing analytics for their students' actions over the lab. From the logs of this evaluation, we noticed that students used ARDLAB for LAB exercises especially during the late hours and, also, that many students repeated the courses more than once, which potentially showed that it might be helpful for them or that they needed some help to complete the tasks. The experience acquired from SYS-STEM led to the design and implementation of a different platform that provides more capabilities and functions toward pervasive and supervised learning, concepts that require more than an RL on which the user only monitors the results of the code uploaded on the experimental microcontroller. A first presentation of the HMU-RLP was issued in the paper [5], where we described the ability of an RL to connect with a learning management system (LMS) and how the experiments it provides can be used as learning objects (LOs) within an LMS course.

In the present paper, we proceed one step further toward pervasive learning, describing how an RL can supervise user actions over its infrastructure for potentially providing more powerful analytics to the instructors and/or intelligent tutoring to the learners. The need for these features was revealed by the outcomes of [4] and is implemented in this paper with the three assessment types we introduced in the HMU-RLP.

This paper also implements a literature search for finding other remote laboratories (RLs) that are used for teaching microcontroller programming of the Arduino family. This search was underlined by our need to understand how education based on modern remote labs is built today and what new features could potentially be added as a complement. The literature search revealed similarities in the architecture of remote laboratories but did not show any type of assessment apart from monitoring the results of the coding of the microcontrollers. The HMU-RLP is a use case for teaching programming in the Arduino ecosystem. The HMU-RLP implements three different types of assessment of the user's use of the RL and the uploaded code, which is used for further processing of adaptive learning concepts and aligns with the concept of pervasive and supervised learning. Furthermore, the experimental microcontroller is connected to many sensors and actuators, which aligns it with the concept of one RL for many experiments.

The literature search also revealed concepts that should be considered for implementation in the HMU-RLP and will be included in future work, such as hosting of the RL on a remote laboratory management system, remote H/W configuration, and user complements with augmented and virtual reality value-added concepts. Also, the need for an extended qualitative review of our RLP from teachers and students.

The paper is organized as follows: Section 1 presents an overview introduction of the paper. Section 2 includes the related literature search. Section 3 presents the architecture of the remote lab. Section 4 presents the remote lab assessment types. Section 5, Experience API Statements and Tools, presents the experience API used for the users' learning analytics reporting, while Section 6 presents the remote lab experiments. Finally, Section 7 presents the conclusion and future work.

2. State of the Art

Remote laboratories seem to be a necessity in recent years because they cover a gap between theory and hands-on laboratories, real on-site laboratories. The users work on real equipment remotely. Under this context, there are advantages and disadvantages. The advantage is that more users can practice on experiments from the comfort of their homes and at any time they wish, 24/7, and RLs promote inclusion and diversity, providing lab access to persons with special needs.

The disadvantages are that users do not acquire hands-on skills, and the experiments are dependent on network communication and the quality of video and audio streaming.

The literature search that was performed revealed the number of papers published per year. Queries were issued on the following databases: Google Scholar, Scopus, Semantic Scholar, and OpenAlex. The keywords used were "Remote Labs", "Remote Laboratories", "Remote experiments", "Online Labs", "Online Laboratories", and "Online Experiments". The keywords were limited to the paper's title only. The initial search returned 13,544 results and after removing duplicates, and again filtering keywords in the title for a second time, the number was reduced to 1615, as can be seen in Table 1.

Table 1. Results of analysis.

Data Source	Initial Search	1st Stage (Identification)	2nd Stage (Screening)
Google Scholar	2267	620	
Scopus	2152	842	
Semantic Scholar	5186	492	
OpenAlex	3939	16	
Total	13,544	1970	1615

In Figure 1, the number of papers related to remote laboratories published per year can be seen. What we can comment on in the graph is that, during the period of 2012–2015, there was a peak in the number of papers published related to remote laboratories. After the period of 2016–2019, there was a decrease in the number of papers on the topic of remote laboratories, which then increased again during the period of 2020–2023, probably due to the COVID-19 pandemic period, where remote laboratories were an important and required solution for teaching, apart from the fact that technologies were developed.

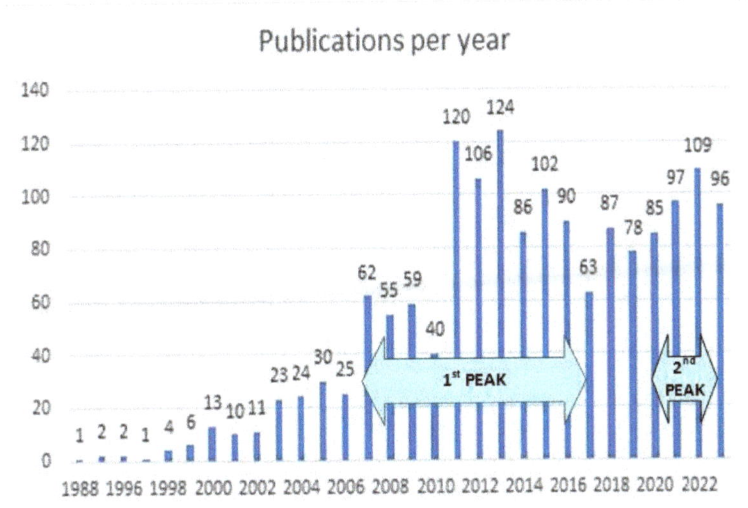

Figure 1. Papers on remote laboratories published per year.

Another literature research was performed to reveal the differences between the proposed RL platform and other relevant use cases. Since the proposed RL is about programming Arduino ecosystem boards, the keyword Arduino was added to the search keywords. For this particular literature research, a search in the Scholar Google database issued the following statement: "Remote Labs" OR "Remote Laboratories" AND Arduino.

The criteria set for the eligibility of a paper were the following:

- Recent papers that were published during the second peak period (see Figure 1), which included also the COVID-19 and post-COVID-19 period. Additionally, it revealed new generation RLs that included new technologies, e.g., complying with IEEE 1876-2019 [1].
- RLs related to teaching microcontroller programming, specifically Arduino, because the proposed RL is about teaching programming of Arduino boards.
- Access to the article using the institutional credentials of Hellenic Mediterranean University.
- Papers written in the English language.

The initial results were 1590 papers and, after applying the selection criteria, the returned number of results was reduced to 625 papers (query performed on 21 January 2024).

The screening was performed on the title and the abstract, and we present the selected papers in Table 2.

By reviewing the related literature listed in Table 2, we saw some common characteristics. These characteristics are the following: RL name, project website, programming board, programming language, RL controller, hosted in platform, direct access, assessment, GUI for access, statistics, monitoring, remote H/W reconfiguration, LMS integration, and logging learning analytics.

Some RLs have a name that can be found in the literature and a website for the project they belong to, for example, VISIR+, etc. The RLs are for programming microcontrollers, so knowing the board that is programmed is important (Arduino, STM, etc.) and also the programming language used. For example, in some cases of Arduino programming, it is C/C++ coding or visual programming.

Table 2. List of reviewed papers.

Reference No	Title	Publication Year
[5]	Integration of a Remote Lab with a Learning System for training on Microcontrollers' programming	2023
[6]	LabsLand Electronics Laboratory: Distributed, Scalable and Reliable Remote Laboratory for Teaching Electronics	2023
[7]	Remote Laboratory for the Development of Customized Low-Power Computing and IoT Systems	2023
[8]	ARM Distributed and Scalable Remote Laboratory for Texas Instruments Launchpad Boards	2023
[9]	Mobile Arduino Robot Programming Using a Remote Laboratory in UNAD: Pedagogic and Technical Aspects: Experience Using a Remote Mobile Robotics	2021
[10]	Learning CAN bus communication with a remote laboratory	2022
[11]	Learning Management Systems as a platform for deployment of remote and virtual laboratory environments	2022
[12]	ERPLab: Remote Laboratory for Teaching Robotics and Programming	2023
[13]	Fpga-based remote laboratory for digital electronics	2020
[14]	Practice Projects for an FPGA-Based Remote Laboratory to Teach and Learn Digital Electronics	2023
[15]	Teaching programming and microcontrollers with an arduino remote laboratory application	2023
[16]	Remote Experimentation Through Arduino-Based Remote Laboratories	2021
[17]	Remote laboratory for microcontroller programming course	2022
[18]	An Implementation of a Web Laboratory Converting Off-Line Experiments into Remotely Accessible Experiments	2022
[4]	Remote Arduino Labs for Teaching Microcontrollers and Internet of Things Programming	2022
[19]	Remote Laboratory Offered as Hardware-as-a-Service Infrastructure	2022
[20]	A Remotely Configurable Hardware/Software Architecture for a Distance IoT Lab	2021
[21]	Internet of things network infrastructure for the educational purpose	2020
[22]	IOT-OPEN.EU: Introduction to the IoT Practical Projects in English–IOT-Open	2024
[23]	µLAB A remote laboratory to teach and learn the ATmega328p µC	2020
[24]	Block. Ino: Remote lab for programming teaching and learning	2020
[25]	Training Laboratories with Online Access on the ITMO. cLAB Platform	2020
[26]	Design and development of remote laboratory system to facilitate online learning in hardware programming subjects	2020
[27]	Implementation of an Arduino remote laboratory with Raspberry Pi	2019
[28]	Improving the scalability and replicability of embedded systems remote laboratories through a cost-effective architecture	2019

RLs are either autonomous, where the user connects directly to the RL, the direct access characteristic, or they are hosted in platforms like remote laboratory management systems (RLMSs), massive online open courses (MOOCs), small private online courses (SPOCs), learning management systems (LMSs), repositories like remote lab management systems (RLMSs), etc., see the hosted in platform characteristic. These platforms handle user authentication and interoperability with the RL. There are also cases where the RLs can work in both cases autonomously or provided via a platform. All of these are suggested in the IEEE 1876-2019 Standard for Networked Smart Learning Objects for Online Laboratories [1], which defines methods for storing, retrieving, and accessing online laboratories. According to the standard, an online laboratory (RL) is defined as a lab as a service (LaaS).

The assessment type characteristic is whether the RL provides an assessment type and reports grading to the user profile, like it is presented in the HMU-RLP in the next sections.

The type of access to the RL is another characteristic, via a web interface, an application, or via remote desktop application. Another characteristic is the monitoring of the RL's experiment and equipment. In most cases, there is a web camera or an RPI camera module connected, but we also found a paper in which there was no monitoring but images of the board were displayed with the current status. In most of the use cases, the H/W experiments are ready connections, but in one case they suggested an online H/W reconfiguration to expand experiment scenarios.

Finally, the last two characteristics are very important in the concept of pervasive and supervised learning. The possibility of integration in a learning management system (LMS) and the logging of learning analytics on the user's learning experience. According to the paper [5], when an RL is integrated into an LMS, as a learning object (LO), then there are learning analytics that can be stored in the LMS user profile, for example, updating the user grade book, marking an LO as completed, etc. The use of these LAs can be used in updating and planning the learning path of a user. For example, imagine a user who cannot complete an experiment that has to turn on some LEDs. Then, the course path in the LMS can be updated to simpler experiments to help the user complete the learning task and acquire the learning skill. On the other hand, when a user archives directly to complete the experiment, then the LMS enables a more complex experiment to be assigned to the user.

All of the characteristics mentioned above are about working in an LMS and the RL is an LO. Now what about when the RL is not integrated into an LMS and works independently. Then, the learning analytics are stored in learning record storage (LRS), the database defined in the xAPI standard. With the use of tools, the instructor may follow the learning path of the user and accordingly assign experiments to the user. This process needs to be developed in such a way as to automate it and is currently under research by the team developing the HMU-RLP. The Table 3, presents the summary of the paper's review characteristics.

Table 3. Summary of the paper's review. Reviewed papers' characteristics.

RL Paper	RL Name	Project Web-site	Programming Board	Programming Language	RL Controller	Hosted in Platform	Direct Access	Assessment	GUI for Access	Statistics	Monitoring	Remote H/W Reconfigu-ration	LMS Integra-tion	Logging Learning Analytics
[5]	HMU-RLP	[29]	Arduino UNO	C/C++	Raspberry Pi 3 B+	No	Yes	Yes	Web In-terface	Yes	RPI Camera Module	No	Yes	Yes
[6]						LabsLand [30]				Yes	Yes		Yes	
[7]	RemoCLEC	[31]	NUCLEO-WB55RG			LabsLand [30]				Yes	Yes			
[8]			Pololu Zumo 32u4 robot with Arduino	C/C++ or Visual programming Blockly		LabsLand [30]				Yes	Yes			
	TIVA Remote Laboratory at the University of Washington	[32]	TIVA			LabsLand [30]	No			Yes	Yes		Yes	
[9]	Robotics Remote Lab of UNAD		ATmega 32U4 compatible with Arduino	C/C++	Raspberry Pi 3 B+	LabsLand [30]				Yes	Yes		Yes	
[10]	Public University of Navarra CAN Bus RL		Arduino Board			Go Lab [33] LabsLand [30]				Number of Accesses and time spent	Yes			
[11]	UbiLAB project					Weblab-Deusto [33]							Yes	
[12]	ERPLab Environment-Robotic-Programmin-Laboratory		Arduino Uno development board and Ethernet Shield W5100	C/C++		No					ESP32CAM		No	
[13]			FPGA Nexys 3								No camera			
[14]	RLAB University of Málaga	[34]	FPGA Nexys 3		Raspberry Pi 4		Yes				No camera			
[15]	RemoLab in Croatian schools		Arduino								WebCam			
[16]	University UNED		Arduino		Raspberry Pi						WebCam			
[17]	Riga Technical University		Texas Instruments (TI) MSP430						Remote Desktop		VLC			

Table 3. *Cont.*

RL Paper	RL Name	Project Website	Programming Board	Programming Language	RL Controller	Hosted in Platform	Direct Access	Assessment	GUI for Access	Statistics	Monitoring	Remote H/W Reconfiguration	LMS Integration	Logging Learning Analytics
[18]	WEB Laboratory at University of Kragujevac		Arduino UNO Arduino DUE						X2GO-Remote Desktop					
[4]	SYS-STEM Hub	[35]	Arduino UNO	C/C++	Raspberry Pi	SYS-STEM Hub [35]	No	No	Web Interface	No	RPI Camera module		No	No
[19]	Wrocław University of Science and Technology WUST		STM32 microcontrollers • Nucleo-L476RG • 32L476GDISCOVERY • STM32F429I-DISC1		Raspberry Pi						RPI Camera module			
[20]			Arduino UNO		Raspberry Pi 3B+							Yes		
[21, 22]	IOT-OPEN.EU VREL	[35]	Arduino Uno ESP 8266 (ESP-12E);	C/C++	Raspberry-Pi 2 & 3	VREL management server			Web Interface		RPI Camera module			
[23]	µLAB-Polytechnic Institute of Porto		ATmega328p		Raspberry Pi		Yes		Web Interface		Webcam			
[24, 36]	Block.Ino -University of Santa Catarina, Brazil		Arduino	visual programming environment		RELLE [37]					Yes			
[25]	ITMO.cLAB-ITMO University		SDK-1.1M STM32F407VG microcontroller TFK-4.0U MA842 analog I/O module											
[26]			MK20DX128VLH5 by NXP Semiconductors						Chrome Remote Desktop					
[28]	ArduinoRL		Arduino			LabsLand [30]	No		Web Interface		Yes			
[27]	UNED		ATmega328p ATmega2560 MKR1000		Raspberry Pi				Web Interface		Yes			

3. The Architecture of the HMU-RLP

The literature search revealed similarities in the architecture of remote laboratories but did not show any type of user assessment apart from monitoring the results of the coding of the microcontrollers. The HMU-RLP attempts to fill this gap by implementing, as we will present, three different types of assessment of the user's interaction with the RL and of the code the user uploads. Furthermore, the experimental microcontroller is connected to many sensors and actuators, which aligns with the concept of one RL for many experiments. The literature search also revealed concepts that should be considered for implementation in the HMU-RLP and will be included in future work, such as hosting of an RL on a remote laboratory management system and user complements with augmented and virtual reality value-added concepts.

The presented platform for remote labs has a similar architecture to most RLs presented in Section 2. There is a main server, which is a Raspberry PI single-board computer, that runs an application developed in the Flask framework. The application provides the user interface and all necessary tasks needed for the function of the RL.

In most of the cases reviewed, the user connects to the RL via the Internet from anywhere the user wishes and by any device s(he) prefers (PC, laptop, tablet, or smartphone), exploiting distance learning positives. Such an architecture is proposed in [28].

The application provides a platform for the users to use the RL. There are tasks granted to administrators and users. Table 4 lists the main tasks of the application for users and administrator groups.

Table 4. Tasks per users and administrators.

Users:	**Administrators:**
• User creation • User profile update • User Arduino sketch storage • User sketch compilation and uploading • Remote lab monitoring • Remote lab user interaction • Activating activities • Monitoring activity status	• Setting configuration parameters • Shadow controller check (Enable/Disable) • Booking system (Enable/Disable) • Clearing users' sessions • Setting user group (Student/Administrator) • Setting xAPI status (enable/disable) • Creating activities

Additionally, the application implements the learning tools interoperability (LTI), which allows the platform to act as an external tool in a learning management system like Moodle.

From the above discussion, the platform can work in two modes. The first mode is "LMS-Free", which handles everything, like users' creation and authentication, security issues, activities, learning analytics, etc. The second mode is the LMS external tool, where the platform is called from an LMS for a particular activity (learning object), and the user executes the activity according to the instructions and finally submits the activity. The performed activity is assessed by the RL and the grade is updated in the LMS users' grade book. Additionally, the users' learning analytics are stored in a learning record store (LRS) as part of the experience API standard.

This is what differentiates the remote lab presented from the other cases mentioned in Section 2. The RL can assess the uploaded code, but on the contrary, the other remote labs only stay in the phase of monitoring the results of the sketch uploaded on the Arduino board.

To summarize the above-mentioned information and visualize it, we can see Figure 2. The user logs in to the HMU-RLP via the Internet using a web browser. The HMU-RLP is located on the campus of the HMU, in the Sensor Networks LAB.

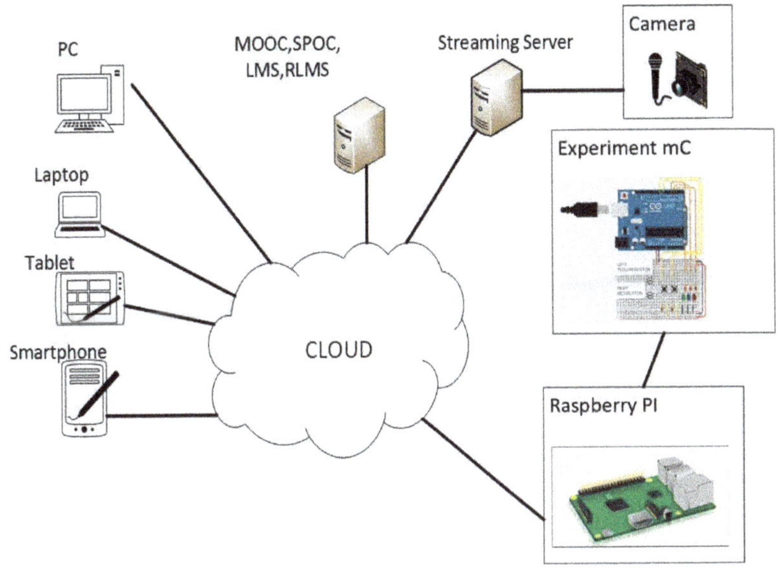

Figure 2. Generic architecture of remote laboratories.

HMU–RLP User Workflow

The user's workflow of the HMU-RLP is presented in Figure 3. After the user logs in to the HMU-RLP, the user then has to activate an activity.

The main object in the RLP is the activity. The activity is the scenario or the experiment that will run on the RLP; everything begins with activity activation. The user activates an activity in the RLP and receives instructions about the scenario and also relevant documentation. After activation, the user creates the code required from the activity's scenario in the Arduino IDE. It must be reassured that the code contains no syntax errors using the Arduino IDE Compiler. After verifying that the code is correct, the code is stored in the RLP users' profile. The code must be compiled and uploaded in the Arduino experimental microcontroller.

When the new code is uploaded, then the user can remotely monitor the execution of the code via the camera/s of the RLP. Additionally, the user can interact with the experiment (turn on/off the lights, the heater, the FAN, or pressing the red or the yellow buttons); in this way, the activity's scenario is tested. After testing the scenario, the user can check that they have followed the correct instructions from the activity status report.

When the user is ready, they can submit the activity for assessment. The activity submission concludes the experiment.

During the whole process, the user's learning analytics profile is updated using xAPI statements.

As we mentioned, there are three types of assessments in the RLP. The logging of the path of the actions assesses the actions instructed to the user. The shadow microcontroller verification checks the code at a hardware level. And last, the machine learning check of the code, which uses a decision tree algorithm to verify that the code adheres to the instructions. All assessment test results are stored in the xAPI LRS as statements.

Finally, instructors can use reporting and visualization tools for further processing, the user's learning path, and adaptive learning.

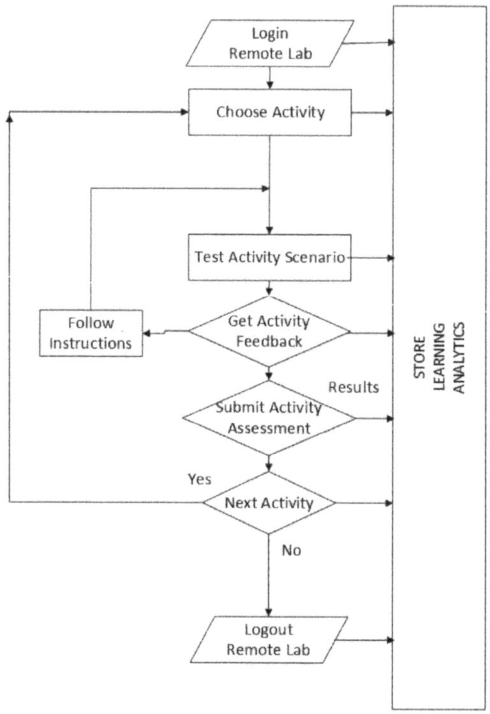

Figure 3. HMU-RLP user workflow.

4. Remote Lab Assessment Types

We will begin by explaining what we mean by the term assessment of experiment in the context of a remote lab. When a user is assigned an activity (experiment), the user needs to have the instructions for the activity, the relevant theory documentation, and the schematic of the connections on the microcontroller. The user has to make the required code and compile it to the Arduino IDE to verify that there are no syntax errors and store it in his profile sketches in the remote laboratory. Consequently, the user has to compile the code and upload it to the microcontroller via the remote laboratory. After this, the user has to check that the code created is doing what it is supposed to do, according to the activity instructions. This is where the assessment starts. How is the instructor informed that the code is correct and aligned with the activity's instructions? One way is to show the teaching assistant the results, but this cannot be performed since we are working remotely and there is not an available assistant. Another way is to send the code to the teaching assistant, which is time-consuming. Another way is to create a report, but also in this case there is no guarantee that the user is the one who performed the experiment or that the user did not copy the report of somebody else. So, the best way is for the code to be assessed automatically by the RL.

The remote lab has three types of assessment that can be implemented. The following subsections will present the HMU-RLP assessment types.

4.1. Actions Assessment

The first one is to monitor the "microactivities", each action that the user performs on the RL, for example, login, logout, create, update, delete, compile, and upload a sketch, and interactions with the experiments, such as turn on/off the lights, the fan, the heating resistance, and change the position of the potentiometer.

Every action performed is logged and compared to the required "microactivities" path of the exercise. This type of assessment shows what exactly the user is doing on the RL. Also, it is very useful in a simulation scenario of an operator. In case there is a need to train an operator, when an alarm occurs indicating what sequence of action must be followed.

All of these "microactivities" are logged via xAPI statements to an LRS for learning analytics. Using xAPI reporting tools, the instructor can obtain conclusions about the user and how they performed in the experiment. For example, the number of experiment attempts, the duration of the experiment session, how many times the code was uploaded, what experiments the user performed, etc.

4.2. Shadow Microcontroller Assessment

The second type is to monitor the experimental microcontroller by another microcontroller called a "shadow" microcontroller. Ideally, the "shadow" microcontroller must be the same as the experimental microcontroller because the pins must be the same in both cases. All pins between the two microcontrollers are connected (exp. Mc A0-sha. Mc A0 ... exp. Mc Pin13-sha. Mc Pin13).

Using the shadow microcontroller, the instructor assesses the activity on a hardware level. The shadow microcontroller monitors what happens on the pins of the experimental microcontroller and reports it to the RPI.

The communication between the RPI and the shadow microcontroller is implemented via the I2C bus. We must mention that for protecting the I2C bus of the RPI, we must use a Bi-Directional Logic Level Converter due to different voltage levels between RPI (3.3 V) and the Arduino UNO (5 V).

There are two ways that we can implement the shadow microcontroller assessment. The "Activity Specific Firmware" and the "General Firmware".

4.2.1. Activity-Specific Firmware

The "shadow" microcontroller is loaded with a specific firmware when an activity is activated. This firmware instructs the shadow microcontroller what to monitor from the experimental microcontroller. If the shadow microcontroller obtains the expected results, it reports to the RPI. The validity of the activity's code is reported by the shadow microcontroller sending "success" or "failure" when it is inquired from the RPI. Then, the RPI sends an xAPI statement of the shadow controller report.

To elaborate on this case, we will present an example. Suppose that the activity instructs the user to create a code that will flash the green LED connected to pin 2 on the experimental microcontroller. As mentioned before, pin 2 of the shadow microcontroller is also connected with pin 2 of the experimental microcontroller (see Figure 4).

The shadow microcontroller is programmed to set pin 2 as the input and monitor the pulses that come to pin 2. The shadow firmware is created and stored from the instructor/creator of the activity (see Figure 5, section C), where the activity creation is enabled only for administrators/instructors.

The user, from their side, is instructed to create a code that flashes the green LED every 1 s. If the code is correct, then the green LED flashes every 1 s and the shadow microcontroller starts counting pulses from its side. Then, the shadow microcontroller reports "success" when it is inquired from the RPI. Now, in case the user made a mistake and instead of flashing the green LED flashes the red, which is connected to pin 4, which is a logical error, then the shadow microcontroller will not count any pulses on pin 2 and then will report "failure" when it is inquired from the RPI. The same result will occur if the user loads a different code, different than the code instructed.

As a second example, we have a scenario where the user is instructed to control the RL environment temperature. The user creates a code that monitors the temperature using pin A5 (see Table 5), and when the temperature exceeds the threshold value, then the user's code must activate the fan connected to pin 11 (see Table 5). The shadow microcontroller firmware will monitor the value of the analog pin A5, and if it exceeds the threshold, pin

11 should be set to HIGH. If these conditions exist, then the shadow microcontroller will set the activity as successful; as long these conditions never occur, then the activity is set as failed.

It must be highlighted that the shadow microcontroller is transparent to the user, that the user does not know of its existence, so it is hidden in the RL. The drawback of this option is that when activating an activity, there is a delay in compiling and uploading the shadow microcontroller firmware, which may confuse the user.

Additionally, it must be clarified that the shadow controller-specific firmware is created by the instructor and creator of the activity. The code is in Arduino programming and it is based on a template of code that at the end sends the report SUCCESS or FAILURE to the RPI when it is requested from the RPI.

Figure 4. Experiment and shadow microcontrollers.

Table 5. Sensors and actuators connected to the experimental microcontroller.

1.	LED	Three LEDs connected to pins 2, 3, and 4
2.	Servomotor	One servo motor connected to pin 5
3.	Buzzer	One buzzer connected to pin 6
4.	Push Buttons	Two push buttons connected to pins 7 and 8
5.	H-Bridge	One HW95 H-Bridge connected on pins 9, 12, and 13
6.	Relay	One relay connected on pin 10, which turns on an LED stripe One relay connected on pin 11, which turns on a fan
7.	LCD Display	One LCD connected to A0 and A1
8.	Potentiometer	One potentiometer connected on pin A3
9.	Photoresistor	One photoresistor connected on pin A4
10.	Temperature sensor LM35	One temperature sensor connected on A5

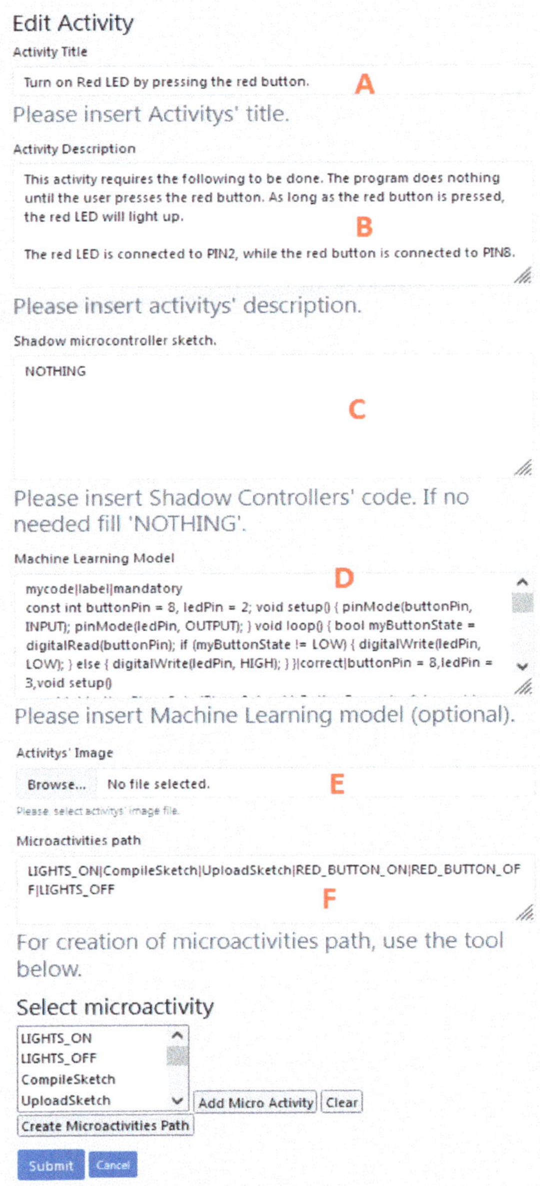

Figure 5. Activity creation/update.

4.2.2. General Firmware

In the case of the general firmware, we upload a general firmware on the shadow microcontroller, which monitors all the pins of the experimental microcontroller. When there is a change in the pin status, the shadow microcontroller sends a report to the RPI of a string with the status of all of the pins (High/Low for digital pins and the value read from analog pins). Then, the RPI sends the pins' status reports as xAPi statements to the

LRS. The instructor, using the xAPI tools and filtering, can view the sequence of pin status in the specific activity and understand that the code is working as it should.

As can be understood, in this case, there is the option for bigger users' learning analytics and reporting. We will use the above two examples to elaborate on this case.

In the first example, where the user is assigned to flash (turn on/off) the LED connected to pin 2, each time that pin 2 changes status, the RPI is triggered and the pin status report is sent. Next, the RPI sends the xAPI statement to the LRS. Then the teacher can filter the users' profile statements for the specific activity and see the status of all of the pins. It will be visible that pin 2 flashes, changing from LOW to HIGH and from HIGH to LOW.

In the second example, where the user has to control the environment temperature, as the temperature increases, the value of pin A5 increases. As long as the value increases, the status of the pins changes, and for each change, the RPI receives the pins' status reports from the shadow microcontroller, which are sent as xAPI statements. When the value of pin A5 reaches the threshold, then pin 11 will be set to HIGH. Filtering the xAPI statements, the instructor will see that the value of pin A5 increases and after the value of A5 exceeds the threshold value, pin 11 will be set to HIGH.

The positive aspect of this method is that the shadow controller is loaded once with the generic firmware and there is no delay in loading the firmware each time an activity is activated. The drawbacks of this method are that there are a large amount of data stored in the LRS and that the instructor must check for the validity of the user's code, which is time-consuming. We are working on automating this process.

The report that the Arduino sends and is received from the RPI is in the following format:

{'digital_pin_statuses': [0, 1, 1, 1, 1, 1, 1, 1, 1, 1, 1, 0], 'analog_pin_values': [1023, 993, 929, 816, 25, 25]}

The drawback in this case is that there are a lot of data sent due to unstable analog pins values, which is why in the code we have implemented a percentage threshold of difference. This method of assessment is a work in progress, and we suggest the use of activity-specific firmware.

4.3. *Artificial Intelligence Assessment*

The third type of assessment in the RL is to use artificial intelligence (AI) to assess the uploaded code. Before we start the presentation, we must highlight that this type of assessment is a broad topic that needs particular research and development. Different algorithms can be implemented, like neural networks, decision trees, etc., or even using Ghat-GPT. For this paper, we created a simple use case for proof of concept, which will be elaborated on in future works.

In this use case, we created an algorithm based on a decision tree. For each activity activated, there are stored defined accepted codes and wrong codes. When the user uploads the required code and submits the activity for assessment, the code is compared with the AI model and replies if it is successful or wrong.

The model is stored during the activity creation, as shown in section D in Figure 5. The model is created by the instructor and can be enriched gradually using users' successful or wrong submitted codes.

In the paper [38], they propose a generic AI-based technique for assessing student performance in conducting online virtual and remotely controlled laboratories. Their suggestion is based on the dynamic use of the mouse, which is different than our suggestion, but has a similarity in breaking an experiment into different stages or user steps, which in our case we call "microactivities".

Based on these three types of assessment, the RL can reply if the code submitted by the user aligns with the scenario of the experiment or not, and reply accordingly, updating the grade book or sending the relevant xAPI statements for the profile of the user.

4.4. Assessment Types Roles

For a better understanding, we must clarify that each assessment type is independent and has a particular role in the assessment process. Also, each activity's assessment type result is logged separately to obtain the final activity grade.

Action assessment: Each activity has a set of paths of actions to be followed. This is used for training purposes in a use case where a scenario has to be followed. For example, a user who receives an alarm or an indication has to perform a specific workflow. In this assessment, it is monitored that the users use the experiment and test it, and we can avoid cases where users log in to the RL and only log their session time.

Shadow microcontroller assessment: The role of the shadow microcontroller is to report the status of the experimental microcontroller. For example, if there is an activity where the user has to log and report some analog values on the experimental microcontroller, who will verify that the values reported are correct and not copied from another report? The comparison of the user's report and the statements of the shadow microcontroller will show the real activity testing.

Artificial intelligence assessment: This is used to verify that the code used by the user is according to the code expected. In this case, if the instructor wants to teach a specific programming method, for example, "while loops", and does not want the use of "for loop", this can be easily traced.

4.5. Activity Creation

In Figure 5, it can be seen how an activity is created in the platform. In the fields from A to F, we can see the information that needs to be filled in by the activity creator.

A. Activity Title
B. Activity Description–User Instructions
C. Shadow Controller Code
D. Machine Learning Models
E. Activity Image
F. Microactivities–User action path

5. Experience API Statements and Tools

In the paper, we have mentioned the experience API (xAPI), formerly known as Tin Can API. It is a specification for learning technology that allows for data collection about a wide range of a person's experiences (both online and offline). Developed by the Advanced Distributed Learning Initiative (ADL), xAPI provides a way to store, manage, and share data about learning experiences in a consistent format and can be used to integrate various learning tools and platforms. It is often seen as a successor to SCORM (Sharable Content Object Reference Model), offering more flexibility and capabilities for tracking learning experiences.

The basic components of xAPI are the following:

1. **Statements**: At the heart of xAPI are "statements" that record what a learner has done. A statement is usually formatted as "I did this" or "[Actor] [Verb] [Object]". For example, "Manos logged in the Remote Lab 1", "Manos activated activity No 1 on Remote Lab 1", or "Manos passed activity No 1 on Remote Lab 1".
2. **Actor**: The individual or group that the statement is about.
3. **Verb**: Describes the action taken by the actor.
4. **Object**: What the action is performed on.
5. **Result**: Additional data about the outcome (optional).
6. **Context**: Additional data to help understand the context in which the action occurred (optional).

xAPI statements can be generated by tools, simulations, quizzes, serious games, or learning environments. Depending on the learning analytics required, the designer decides what kind of user activities need to be stored. The xAPI statements are stored in and retrieved from the learning record store (LRS).

The project uses an LRS, which is a learning locker [39] hosted in the site https://lrs.nile.hmu.gr/ (accessed on 28 March 2024) and provided by the Natural Interactive Learning Games and Environments Lab (NILE) of the HMU. While the learning locker primarily serves as an LRS, it also offers some basic reporting and visualization features. Users can build custom dashboards and reports using the built-in reporting engine or by integrating with external visualization tools (see Figures 6 and 7).

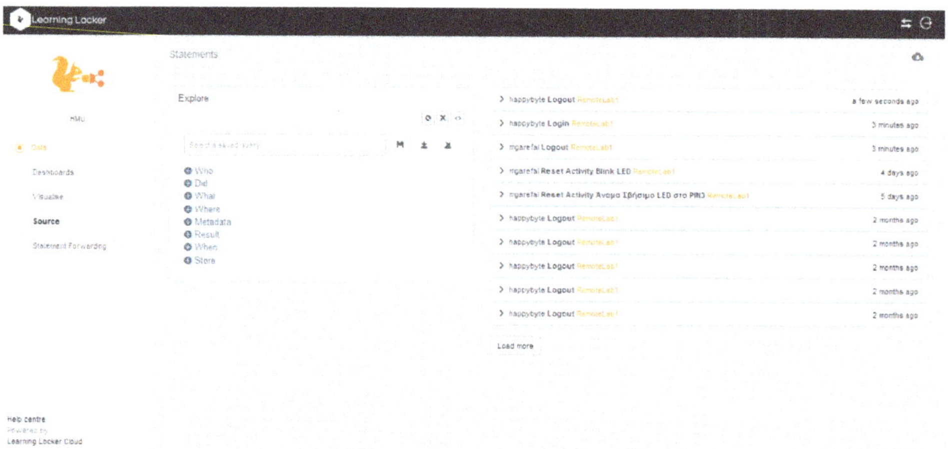

Figure 6. Learning locker reporting features.

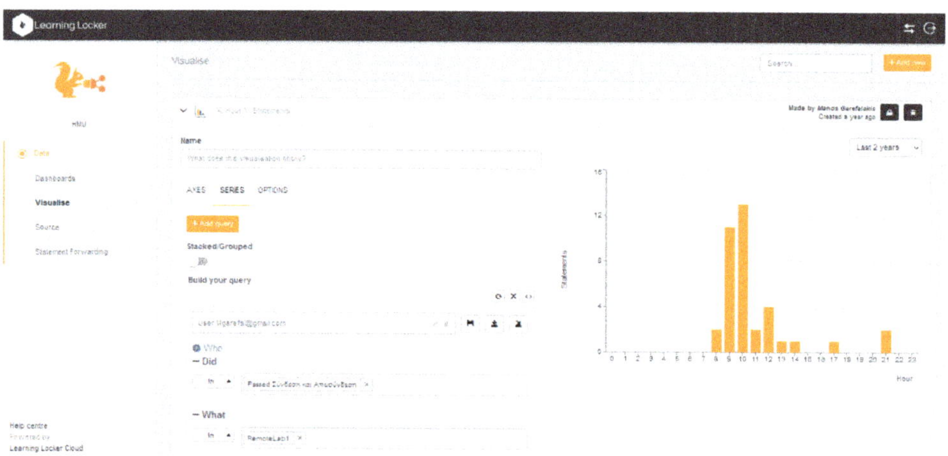

Figure 7. Learning locker visualization features.

During our research, we also found other commercial LRS reporting and visualizing tools, like Watershed LRS, Rustici LRS, etc.

Using the experience works of the HMU in the papers [40,41], we created a use case webpage (https://garefalakis.eu/xAPI-Dashboard/examples/verbs2.html (accessed on 28 March 2024)) where a graph of the statements stored in the LRS of the RL can be viewed (Figure 8).

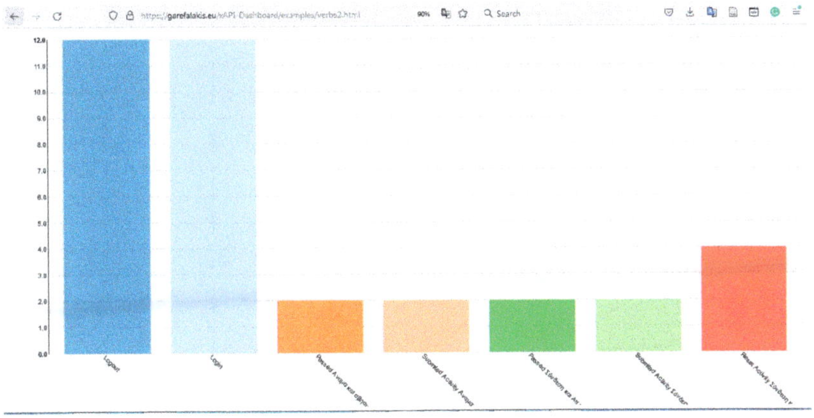

Figure 8. xAPI graph representation.

Additionally, the xAPI statement viewer can be seen in the webpage (https://garefalakis.eu/xAPI/1.0/original_prototypes/StatementViewer/ (accessed on 28 March 2024)), which can show all of the statements stored in the LRS (Figure 9). The two web pages were created for demonstration purposes and will be developed shortly on the project's site (https://iot.hmu.gr (accessed on 28 March 2024)).

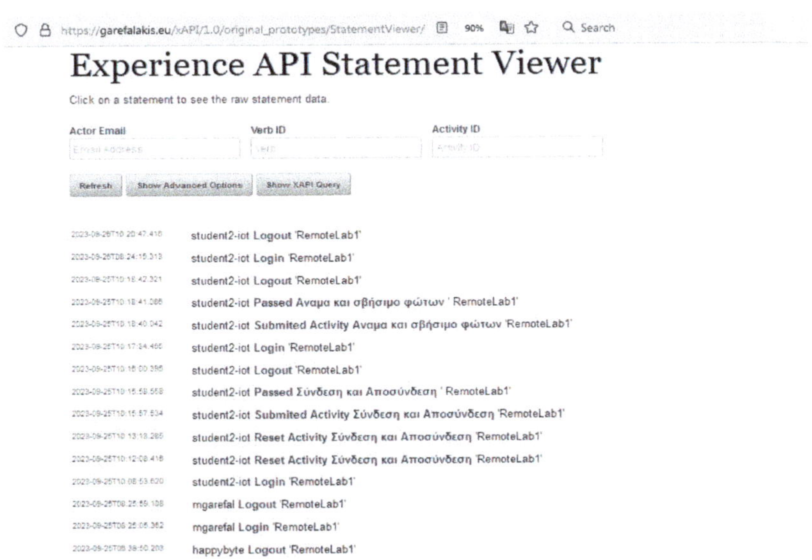

Figure 9. xAPI statement viewer.

6. Remote Lab Experiments

To give RL flexibility, the experiment controller is connected to many sensors and actuators. This gives the flexibility to create different activities and scenarios for the users' training. In Table 5, the list of electronic parts and the pins they are connected to can be seen. In Figure 10, is depicted the schematic, of all the components connected to the experiment controller, and in Figure 11, are depicted all the components connected.

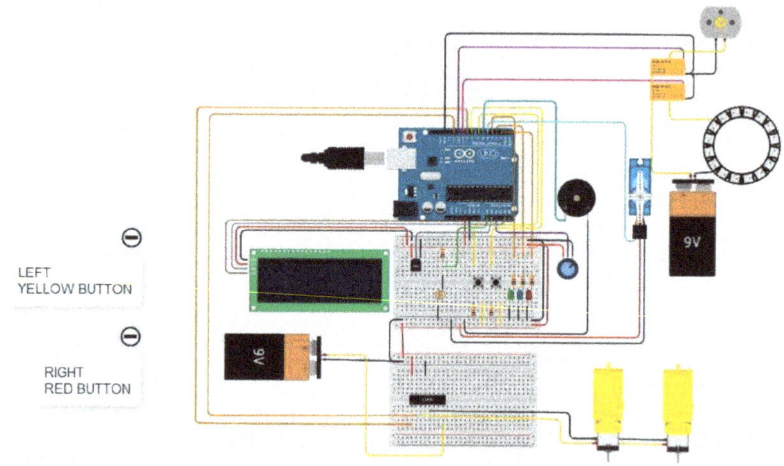

Figure 10. Tinkercad sensors and actuators connected to the experimental microcontroller.

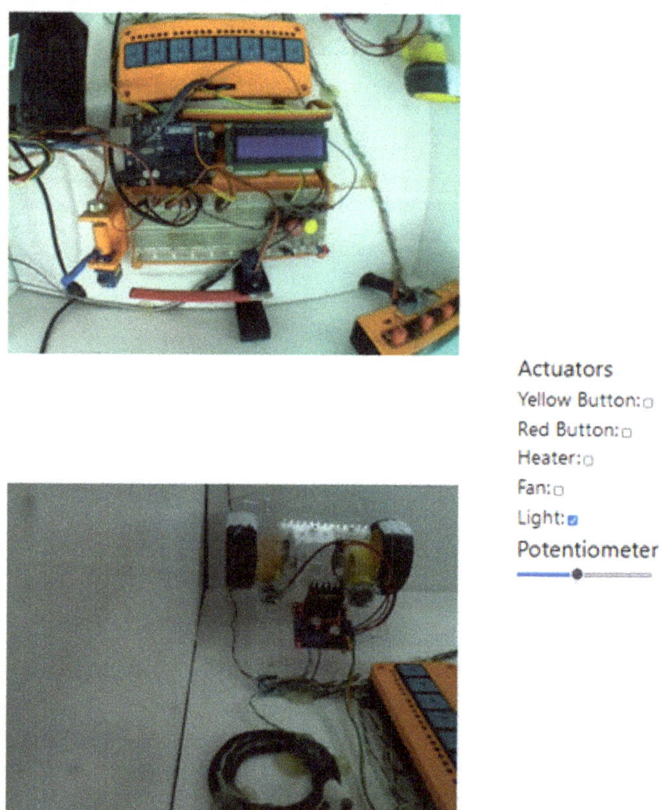

Figure 11. Experimental microcontrollers with connected sensors and actuators.

Apart from the components connected to the experimental microcontroller, other parts are connected to the RPI. These parts are activated by the user from the user interface. The user, by using these parts, changes some conditions in the RL, and the sensors sense the changes. For example, an experiment is to write a code that the microcontroller monitors when the environmental light of the RL is turned off and the microcontroller turns on the LED stripe connected to pin 10. The list of these parts is given in Table 6.

Table 6. External parts.

Relay 1–Activates RL lights
Relay 2–Pushes red button
Relay 3–Pushes yellow button
Relay 4–Activates heating resistor
Relay 5–Activates fan
Relay 6–Future use
Relay 7–Future use
Relay 8–Future use

7. Discussion and Conclusions

In this paper, we presented the related literature on the subject of remote laboratories, specifically on the RLs that aim to teach microcontroller programming of the Arduino family. The literature revealed the remote laboratory management systems (RLMSs) that exist and host a great number of such RLs, some of which are Labsland, GO-LAB, and RELLE. Also in this review, it was found that the RLP we have created in the Sensor Network LAB of the HMU shares the same architecture with a great number of RLs found in the literature. The different features of the HMU-RLP that were not found in the literature relate to the fact that, in our platform, the assessment of the experiments is not performed only by the users. In most RL cases, the user uploads his code to the microcontroller and then monitors the results on the streaming video and also interacts with the RL to verify that the code is doing what it is supposed to do.

In the HMU-RLP, we have introduced three types of user assessment: The first type monitors each action users perform over the web page offered by the RL. The second type monitors the activities of users at the hardware level. To this end, a shadow microcontroller is used that monitors the pins of the microcontroller programmed by the users. The third type automatically assesses the code uploaded by the users, checking its similarity with the prototype code uploaded by the instructors. A trained AI model is used to this end. For the assessments provided by the HMU-RLP, the experience API (xAPI) standard is exploited to store users' learning analytics (LAs). The LAs can be processed by the instructors for the students' evaluation and personalized learning. These assessment types work together toward the pervasive and supervised learning with which the HMU-RLP project intends to comply.

Future work planned for development of the HMU-RLP includes the following topics:

- We intend to further develop the features of our platform that exploit xAPI statements with users' learning analytics data to create personalized learning paths according to the adaptive and pervasive learning paradigm. We intend to follow a hybrid approach so that this process works both manually, under the supervision of the instructors with decision support offered by our system, and automatedly, so users are automatically assessed and tutored by the system.
- Further development of the AI type of assessment offered by our platform for automatically checking user coding. To this end, we plan to fine-tune a pretrained open-source large language model (e.g., Llama2) to assist the user with the coding actions that must be followed for a specific activity or provide feedback on a sketch that is not aligned with the activity scenario.

- The literature revealed that many RLs are hosted in remote laboratory management systems (RLMSs). Although the HMU-RLP can currently accommodate several RLs, it cannot be considered an RLMS. Such an option will help in RL sharing and dissemination via the HMU-RLP.
- Creation of an RL that will be used in teaching microcontroller programming for the Internet of Things. There will be two remote laboratories, one of which will have an experimental microcontroller connected with sensors and actuators, like ESP32, that will communicate with an MQTT server, and one software RL that will host a Node-RED server where the user will develop an application that will interact with the MQTT server and will display in the user interface controls and charts.
- Thorough evaluation and testing of the HMU-RLP into real training and learning environments so we can scale its readiness level from an experimental proof-of-concept platform to a production-ready toolkit. The HMU-RLP has already been presented to teachers of Greek secondary education and we will soon have their opinions and evaluation. Next, it is planned for the teachers to use the HMU-RLP in their classes for teaching Arduino and IoT programming, and students will be able to further evaluate the HMU-RLP.
- The implementation of the H/W configuration of the RL by the user remotely, as is described in the paper [20]. The user will be able to use more components connected to the Arduino board by switching and enabling different connections to new circuits remotely, using relay matrixes.
- Implementation of augmented and virtual reality applications for the user to see the experiments working and interact with them using AR and VR technology.

Author Contributions: Conceptualization, M.G., Z.K. and S.P.; methodology, M.G. and Z.K.; validation, M.G. and Z.K.; formal analysis, M.G. and Z.K.; investigation, M.G. and Z.K.; resources, M.G. and Z.K.; data curation, M.G. and Z.K.; writing—original draft preparation, M.G.; writing—review and editing, M.G. and Z.K.; visualization, M.G. and Z.K. All authors have read and agreed to the published version of the manuscript.

Funding: This research received no external funding.

Institutional Review Board Statement: Not applicable.

Informed Consent Statement: Not applicable.

Data Availability Statement: Data are contained within the article.

Conflicts of Interest: The authors declare no conflict of interest.

References

1. 1876–2019—IEEE Standard for Networked Smart Learning Objects for Online Laboratories | IEEE Standard | IEEE Xplore. Available online: https://ieeexplore.ieee.org/document/8723446 (accessed on 30 August 2023).
2. Zutin, D.G.; Auer, M.E.; Maier, C.; Niederstätter, M. Lab2go—A repository to locate educational online laboratories. In Proceedings of the IEEE EDUCON 2010 Conference, Madrid, Spain, 14–16 April 2010; pp. 1741–1746. [CrossRef]
3. Tinkercad—From Mind to Design in Minutes. Tinkercad. Available online: https://www.tinkercad.com/ (accessed on 13 March 2024).
4. Panagiotakis, S.; Karampidis, K.; Garefalakis, M.; Tsironi-Lamari, A.; Rallis, I.; Kamarianakis, Z.; Papadourakis, G. Remote Arduino Labs for Teaching Microcontrollers and Internet of Things Programming. In Proceedings of the 2022 31st Annual Conference of the European Association for Education in Electrical and Information Engineering (EAEEIE), Coimbra, Portugal, 29 June–1 July 2022.
5. Garefalakis, M.; Panagiotakis, S. Integration of a Remote Lab with a Learning System for training on Microcontrollers' programming. In Proceedings of the 27th PanHellenic Conference on Progress in Computing and Informatics—PCI 2023, Lamia, Greece, 24–26 November 2023.
6. Villar-Martinez, A.; Ortiz-de-Zarate, L.; Rodriguez-Gil, L.; Hernandez-Jayo, U.; Garcia-Zubia, J.; Angulo, I.; Terkowsky, C.; Ortelt, T.R.; Wilkesmann, U.; Nowak, R.; et al. LabsLand Electronics Laboratory: Distributed, Scalable and Reliable Remote Laboratory for Teaching Electronics. In *Open Science in Engineering*; Auer, M.E., Langmann, R., Tsiatsos, T., Eds.; Lecture Notes in Networks and Systems; Springer Nature: Cham, Switzerland, 2023; Volume 763, pp. 261–272. [CrossRef]

7. de Zarate, L.O.; Angulo, I.; Villar-Martínez, A.; Rodriguez-Gil, L.; García-Zubía, J. Remote Laboratory for the Development of Customized Low-Power Computing and IoT Systems. *Lect. Notes Netw. Syst.* **2023**, *763*, 249–260. [CrossRef]
8. Villar-Martinez, A.; Rodriguez-Gil, L.; Ortiz-de-Zarate, L.; Hussein, R.; Orduña, P. ARM Distributed and Scalable Remote Laboratory for Texas Instruments Launchpad Boards. *Lect. Notes Netw. Syst.* **2023**, *763*, 177–186. [CrossRef]
9. Buitrago, P.A.; Camacho, R.; Pérez, H.E.; Jaramillo, O.; Villar-Martinez, A.; Rodríguez-Gil, L.; Orduna, P. Mobile Arduino Robot Programming Using a Remote Laboratory in UNAD: Pedagogic and Technical Aspects: Experience Using a Remote Mobile Robotics Laboratory at UNAD. *Adv. Intell. Syst. Comput.* **2021**, *1231*, 171–183. [CrossRef]
10. Del Villar, I.; Rodriguez-Gil, L.; Orduña, P. Learning CAN Bus Communication with a Remote Laboratory. 2022. Available online: https://ieeexplore.ieee.org/abstract/document/9766633/ (accessed on 28 March 2024).
11. Sapeha, A.; Zlatkova, A.; Poposka, M.; Donchevski, F.; Karpov, K.B.; Todorov, Z.; Efnusheva, D.; Kokolanski, Z.; Sarjas, A.; Gleich, D.; et al. Learning Management Systems as a Platform for Deployment of Remote and Virtual Laboratory Environments. 2022. Available online: https://repo.bibliothek.uni-halle.de/handle/1981185920/78898 (accessed on 28 March 2024).
12. da Silva, R.C.; de Magalhães Netto, J.F.; Lopes, A.M.M.; de Menezes, M.F.; Menezes, R.A. ERPLab: Remote Laboratory for Teaching Robotics and Programming. 2023. Available online: https://ieeexplore.ieee.org/abstract/document/10343379/ (accessed on 28 March 2024).
13. Oballe-Peinado, Ó.; Castellanos-Ramos, J.; Sánchez-Durán, J.A.; Navas-González, R.; Daza-Márquez, A.; Botín-Córdoba, J.A. Fpga-Based Remote Laboratory for Digital Electronics. 2020. Available online: https://ieeexplore.ieee.org/abstract/document/9163676/ (accessed on 28 March 2024).
14. Navas-González, R.; Oballe-Peinado, Ó.; Castellanos-Ramos, J.; Rosas-Cervantes, D.; Sánchez-Durán, J.A. Practice Projects for an FPGA-Based Remote Laboratory to Teach and Learn Digital Electronics. *Information* **2023**, *14*, 558. [CrossRef]
15. Bukovac, A.; Pleše, E.; Maravić, U.; Petrović, P.; Jagušt, T. Teaching Programming and Microcontrollers with an Arduino Remote Laboratory Application. 2023. Available online: https://ieeexplore.ieee.org/abstract/document/10159730/ (accessed on 28 March 2024).
16. Martin, S.; Fernandez-Pacheco, A.; Ruipérez-Valiente, J.A.; Carro, G.; Castro, M. Remote experimentation through Arduino-based Remote Laboratories. *IEEE Rev. Iberoam. De Tecnol. Del Aprendiz.* **2021**, *16*, 180–186. [CrossRef]
17. Terauds, M.; Smolaninovs, V. Remote Laboratory for Microcontroller Programming Course. 2022. Available online: https://ieeexplore.ieee.org/abstract/document/9978868/ (accessed on 28 March 2024).
18. Seničić, Đ.; Matijević, M.; Tanasković, M.; De La Torre, L. An Implementation of a Web Laboratory Converting Off-Line Experiments into Remotely Accessible Experiments. In Proceedings of the Sinteza 2022—International Scientific Conference on Information Technology and Data Related Research, Belgrade, Serbia, 16 April 2022. [CrossRef]
19. Domski, W. Remote Laboratory Offered as Hardware-as-a-Service Infrastructure. *Electronics* **2022**, *11*, 1568. [CrossRef]
20. Scaffidi, C.; Distefano, S. A Remotely Configurable Hardware/Software Architecture for a Distance IoT Lab. 2021. Available online: https://ieeexplore.ieee.org/abstract/document/9556236/ (accessed on 28 March 2024).
21. Tokarz, K.; Czekalski, P.; Drabik, G.; Paduch, J.; Distefano, S.; Di Pietro, R.; Merlino, G.; Scaffidi, C.; Sell, R.; Kuaban, G.S. Internet of Things Network Infrastructure for the Educational Purpose. 2020. Available online: https://ieeexplore.ieee.org/abstract/document/9274040/ (accessed on 28 March 2024).
22. Admin. IOT-OPEN.EU: Introduction to the IoT Practical Projects in English—IOT-Open. Available online: https://iot-open.eu/download/iot-open-eu-introduction-to-the-iot-practical-projects-in-english/ (accessed on 23 January 2024).
23. Costa, R.; Pérola, F.; Felgueiras, C. µLAB A Remote Laboratory to Teach and Learn the ATmega328p µC. 2020. Available online: https://ieeexplore.ieee.org/abstract/document/9125336/ (accessed on 28 March 2024).
24. Da Silva, J.B.; De Oliveira, G.; Da Silva, I.N.; Mafra, P.M.; Meister, S.; Bilessimo, S. Block. Ino: Remote lab for programming teaching and learning. *Int. J. Adv. Eng. Res. Sci.* **2020**, *7*, 41–47. [CrossRef]
25. Platunov, A.; Kluchev, A.; Pinkevich, V.; Kluchev, V.; Kolchurin, M. Training Laboratories with Online Access on the ITMO. cLAB Platform. 2020. Available online: http://ceur-ws.org/Vol-2893/paper_12.pdf (accessed on 28 March 2024).
26. Jo, H.S.; Jo, R.S. Design and development of remote laboratory system to facilitate online learning in hardware programming subjects. In Proceedings of the 2020 13th International UNIMAS Engineering Conference (EnCon) 2020, Kota Samarahan, Malaysia, 27–28 October 2020. [CrossRef]
27. Fernández-Pacheco, A.; Martin, S.; Castro, M. Implementation of an Arduino remote laboratory with raspberry Pi. In Proceedings of the 2019 IEEE Global Engineering Education Conference (EDUCON), Dubai, United Arab Emirates, 8–11 April 2019; pp. 1415–1418. Available online: https://ieeexplore.ieee.org/abstract/document/8725030/ (accessed on 23 January 2024).
28. Villar-Martinez, A.; Rodriguez-Gil, L.; Angulo, I.; Orduña, P.; García-Zubía, J.; López-De-Ipiña, D. Improving the scalability and replicability of embedded systems remote laboratories through a cost-effective architecture. *IEEE Access* **2019**, *7*, 164164–164185. [CrossRef]
29. HMU-RLP Helenic Mediterranean University Remote Laboratory Platform. Available online: http://rlp.hmu.gr:5000/ (accessed on 11 March 2024).
30. LabsLand—Home. Available online: https://labsland.com/en (accessed on 28 February 2024).
31. The REMOCLEC Project. Available online: https://remoclec.eu/ (accessed on 13 March 2024).
32. The Remote Hub Lab. Remote Hub Lab. Available online: https://rhlab.ece.uw.edu/ (accessed on 14 March 2024).
33. Home | Golabz. Available online: https://www.golabz.eu/ (accessed on 28 February 2024).

34. Inicio | Laboratorio Remoto de Electrónica Digital. Available online: https://fpga-lab.uma.es/ (accessed on 15 March 2024).
35. Home. IOT-Open. Available online: https://iot-open.eu/ (accessed on 16 March 2024).
36. de Lima, J.P.C.; Carlos, L.M.; Simão, J.P.S.; Pereira, J.; Mafra, P.M.; da Silva, J.B. Design and implementation of a remote lab for teaching programming and robotics. *IFAC-PapersOnLine* **2016**, *49*, 86–91. [CrossRef]
37. Labs | RELLE—Remote Labs Learning Environment. Available online: http://relle.ufsc.br/ (accessed on 28 February 2024).
38. Abd El-Haleem, A.M.; Eid, M.M.; Elmesalawy, M.M.; Hosny, H.A.H. A Generic AI-Based Technique for Assessing Student Performance in Conducting Online Virtual and Remote Controlled Laboratories. Available online: https://ieeexplore.ieee.org/abstract/document/9973300/ (accessed on 28 March 2024).
39. "Home", Learning Locker. Available online: https://www.learninglocker.co.uk/ (accessed on 28 March 2024).
40. Arvaniti, D. Tracking learning with Experience API. March 2023. Available online: https://apothesis.lib.hmu.gr/handle/20.500.12688/10503 (accessed on 7 February 2024).
41. Papadokostaki, K. Ubiquitous learning with Experience API. December 2017. Available online: https://apothesis.lib.hmu.gr/handle/20.500.12688/8505 (accessed on 7 February 2024).

information

Article

An Edge Device Framework in SEMAR IoT Application Server Platform

Yohanes Yohanie Fridelin Panduman [1], Nobuo Funabiki [1,*], Sho Ito [1], Radhiatul Husna [1], Minoru Kuribayashi [1], Mitsuhiro Okayasu [1], Junya Shimazu [1] and Sritrusta Sukaridhoto [2]

[1] Graduate School of Natural Science and Technology, Okayama University, Okayama 700-8530, Japan; p8f01q6f@s.okayama-u.ac.jp (Y.Y.F.P.); p3ti3fqh@s.okayama-u.ac.jp (S.I.); pwmn7i7q@s.okayama-u.ac.jp (R.H.); kminoru@okayama-u.ac.jp (M.K.); mitsuhiro.okayasu@utoronto.ca (M.O.); p5835ic4@s.okayama-u.ac.jp (J.S.)

[2] Department of Informatic and Computer, Politeknik Elektronika Negeri Surabaya, Surabaya 60111, Indonesia; dhoto@pens.ac.id

* Correspondence: funabiki@okayama-u.ac.jp

Abstract: Nowadays, the *Internet of Things (IoT)* has become widely used at various places and for various applications. To facilitate this trend, we have developed the IoT application server platform called *SEMAR (Smart Environmental Monitoring and Analytical in Real-Time)*, which offers standard features for collecting, displaying, and analyzing sensor data. An *edge device* is usually installed to connect sensors with the server, where the interface configuration, the data processing, the communication protocol, and the transmission interval need to be defined by the user. In this paper, we proposed an *edge device framework* for *SEMAR* to remotely optimize the edge device utilization with three phases. In the *initialization phase*, it automatically downloads the configuration file to the device through *HTTP* communications. In the *service phase*, it converts data from various sensors into the standard data format and sends it to the server periodically. In the *update phase*, it remotely updates the configuration through *MQTT* communications. For evaluations, we applied the proposal to the *fingerprint-based indoor localization system (FILS15.4)* and the *data logging system*. The results confirm the effectiveness in utilizing *SEMAR* to develop IoT application systems.

Keywords: Internet of Things; edge device; framework; application server platform; SEMAR

Citation: Panduman, Y.Y.F.; Funabiki, N.; Ito, S.; Husna R.; Kuribayashi, M.; Okayasu, M.; Shimazu, J.; Sukaridhoto, S. An Edge Device Framework in SEMAR IoT Application Server Platform. *Information* **2023**, *14*, 312. https://doi.org/10.3390/info14060312

Academic Editors: Spyros Panagiotakis and Evangelos K. Markakis

Received: 17 April 2023
Revised: 26 May 2023
Accepted: 27 May 2023
Published: 29 May 2023

1. Introduction

Currently, the *Internet of Things (IoT)* is receiving much attention from both industries and academics as an emerging technology that uses the Internet infrastructure to connect physical worlds to cyberspaces [1]. The IoT application infrastructure is continuously being extended to become more ubiquitous around the world and is composed of numerous physical devices distributed across multiple domains [2]. In this context, the success of an IoT application system depends on the ability to collect, manage, and analyze the data easily and flexibly, as well as to distribute it to users and other systems efficiently [3,4]. Nowadays, the amount of data generated by sensor devices is increasing rapidly with the availability of diverse network connectivity and various protocol services; IoT application system developers should design and build these systems considering standardizations with heterogeneous device management.

In an IoT application system, *edge computing* is often adopted to bring computing capabilities for data processing to locations closer to sensors or target devices [5]. Some IoT applications may require low latency and real-time data processing, which cloud servers cannot provide [6,7]. Due to the diversity of sensor resources, the introduction of *edge computing devices* has become a valuable solution to reducing the computational complexity of data processing in cloud servers [8]. Edge computing devices enable various functions at the edges of networks before sending data to the server and can increase the efficiency of data processing [9]. It also offers the data conversion capability to convert raw data to the standard data format. It is expected that the *edge device framework* was introduced to facilitate application developments in edge computing devices [10]. The framework interacts with devices in the physical world that may change over time [11]. Therefore, it should support the dynamic development of edge systems.

Recently, cloud-based solutions have been widely used for IoT application systems [12]. Instead of focusing on the implementation details, the prepared tools allow developers to focus on the implementation of logic by using functions that efficiently support the design and implementation of IoT applications [13]. However, most of the existing cloud-based solutions did not support effective and efficient developments at the edge devices level, and their technologies have often limited the interoperability with third parties.

Previously, we designed and implemented the IoT application server platform as a cloud-based solution for integrating various IoT application systems, called *SEMAR (Smart Environmental Monitoring and Analytical in Real-Time)* [14]. *SEMAR* provides standard features for collecting, displaying, processing, and analyzing sensor data from different domains. It offers *built-in* functions for data synchronizations, aggregations, and classifications with machine learning in *Big Data* environments, and *plug-in* functions for allowing other systems to access the data through the *Representational State Transfer Application Programming Interface (REST API)*.

Unfortunately, the current implementation of *SEMAR* does not facilitate deployments and implementations of *edge devices* within the context of IoT ecosystem application deployments. As an effective IoT application server platform, *SEMAR* should be able to control and manage various IoT devices remotely. It must be capable of reconfiguring IoT devices to improve their performance and utilization.

In this paper, we proposed an *edge device framework* and its implementation for *SEMAR* to facilitate the development of edge devices for IoT applications. As a popular edge device, the *Raspberry Pi* was selected for this implementation, and the image was created in the *SEMAR* server. This framework can remotely optimize the utilization of this edge device by configuring the connectivity of sensor interfaces, a data conversion approach, a data model, transmitted data, local data storage, local visualization, and the data transmission interval on the server. Actually, it provides features for downloading configuration files to the devices using *HTTP* communications, converting data from diverse sensor resources into standard data formats before delivering them to *SEMAR*, processing data using rules and filter functions, offering multiple output components for utilizing the acquired data, and enabling remote configuration updates using *Message Queue Telemetry Transport (MQTT)* services [15].

For evaluations of the proposal, we applied the edge device framework to the *fingerprint-based indoor localization system (FILS15.4)* [16,17] and the *data logging system*. These integrated systems were deployed in #1 and #2 Engineering Buildings at Okayama University, Japan. In addition, we evaluated the effectiveness of the edge device framework by investigating its computing performance and comparing it with similar research works. The results confirm the feasibility of utilizing the edge device framework in developing IoT application systems with *SEMAR*.

The rest of this paper is organized as follows: Section 2 presents related works. Section 3 describes the IoT application system architecture. Section 4 briefly reviews our previous works on *SEMAR*. Section 5 presents the design and implementation of the edge device framework. Sections 6 and 7 briefly describe the implementation in two IoT application systems. Section 8 presents comprehensive performance evaluations and a comparative analysis with similar related work. Finally, Section 9 concludes this paper with future works.

2. Related Works

In [18], Mahmood et al. presented a simulation of an edge computing implementation for resource allocation in IoT applications for smart cities. The result shows the effectiveness of the edge computing layer in reducing the energy and computational resources for IoT networks.

In [19], Sarangi et al. proposed IoT applications for digital farming by using a microcontroller that connects soil moisture sensors with the mobile system as edge gateways. The edge device captures and transmits sensor data to the mobile system through Wi-Fi communications. Then, the mobile system processes the data and sends them to the cloud server. This approach presented the utilization of the mobile system for collecting and processing information at edges to reduce computational processes at the cloud level.

In [20], Oueida et al. proposed an integration of the edge computing device and the cloud service in the smart healthcare system. Edge computing was used to gather information from smart devices, process it to obtain the necessary data, and transmit it to the cloud server. The proposed system was suitable for emergency departments and other types of queuing systems.

In [21], Mach et al. proposed the concept of the mobile edge computing, which enables IoT applications to perform massive data processing at the device level. However, developers should consider three key aspects, namely, the computation decision, the resource allocation for computational processes, and the mobility management. This approach can reduce the latency of the network in IoT application systems.

In [22], Yousafzai et al. introduced a light-effect migration-based paradigm for managing computational offloading in edge networks in mobile edge computing. They investigated the impacts of edge networks on IoT applications. The evaluation results showed that the execution time for data processing and the amount of transmitted data should be considered to optimize the utilization of edge devices.

In [13], Berta et al. proposed a general end-to-end IoT platform that is composed of the cloud-based service for managing sensor data and devices of IoT applications called *Measurify*, and the tool for facilitating the construction of edge devices called *Edgine*. *Edgine* requests the local configuration and executable scripts. Then, it collects data from the sensors, processes them using downloadable scripts and stores it in the cloud. The proposed system has been installed and used for several IoT application systems. The results demonstrated the efficiency of the system by enabling developers to focus on application requirements and design decisions to define the edge system rather than on implementations.

In [23], Yang et al. proposed an edge computing framework suitable for IoT device development. This framework provides functions to configure the module hardware security, the data conversion, control, and communication to the server. It also offers advanced data processing capabilities at the edge computing level, including rule engines, data analysis, and application integration. By accessing the cloud service, this framework allows users to update the configuration through *MQTT* communications. This approach is similar to our method for updating the configuration remotely.

In [24], Kim et al. proposed plug-and-play in IoT platforms, using a web page to manage IoT devices. They utilized *Arduino* boards as edge devices that were connected to the sensors and actuators. The proposed system allows configuring the device for data collection or control actions by accessing the platform website. The implementation results indicate that the system was able to reduce the deployment complexity and increase the IoT environment dynamicity. However, they only considered the device layer and did not address the data visualization and analysis at the cloud level.

In [25], Iera et al. introduced the *Social Internet of Things (SIoT)* architecture paradigm. This architecture comprises IoT applications in objects that are registered on a social networking platform, where each object collaborates and interacts with other objects to provide specialized services. The architecture includes three elements: objects, gateway, and an *SIoT* server. Each component may consist of three layers: sensing, network, and application. It enables IoT objects to conduct high-computational processes, in contrast to only the server performing these tasks. As a common IoT architecture, the network layer is only used to connect the server and the objects. However, this architecture allows the integration between IoT objects and provides interfaces for IoT objects and humans through network layers. Thus, it provides the development of IoT applications that interact with one another. This architecture can be considered a reference with which to improve the design of the IoT application system architecture proposed in this paper.

In [26], Cauteruccio et al. proposed the *Multi-Internet of Things (MIoT)* architecture to improve object communication in the *SIoT* architecture. In the *SIoT* architecture, IoT objects connect and collaborate with one another. It makes the complexity of data transfer increase. Thus, MIoT architecture solves this issue by considering data-driven and semantics-based aspects of data exchange between objects. Unfortunately, the proposed communication model is not suitable for dynamic IoT application scenarios, where IoT devices are dynamically added and removed.

3. Design of The IoT Application System Architecture

3.1. System Overview

In this section, we describe the design of the IoT application system architecture for generalization. Currently, there are many IoT architecture references that can be considered for developing IoT application systems. However, each IoT application system has unique designs and requirements. The common IoT application system architecture consists of three layers. The *perception layer* represents the physical devices for sensing and actuating that interact with the environment. The *network layer* represents the transport layer for data communications between layers. The *application layer* represents the application software to offer specific services for data processing [27]. There are many IoT application system architectures that need to be addressed to enhance the development of IoT applications and platforms.

In [28], Lombardi et al. presented commonly used IoT architectures such as *cloud-based* architecture, *edge-computing-based* architecture, and *Social Internet of Things (SIoT)* architecture. *Cloud-based* architecture utilizes services deployed on a cloud server to generate, process, and visualize large amounts of data for users. This architecture allows users and other services to access data at any time. *Edge-computing-based* architecture offers computational services close to the device layer by offering data processing, storage, and control capabilities. It is frequently used for industrial devices and IoT application systems that demand a quick response as a result of data processing.

In *SIoT* architecture, IoT applications are comprised of objects registered on a social networking platform, where each object collaborates and interacts with other objects to provide specific services [25]. This architecture enables IoT objects to conduct high-computational processes, as opposed to only the server performing these tasks. It enables the development of IoT applications that interact with one another. In addition, the *MIoT* architecture has been added to the *SIoT* architecture. In order to reduce the complexity of the *SIoT* architecture system, the *MIoT* architecture considers data-driven and semantics-based aspects for data exchange between objects [26].

In this paper, the concept of the IoT application system architecture was based on these references. Figure 1 illustrates the proposed architecture. It is composed of the *sensors and actuators*, *edge*, and *cloud* layers.

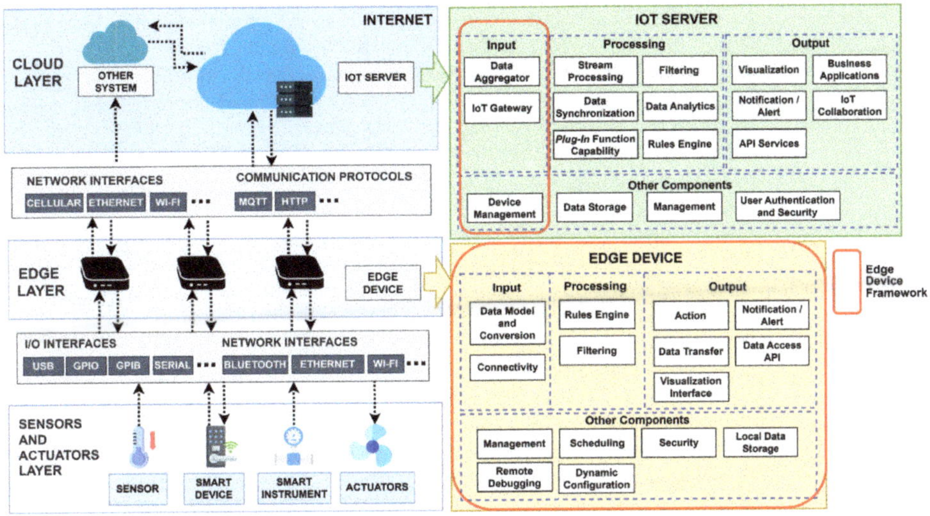

Figure 1. Design overview of general IoT application system architecture.

3.2. *Sensor and Actuators Layer*

In the context of the IoT application system, perception devices as IoT objects are sensors and actuators connected to a controller. Sensors are primarily used to monitor the environment by converting physical parameters into measurable electrical quantities (often voltage), while actuators provide physical actions when presented with an electrical quantity. However, with the rapid development of technologies, Internet-connected devices have become common and diverse in their application purposes.

For instance, in smart homes, developers have often utilized smart devices to improve living experiences and reduce energy consumption. These smart devices are controlled by smartphones and are integrated with cloud services through wireless networks.

The *Industrial Internet of Things (IIoT)* has been presented to connect IoT technologies to industrial machines or instruments to analyze the obtained data and optimize existing industrial processes [29]. It uses smart instrument devices for automatic data collection to enhance the condition monitoring of industrial instruments. Recently, industrial devices in the market have contained features to enable Internet-based data access to central operation management systems through Ethernet and wireless technology. In this paper, we considered smart devices and smart instruments as components in the sensor and actuator layers of the proposed architecture.

3.3. *Edge Layer*

The *edge* layer addresses the issue of the growing data volume in an IoT application system by utilizing computing capabilities of edge devices. In this section, we explain the components of the edge device—*input*, *processing*, *output*, and *other components*.

3.3.1. Input Components

Input components should consider the connectivity of IoT devices and the method for collecting valuable data from them. The connectivity component refers to the *input/output (I/O)* and the network interfaces of the IoT device for data communications. Currently, a single-board computer, such as *Raspberry Pi*, has enabled various interfaces to accept data from a variety of devices. Among them, *General Purpose Input Output (GPIO)* is the standard interface for receiving and sending commands to/from IoT sensors and actuators. *General Purpose Interface Bus (GPIB)* is the I/O interface included in the IEEE-488 standard for industrial instrumentation data. While *GPIO* only transmits data in signals, *GPIB* is able to handle both text data and numeric expressions.

In the context of IoT data communications, serial communication protocols are often used to transfer data among IoT devices. Each device may support different serial interfaces based on its hardware specifications. These include the *RS-232 protocol*, *Universal Serial Bus (USB)*, the *Serial Peripheral Interface (SPI)*, the *Universal Asynchronous Receiver Transmitter (UART)*, and the *Inter-Integrated Circuit (I2C)*. It is necessary to build an edge system that is able to handle different interfaces.

Various network interfaces, including Bluetooth, Ethernet, and Wi-Fi, have been introduced to connect IoT objects and edge computing devices. *Bluetooth* is widely applicable in smart devices due to its capability of low-power communications. *Ethernet* provides stability and security by wired connectivity. However, it is difficult to communicate over long distances. In IoT application systems, *IEEE 802.11 wireless LAN (Wi-Fi)* is the most popular network interface used by current smart devices and smart instruments.

Sensor devices usually generate data in different and non-standard formats. It is challenging to enable the interoperability among sensors from different companies that have different communication technologies. Therefore, the edge device requires the data conversion component to generate data in the standard format from various sensor devices. This component represents the translation process of sensor data. It requires a data model to define the valuable data structures of sensor data that are used for further processing in the edge system. *JavaScript Object Notation (JSON)* format data are frequently used for this purpose.

3.3.2. Processing Components

As an extension of cloud services, edge computing has similar characteristics to cloud computing. Edge computing is able to perform local data processing with minimal computational resources. Processing components in the edge layer are designed to optimize data collections and enable immediate analysis and decision-making. The filtering and the rules engine are included in these components. The *filtering* component reduces data noise and inaccuracies by applying digital filters to sensor data. Several sensors, such as the accelerometer and the gyroscope, may produce noisy data. It is necessary to reduce noises before transmitting data to a cloud server.

The *rules engine* component makes data-driven decisions in real-time. It applies various output services when rule patterns are matched. They include delivering notification messages to users and issuing action commands to actuators. The rules contain basic operations in the format of "if the specific conditions are fulfilled, then trigger the specific actions" or defined as *IF-THIS-THEN-THAT* form—for example, in an IoT application system for smart homes, "if the temperature is higher than 30 °C, then turn on the air conditioner". The rules engine in the edge layer can reduce the time required to generate the response action, compared to waiting for the server response. However, it should avoid complex rule models due to the limited computational resources of edge devices.

3.3.3. Output Components

The output components concern the ability of edge devices to utilize the collected data and transmit it to the cloud server or other systems. Several output components, such as the visualization interface, notification/alert, data transfer, trigger action operations, and data access API, should be considered for this purpose. The *visualization interface* component provides web-based user interfaces to monitor IoT data at the edge continuously. The *notification/alert* component communicates with users through email or push notification services.

The *data transfer* component represents the ability of the edge device to send data across different networks to its cloud server or other systems. Network interfaces of the edge device and communication protocols need to be considered. The edge device, such as *Raspberry Pi*, has enabled diverse network interfaces. Wi-Fi, Ethernet, and 5G cellular are standard network interfaces used to connect edge devices to cloud servers. Communication protocol services consist of the *publish–subscribe* and *request–response* messaging models. *MQTT* communication is the most popular *publish–subscribe* protocol for IoT application systems. It can operate on an edge device with limited processing power and memory. *HTTP* communication is often used for the *request–response* messaging model. In addition, the standardization format of data transfer should be addressed for this component. In this case, the JSON format is utilized.

The *action* component consists of functions that send commands to actuators through connectivity interfaces. Due to the complexity of action functions becoming more diverse, it should be able to execute different action functions in parallel or sequentially. The *data access API* is another output component that should be considered in the edge layer. It provides a function to allow external systems to access local data through *HTTP* communication, which is relevant to the current IoT trends of cross-vendor capabilities and interoperability. Thus, it enables the development of complex IoT systems that utilize multiple vendor services simultaneously.

3.3.4. Other Components

For developing the edge device, we should consider additional components that are not included in the *input*, *processing*, and *output* components. These components are management, scheduling, security, local data storage, remote debugging, and dynamic configuration. The *management* component controls and monitors the lifecycle of the edge device. The *scheduling* component controls the time cycle for executing data streams in the edge device. The *security* component provides privacy and security capabilities of the edge device.

When sensor data cannot be transmitted to the server, the system must provide the reliable local data storage service to archive sensor data records. The local data storage component should consider the battery consumption, latency, and CPU utilization. The lightweight embedded database engine, such as *SQLite*, can be the suitable database option with which to develop this component.

Currently, the edge management system provides dynamic configuration capabilities. It allows users to modify edge system parameter settings by changing environments. Parameter settings include connected sensors and actuators, data processing methods, and data transmission services. However, this component may cause problems and errors if the configuration does not match the current environment of the edge device. Therefore, the remote debugging component will be the solution. It allows running and verifying the new device configuration without affecting the existing system running on the edge device.

3.4. Cloud Layer

The cloud layer components are responsible for processing, analyzing, managing, storing, and visualizing IoT data using cloud-based services. These components perform computations that are not feasible on edge devices. In this paper, we present the cloud layer components in Figure 1. We organized them into *input*, *processing*, *output*, and *other* components.

The input components provide the services to receive sensor data from different devices using different communication protocols. It consists of the IoT gateway and the data aggregator. The components contain a variety of data processing functions for IoT data stream processing, filtering, rules engine, data synchronization, and analytics, with plug-in function capabilities, where each function should be implemented as a standalone one to prevent system failures. The output part concerns the ability of the cloud system to provide capabilities for users or other systems to access IoT data. The output components may include visualization functions, notification/alert functions, REST API services, business application integrations, and IoT collaboration capabilities. The other components provide additional components that will support the main services of the cloud server.

They include management, data storage, device management, user authentication, and security components.

Figure 1 shows the components of the cloud layer in the edge device framework. It consists of the IoT gateway, the data aggregation, and the device management component. The *IoT gateway* provides communication protocol services such as *HTTP* and *MQTT* to receive data from edge devices. The *MQTT* broker service was implemented to enable *MQTT* communication. The *REST API* service was developed for accepting sensor data through *HTTP POST* communications. Additionally, the IoT gateway component should consider potential utilizations of communication protocols provided by other cloud service providers.

The data input process at the cloud layer usually starts when the *IoT gateway* receives sensor data. It will be followed by *data aggregation*. Then, data will be forwarded to data processing functions and be stored in the data storage. The *data aggregation* component collects data from several data sources, applies data processing, and reassembles data in a usable format.

In this paper, we emphasized the importance of the *device management* component in the development of the edge device framework. The component manages the devices in the cloud system. It identifies device specifications, such as sensors that are connected to the edge device, and handles the integration between edge devices and the cloud server. It allows the dynamic configuration component of the edge to be triggered remotely from the cloud server. The device management data are stored in cloud server data storage.

4. *SEMAR* IoT Application Server Platform

In this section, we introduce *SEMAR* as an IoT application server platform to facilitate the development of a cloud layer system. In previous studies, we designed and implemented the *SEMAR* IoT application server platform in consideration of the cloud layer for the general IoT architecture described in Section 3. The current implementation of *SEMAR* has been used in several IoT application systems [30]. It provides the integration functions of collecting, displaying, processing, and analyzing sensor data, including *built-in* and *plug-in* functions. Figure 2 shows the system overview of the *SEMAR*.

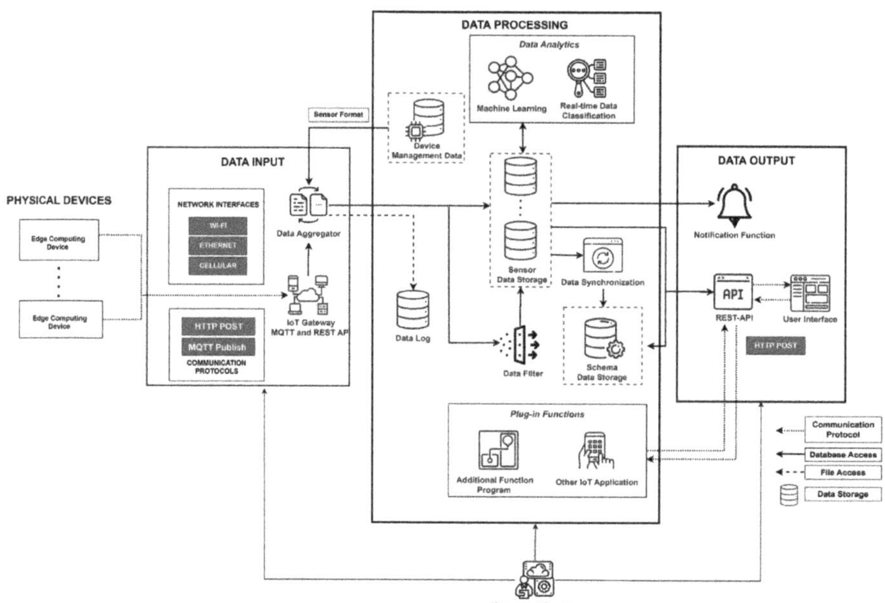

Figure 2. Design overview of *SEMAR* IoT application server platform.

The *built-in* function allows the use of new functions without implementing or modifying the original source codes. The components of the *built-in* function are grouped according to *data input, data processing,* and *data output* that are controlled by the *management* system.

The *data input* provides components for gathering sensor data from various IoT resources that accept connection through network interfaces and communication protocols. It consists of the IoT cloud gateway for communication services through *HTTP POST* and *MQTT* communication protocols, and the *data aggregator* for gathering and processing sensor data with the consumable format based on the sensor format stored in the device management data. It transmits the results to the *data processing* component and stores them in the *MongoDB* data storage [31].

The *data processing* components consist of the *data filter* for reducing noises and inaccuracies in the data obtained, the *data synchronization* for synchronizing the data from different devices and storing it in the dynamic database called the *schema data storage*, and the *data analytics* for analyzing large amounts of data.

The last component employs machine learning techniques and real-time data classification services. The machine learning techniques enable the user to construct a data model for the real-time data categorization feature using sample data from the data storage. In addition, the *SEMAR* IoT application server platform enables *plug-in* functions that can be implemented as system extensions or as the other IoT application systems to access the data through *REST API* services.

The *SEMAR* IoT application server platform includes several output components. Users can access the sensor and synchronized data through the user interface based on a website. It enables the data export function to download sensor data in CSV, JSON, Excel, or text format at a specific time by accessing the user interfaces. The notification function enables the user to set the threshold for each sensor data point as the message notification trigger. If the value fulfills the threshold, the system will generate and send an alert to users. In accordance with the current trend of IoT platforms, we implemented capabilities that enable IoT collaboration, which allows for the connections and integrations of other systems. We used the *REST API* service for data integrations and exchanges through *HTTP POST* communications using the JSON format. The *REST API* retrieved data from storage and translated it into the JSON format.

The *management service* has been implemented in the *SEMAR* to manage user authentications, devices data, and its communication protocol. In this paper, we improved the device management feature by adding the function to create, update, and delete the edge configuration file for the edge device. Users are able to operate and monitor the device remotely. Users can access this service through the user interface.

The procedure for integrating the *SEMAR* platform with a new IoT application system is described as follows:

- The user registers the devices and the sensors of the IoT application system on the *SEMAR* platform;
- The system prepares the IoT cloud gateway services, including the *HTTP POST* and *MQTT* communication protocols, to receive data;
- The device sends data to the server through the defined communication service in JSON format;
- The data are received by the IoT cloud gateway, processed by the *data aggregator* based on the registered sensors, and are stored in the *sensor data storage*.
- The *SEMAR* provides the capability of synchronizing data from several devices by accessing the *sensor data storage* and storing the results in the *schema data storage*;
- The user interface of *SEMAR* displays the data. The user can integrate their programs as *plug-in* functions by utilizing the *REST API*.

5. Design and Implementation of Edge Device Framework

In this section, we present the design and implementation of the edge device framework.

5.1. System Overview

The following section presents the edge device framework as a collection of tools that will make it easier to create edge computing systems. Figure 3 provides the overview of the integrated system of the edge device framework in *SEMAR*. It functions in three phases. In the *initialization phase*, it offers web services that enable the automatic downloading of the configuration file to the device via *HTTP* communications. In the *service phase*, it transforms data from various sensors into the standard data format and periodically transmits them to the server. In the *update phase*, it remotely updates the configuration through *MQTT* communications.

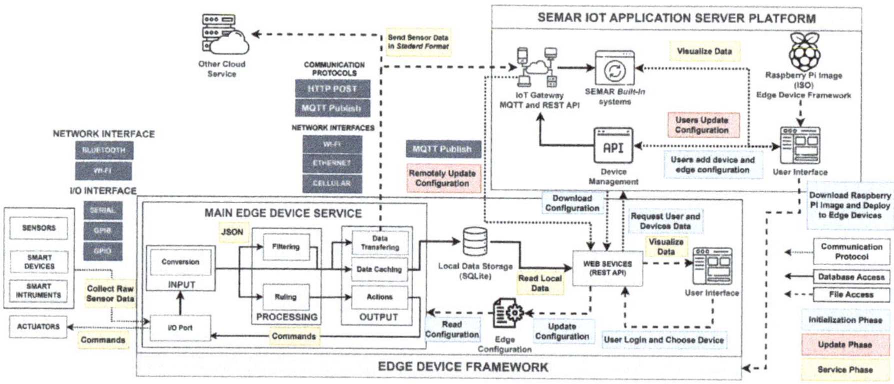

Figure 3. Design overview of the edge device Framework.

5.2. Initialization Phase

In the *initialization* phase, the framework is installed on the edge device, and the initial connection is established between the edge device and the *SEMAR* platform. First, the user registers a new device and configures the edge device on the *SEMAR* platform via the user interface. Then, the user downloads the *Raspberry Pi* image from the *SEMAR* platform and deploys it to the edge devices. The user needs to ensure that the devices are connected to the Internet. Next, the user accesses the web services of the edge device framework through the user interface. The system verifies the user account by accessing the *REST API* services of the *SEMAR* platform. If the user account is authenticated, the system retrieves all the device data of the user from the *SEMAR* platform, generates the *edge ID* of the device, and grants the access to the web services.

In the *initialization* phase, the user needs to choose the data to be applied to the edge device from the user interface. Then, the system downloads the edge configuration, saves it to the JSON file, and runs the *main service* program. Algorithm 1 illustrates the process flow of this program for both the *initialization* and *update* phases. Figure 4 shows the sample edge configuration file used in the framework. It includes the device, the device identity, and the configuration parameters such as the sensor interface, the data conversion method, the data model, transmitted data, the local data storage, and the local visualization. The required libraries to run the system have been installed in the edge device framework.

Algorithm 1 Edge configuration service.

```
Input  : Edge ID (edgeID)
Output : Edge configuration file (EdgeConfig)
begin
    Set EdgeConfig ← read EdgeConfig from the "config.json"
    if EdgeConfig not NULL then
        Run Main Service program(EdgeConfig)
        Connect to the MQTT broker in SEMAR
        Subscribe for the "edgeID" MQTT topic
        while true do
            if Message ← receive data from server through MQTT communication then
                Set EdgeConfig ← convert Message to JSON format
                Save EdgeConfig to the "config.json"
                Restart Main Service program(EdgeConfig)
            end
        end
    end
end
```


```json
{
    "device_code": "ih37",
    "configuration_code": "cn37",
    "resource": "Data Logger [GL240]",
    "interface": [{
        "type": "wlan",
        "config":{...},
        "method": "web_scrapping",
        ...
        "object_used": {"CH 1": "CH 1", "CH 2": "CH 2"}
    }],
    "data_transmitted":{"ch_1": "CH 1","ch_2": "CH 2"},
    "time_interval": 5,
    "communication_protocol": {"mqtt": {"server": "103.106.72.181", "port": "1883", "topic": "sensor/logger"}},
    "local_data": { "ch_1": ["CH 1", "real"],"ch_2": ["CH 2", "real"]},
    "visualization": {
        "table": ["ch_1", "ch_2"],
        "graph": [{"value": [{"title": "Channel 1","field": "ch_1"},{"title": "Channel 2","field": "ch_2"}}]
    }
}
```


Figure 4. A sample of the edge configuration file in JSON format.

5.3. *Service Phase*

In the *service phase*, which is the primary phase of the edge device framework, the framework collects and transmits sensor data to *SEMAR*. Figure 3 illustrates the lifecycle of the edge device framework for this purpose. Based on the general IoT application architecture illustrated in Figure 1, the functions of the main edge framework services are classified into *data input*, *data processing*, and *data output*. Algorithm 2 describes the program flow. To collect the raw sensor data, the edge device must be connected to the sensor or device. The service program then reads the edge configuration file, which was downloaded by the edge configuration services. The program can process the raw sensor data by converting them to the standard data format, reducing inaccuracies in the data using the *filtering* function, generating the decisions based on predefined rule models using the *ruling* function, saving it to local data storage, and sending it to the server in JSON format using a defined communication protocol. The communication protocol can be either *MQTT* or *HTTP POST*. The *SEMAR* platform receives, processes, and analyzes the sensor data using built-in systems on the server and displays the sensor data as output in the user interface. Additionally, the system can send notifications/alerts to the user and trigger actuators based on the rule model results. The program runs periodically at specific intervals and only transmits the sensor item values defined in the configuration file. Therefore, the framework enables the user to manage edge devices and optimize their performance by defining edge configuration files.

Algorithm 2 Service phase.

Input : Edge configuration($EdgeConfig$)
begin
  Set $TimeInterval$, $CommService$, $Interface$, $TransmitData$, $FilterModel$, $RuleModels$, $LocalData$, $ActionModels$ ← read the configuration of time interval, communication service, resource interface, transmitted data from $EdgeConfig$
  Set $SensorResource$ ← connect to the network interface of sensor device($Interface$)
  while *true* **do**
    Set $RawSensor$ ← read raw data of sensor from $SensorResource$
    Set $ConvertData$ ← convert raw data of sensor to the standard format($RawSensor$, $Interface$)
    if *FilterModel not empty* **then**
      Set $ConvertData$ ← procces sensor data using digital filter($ConvertData$,$FilterModel$)
    end
    if *RuleModels not empty* **then**
      Set $RullingResults$ ← applying rule models($ConvertData$,$RuleModels$)
    end
    Save sensor data to the local storage($ConvertData$,$LocalData$)
    Set $Data$ ← select transmitted sensor data($ConvertData$, $TransmitData$)
    Send transmitted data to the server through communication service ($Data$, $CommService$)
    if *RullingResults not empty* **then**
      Send commands to control actuators($RullingResults$,$ActionModels$)
    end
    sleep($TimeInterval$)
  end
end

One difficulty in inputting data into the edge device framework involves the connectivity of the sensor interface. The aim of the edge device framework is to create a versatile edge computing device that can automatically gather and transmit sensor data to the server. Therefore, it is essential to establish connectivity services and data models that can support multiple sensors. Currently, our system can capture and transform sensor data through the GPIO, USB serial, and wireless interfaces. We have created multiple functions with which to collect data from the GPIO interfaces. To use the system, the user must first specify the GPIO ports and modes in the configuration file. Then, the system periodically reads the port value, converts it into a JSON object based on the configuration file, and returns the results to the data processing components.

To use the USB serial interface, the user needs to specify the serial port, the timeout time, and the baud rate that determines the data transmission speed. The user also needs to define the delimiter that the system will use to extract the relevant information when it receives a line of serial communication data. Algorithm 3 shows the data conversion process for serial communications.

Algorithm 3 Data conversion procedure for serial communication.

Input : Raw sensor data (*RawSensor*), Edge configuration(*EdgeConfig*)
Output: Converted sensor data (*ConvertData*)

```
begin
    Set Delimeter, ObjectUsed ← read configuration of delimiter and object used, from
    EdgeConfig
    Initialize ConvertedData, Result ← empty JSON object
    Set DataList ← SPLIT(RawSensor,Delimeter[0])
    for each item in DataList do
        Set Buffer ← SPLIT(item,Delimeter[1])
        Set Result[Buffer[0]] ← Buffer[1]
    end
    for each sensor in ObjectUsed do
        if sensor in Result then
            Set ConvertData[sensor] ← Result[sensor]
        end
    end
    return ConvertData
end
```

To use the wireless interface, the user needs to provide the URL of the web service to receive the HTML data through *HTTP GET* communications. The web scraping technique is used to extract the necessary information from the HTML data and to transform it into an array format. The user needs to define the index array that includes the channel name and sensor value. The data conversion process for the wireless interface data is shown in Algorithm 4, which illustrates the data conversion procedure for the wireless interface data.

Algorithm 4 Data conversion procedure for wireless interface.

Input : Raw sensor data in HTML format (*RawSensor*), Edge configuration(*EdgeConfig*)
Output: Converted sensor data (*ConvertData*)

```
begin
    Set ChannelIndex, ValueIndex, MaxSequence, ObjectUsed ← read configuration from
    EdgeConfig
    Initialize ConvertedData, Result ← empty JSON object
    Set DataList ← WEBSCRAPING(RawSensor)
    for i ← 0 to length(DataList) do
        if i % MaxSequence == ChannelIndex then
            Set ChannelName ← DataList[i]
        end
        if i % MaxSequence == ValueIndex then
            Set SensorValue ← DataList[i]
        end
        if i % MaxSequence == ( Maxsequnce - 1 ) then
            Set Result[ChannelName] ← SensorValue
        end
    end
    for each sensor in ObjectUsed do
        if sensor in Result then
            Set ConvertData[sensor] ← Result[sensor]
        end
    end
    return ConvertData
end
```

The data transfer system has been implemented to enable the transmissions of sensor data to not only the *SEMAR* platform but also to any other IoT gateway service the user prefers for the cross-vendor capability in edge computing. It currently supports *HTTP POST* and *MQTT* communications using the standard JSON format. The data transfer function uses the *"time_interval"* configuration to regulate the data transfer frequency, the *"data_transmitted"* configuration to determine the output data to be transferred, and the *"communication_protocol"* configuration to describe the destination and communication service. While developing edge devices, the communication network is the critical factor for avoiding the unsuccessful data transfer. The data caching function is implemented by using SQLite and Python to store sensor data locally, with the *"local_data"* configuration specifying which data are saved in the local data storage.

The current implementation allows the user to visualize data in the forms of tables and graphs. It is accomplished using the *"visualization"* setting, which retrieves sensor data from an *SQLite* database. The user can access these data through the web interface or the *REST API* service. To make IoT application system developments more flexible, we suggest the use of the *REST API* service at the edge layer to integrate edge device frameworks with other systems.

5.4. Update Phase

In the *update phase*, the user has the ability to remotely modify the edge configuration file on the edge device using the *SEMAR* user interface. This process involves modifying the edge configuration and utilizing the deploy button to initiate the remote update function. The *device management* service transmits the updated edge configuration in the JSON format to the relevant edge device using *MQTT* communications with the edge ID as the topic. The edge configuration service connects to the *MQTT* broker within *SEMAR* and subscribes to the same topic with the edge ID. After receiving the new edge configuration through *MQTT* communications, the service saves it in the designated folder and triggers the function to restart the service program. As a result, the user can easily add new sensor devices or modify device configurations by making adjustments through the user interface. Figure 5 illustrates the flow process of the *update phase*.

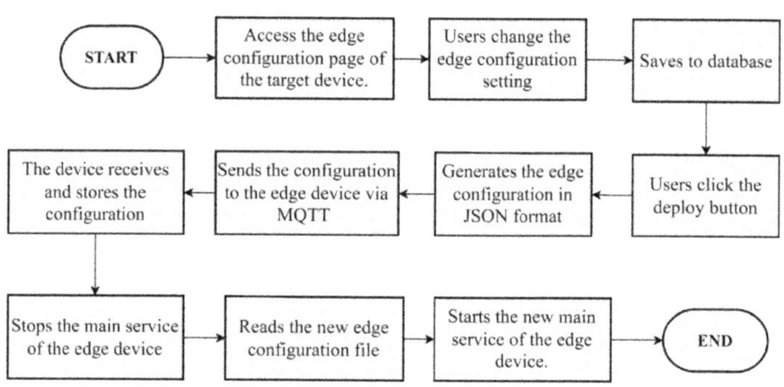

Figure 5. Flow diagram of *update phase*.

6. Application for Fingerprint-Based Indoor Localization System

As the first application, we integrated the *FILS15.4* into the *SEMAR* IoT application platform [30]. This system is used to detect the user locations in indoor environments based on the fingerprints of the target location. The procedure consists of a *calibration phase* and a *detection phase* [16,17].

6.1. System Architecture

Figure 6 illustrates the overview of the *FILS15.4* architecture. The *FILS15.4* system utilizes the transmitting and receiving devices produced by *Mono Wireless* that operate on the *IEEE802.15.4* standard at 2.4 GHz [32]. The transmitter *Twelite 2525* has the dimensions of 2.5 × 2.5 cm, and is powered for a long time by a coin battery. The receiver *Mono Stick* is connected to the *Raspberry Pi* through a USB connection.

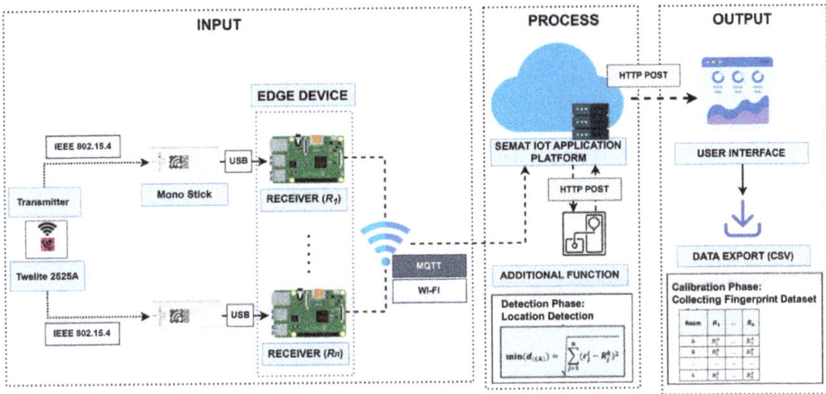

Figure 6. System overview of *FILS15.4*.

Raspberry Pi collects the data from a transmitter by receiving data at the *Mono Stick* through USB serial communications. It determines the *link quality indication (LQI)* for each transmitter and sends the data consisting of the LQI value and the transmitter ID to the server through the *MQTT* communication protocol. The server receives the data, synchronizes the data from all the receivers by calculating the average LQI with the same transmitter ID, and stores the results in one record in the database.

6.2. Calibration Phase

The *calibration phase* produces and records the fingerprint dataset. Each fingerprint consists of n LQI values, where n represents the number of receivers. It indicates the features of LQI values when a transmitter is placed at the specified location (room in *FILS15.4*).

6.3. Detection Phase

The *detection phase* identifies the current location of the transmitters by measuring the *Euclidean distance* between the current LQI values from the receivers and the fingerprint dataset for each room stored in the database and selecting the fingerprint with the shortest distance.

6.4. Evaluation of Implementation

The implemented edge device framework for *FILS15.4* was deployed on two floors in the #2 Engineering Building at Okayama University for evaluations. Our evaluations intended to verify the adaptability and the validity of the edge device framework in *SEMAR*. Table 1 presents the device and software specifications for this evaluation.

Table 1. Device and software specifications of *FILS15.4*.

Components	Items	Specifications
Edge Device	Model Operating System	Raspberry Pi 4B Linux Raspbian
Sensor Device	Model Sensor Interface Communication Method Collected Data	Twelite Mono Stick USB Serial Communication *id, lqi, accelero x, y and z*

We evaluated the ability of the edge device framework to automatically install the edge configuration built on *SEMAR* to the edge device, collect sensor data, convert them, and send them to the server by following the configuration file. In addition, we evaluated the configuration update feature by modifying the edge configuration setting and remotely deploying it to the edge device through the user interface of *SEMAR*.

Figure 7 shows the initial configuration file for *FILS15.4*. The interface includes the configuration of the serial communication for collecting data from a USB receiver and the parameter for obtaining the necessary data by converting them to the standard format. According to the edge configuration, the system sends sensor data that consist of *ID*, *LQI*, and *accelerometer* x, y, z, to the server at every 0.5 s (500 ms) through the *MQTT* communication.

```
{   "resource": "USB Mono Stick",
    "interface": [{
        "type": "usb_serial",
        "config": {"port": "/dev/ttyUSB0","baudrate": 115200,"timeout": 1},
        "string_pattern": "rc=[rc-value]:lq=[lq-value]:ct=[ct-value]:ed=[ed-value]:id=[id-value]:ba=[ba-value]
        :a1=[a1-value]:a2=[a2-value]:x=[x-value]:y=[y-value]:z=[z-value]",
        "delimeter": [":", "="],
        "object_used": {"id": "id","lq": "lq","x": "x","y": "y","z": "z"}
    }],
    "data_transmitted": {"id": "id","lqi": "lq","x": "x","y": "y","z": "z"},
    "time_interval": 0.5,
    "communication_protocol": {"mqtt": {"server": "103.106.72.181","port": "1883","topic": "sensor/edge/fils"}}
}
```

Figure 7. Edge configuration for receiver device of *FILS15.4*.

Figure 8 illustrates the updated edge configuration for *FILS15.4*. It is changed from the initial configuration. The configuration was modified by removing the accelerometer data from the result of the data converter process, and only transmitting *ID* and *LQI* data to the server. The data transmission interval is similar to the previous configuration.

```
{   "resource": "USB Mono Stick",
    "interface": [{
        ....
        "object_used": {"id": "id","lq": "lq"}
    }],
    "data_transmitted": {"id": "id","lqi": "lq"},
    ...
}
```

Figure 8. Updated edge configuration for receiver device of *FILS15.4*.

Figure 9 shows the data visualization of *FILS15.4* through the *SEMAR* user interface. The initial configuration part indicates that the edge device can collect data from the USB receiver, convert them, and send them to the server by following the initial configuration in Figure 7. The updated configuration part represents the edge device when it collects, processes, and transmits data by following the updated configuration in Figure 8.

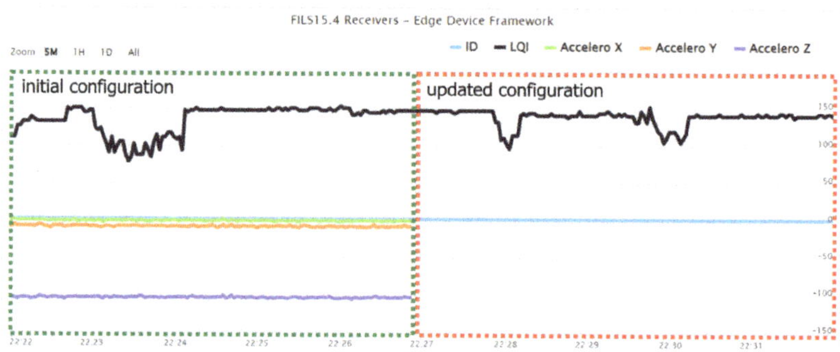

Figure 9. Data visualization of the *FILS15.4* receiver device.

7. Application for Data Logging System

As the second application, the data logging system is integrated to enable real-time monitoring of the temperature data of some materials during the quenching heat treatment process.

7.1. System Overview

Figure 10 illustrates the overview of the *data logging system* architecture. This system uses *midi Logger GL240* with *WLAN B-568* that is provided by *Graphtec* [33] to capture the temperature data during the quenching heat treatment process by attaching the sensor to the material. The treatment process is used for hardening steel by putting the material into the heater machine to improve metal performances. *WLAN B-568* provides the *HTML* web service for displaying the data collected by the data logger. The integration of the data logger with the IoT application server platform is as follows:

- The edge device for the data logging system captures raw sensor data in the HTML format by accessing the data logger web services through wireless communications;
- It reads the input HTML data, extracts the temperature value using web scraping techniques, and transforms it into JSON format;
- It transmits the JSON data to the *SEMAR* platform through the *MQTT* communication protocol;
- The *SEMAR* platform receives, processes, and saves the sensor data in the database;
- The *SEMAR* platform displays the sensor data through the user interfaces.

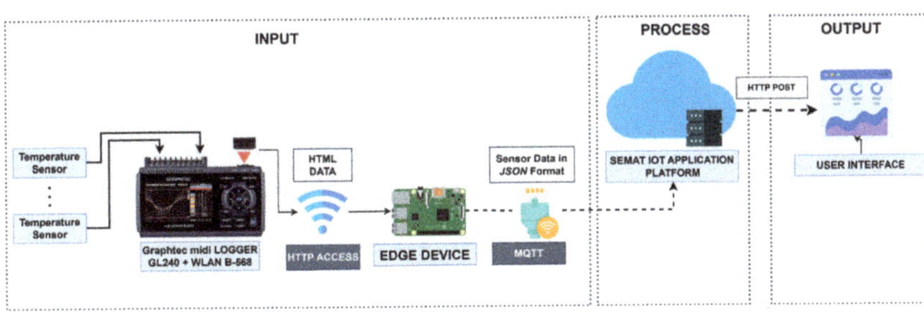

Figure 10. System overview of *data logging system*.

7.2. Evaluation of Implementation

We evaluated the implementation of the edge device framework for the *data logging system* by running it in the #1 Engineering Building at Okayama University. Our evaluations intended to verify the adaptability and the validity of the edge device framework in *SEMAR*. Table 2 presents the device and software specifications for this evaluation.

Table 2. Device and software specifications for *data logging system*.

Components	Items	Specifications
Edge Device	Model	Raspberry Pi 4B
	Operating System	Linux Raspbian
Sensor Device	Model	midi Logger GL240 and WLAN B-568
	Sensor Interface	Wireless Connection
	Wireless LAN Mode	Access Point
	Wireless LAN IP	*192.168.230.1*
	Web Services URL	http://192.168.230.1/digital.cgi?chgrp=0 (accessed on 19 December 2022)
	Communication Method	*HTTP* Communication
	Collected Data	*temperature*

Figure 11 shows the initial configuration file for the *data logging system*. The edge configuration indicates that the edge device collects data from the data logger through the wireless network. It transmits the measured temperature data from *channels* 1 and 5 to the server every five seconds through the *MQTT* communication.

```
{   "resource": "Data Logger",
    "interface": [{
        "type": "wlan",
        "config": {"url": "http://192.168.230.1/digital.cgi?chgrp=0","timeout": 8},
        "method": "web_scrapping",
        "index_name": 0,
        "index_value": 1,
        "max_sequence": 3,
        "object_used": {"ch_1": "CH 1","ch_2": "CH 2","ch_3": "CH 3","ch_4": "CH 4","ch_5": "CH 5","ch_6": "CH
        6","ch_7": "CH 7","ch_8": "CH 8","ch_9": "CH 9","ch_10": "CH 10"
        }
    }],
    "data_transmitted": {"ch_1": "ch_1","ch_5": "ch_5"},
    "time_interval": 5,
    "communication_protocol": {"mqtt": {"server": "103.106.72.181","port": "1883","topic": "sensor/logger2"}}
}
```

Figure 11. Edge configuration for edge device in the data logging system.

Figure 12 shows the updated edge configuration of the data logger monitoring system. It was modified from the initial configuration. In this configuration, the transmitted data were changed by only sending the temperature data from *channel*1 every 2 s through the *MQTT* communication.

```
{   "resource": "Data Logger",
    "interface": ....,
    "data_transmitted": {"ch_1": "ch_1"},
    "time_interval": 2,
    "communication_protocol": ....
}
```

Figure 12. Updated edge configuration for edge device of data logging system.

Figure 13 illustrates the data visualization of the data logging system. The initial configuration part represents the edge device for collecting, processing, and transmitting data according to the initial configuration in Figure 11. Additionally, the updated configuration part shows the data sent by the edge device when the configuration is modified according to Figure 12.

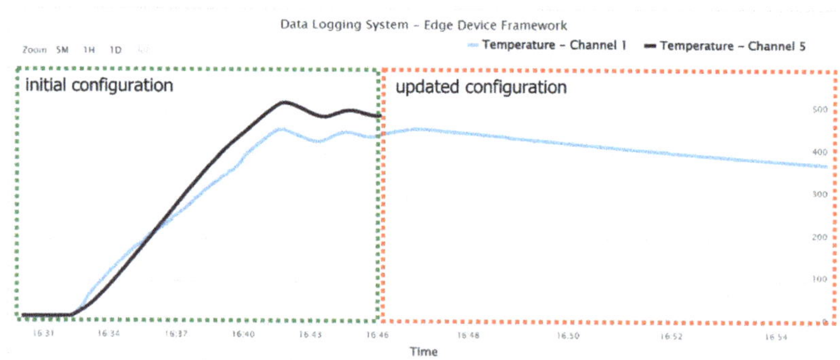

Figure 13. Data visualization of the data logging system.

8. Evaluations

In this section, we evaluated the implementation of the *SEMAR* IoT server platform.

8.1. Performance Analysis

The first evaluation of the edge device framework's performance involved investigating the average CPU and memory usage of the main service program while collecting and transmitting sensor data at various time intervals. This evaluation was crucial for assessing the computational performance of the framework during the main phase. To carry out this evaluation, we employed the data logging system application and measured the average memory and CPU usage during the experiment time as shown in Figures 14 and 15. We tested different time intervals ranging from 0.1 s to 10 s for three minutes each and utilized the feature described in Section 5.4 to modify the time interval configuration.

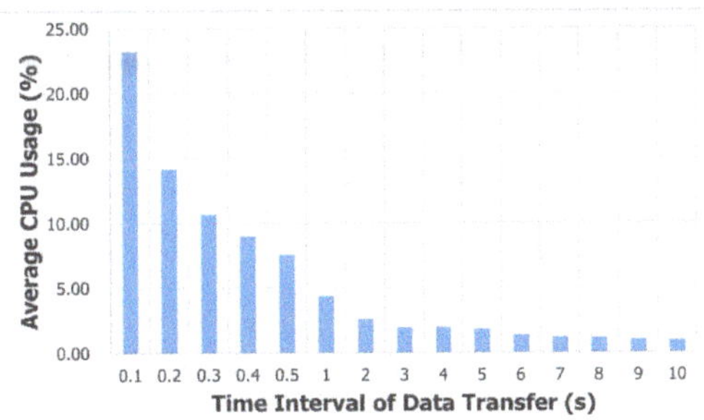

Figure 14. Average CPU usage rate of main services with different time intervals.

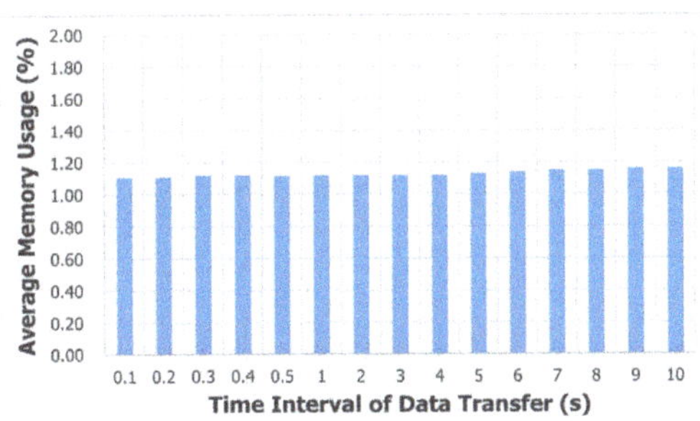

Figure 15. Average memory usage of main services with different time intervals.

The second evaluation involved examining the average response time of the web services when accessed by multiple users simultaneously via *HTTP POST* communications. The edge device framework was installed on a *Raspberry Pi*, and a considerable amount of sensor data were stored. To simulate multiple users, we developed a simulation program that generates virtual users, and ran it on personal computers connected to the *Raspberry Pi* via Ethernet in the local area network. During the experiments, we increased the number of user accesses from 5 to 150, with each virtual user representing an actual user or system using the device data. All the virtual users used similar parameter requests to access sensor data stored in local data storage.

To measure the response time, we calculated the time difference between the case where a virtual user sends a request to the web services and the case where it receives the response message. The response message is the 56 KB JSON message containing 500 records of data. During the experiment, we also evaluated the throughput of web services, which was 2.3 MB/s. It can handle 41 requests per second. Figures 16 and 17 illustrate the average response time and the average CPU usage rate, respectively, when the number of virtual users increased from 5 to 150.

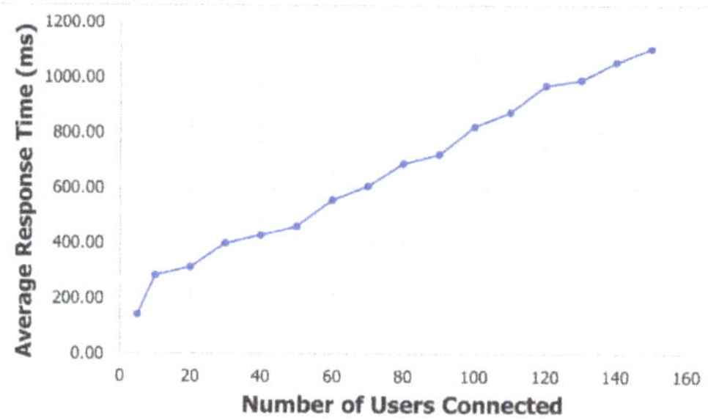

Figure 16. Average response time of web services with different numbers of users connected.

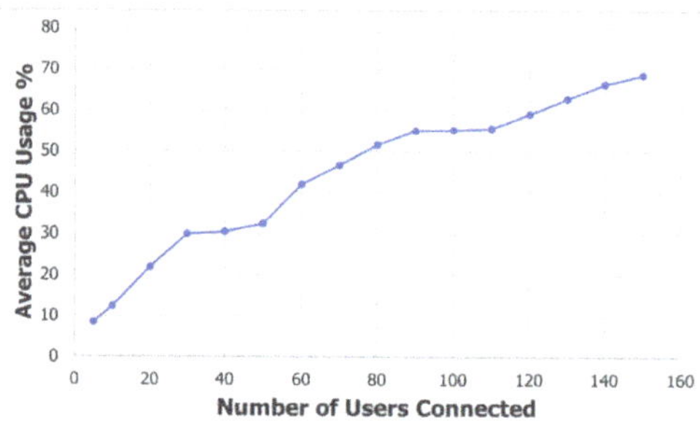

Figure 17. Average CPU usage rate of web services with different numbers of users connected.

8.2. Comparative Analysis

We compared features of the edge device framework with eight research works taking similar approaches in the literature. We compiled a list of features to be considered for comparing different edge computing systems frameworks. They were used to characterize each proposal, and included the following:

- The *main purpose* was to identify the issue that the proposed system intends to address and the key reason for selecting it to run edge IoT applications.
- *Edge devices* represent devices that installed an edge computing framework system.
- *Dynamic deployment* shows the ability to allow users to dynamically configure the flow system to run their own edge applications based on hardware and process requirements (Yes or No).
- *Remotely update* indicates the capability to remotely update the system (Yes or No).
- *Data conversion* implies the capability to preprocess data across several devices into a standard format (Yes or No).
- *Scalability* demonstrates the ability to expand their applications and to execute the number of data processing requests simultaneously (Yes or No).
- *Interoperability* indicates the capability to connect through several widely adopted and supported protocols provided by multiple devices (Yes or No).
- *Cross-vendor capabilities* illustrate the capacity of edge computing to collaborate with multiple vendors to develop complex IoT application platforms (Yes or No).

Table 3 compares the fulfillment of the eight features among the eight related works and our proposed edge devices framework.

Table 3. The comparative evaluation between the proposed framework and the existing related studies.

Work Reference	Main Purpose	Edge Devices	Dynamic Deployment	Remotely Update	Data Conversion	Scalability	Interoperability	Cross-vendor Capabilities
[34]	Data stream processing and task management	Wi-Fi Home Gateway	✓	✗	✗	✓	✓	✗
[35]	Edge devices gateways and support tool	Personal Computer and Server	✓	✓	✗	✓	✓	✓
[36]	Edge devices for smart manufacturing	Single-Board Computer	✓	✓	✓	✓	✓	✗
[37]	Edge framework for smart farming	Personal Computer	✓	✗	✗	✓	✓	✗
[38]	Edge computing gateways	Server	✓	✗	✗	✓	✓	✓
[39]	Edge computing framework	Personal Computer	✓	✓	✓	✓	✓	✗
[40]	Edge devices for smart home	Personal Computer	✓	✗	✗	✓	✓	✗
[13]	Edge computing framework	Single-Board Computer	✓	✓	✗	✓	✓	✗
Our Proposal	General edge computing framework	Single-Board Computer	✓	✓	✓	✓	✓	✓

8.2.1. Overview

Sajjad et al. in [34], Banerjee et al. in [35], and Ullah et al. in [38] developed systems that are consistent with the main objective of the edge computing framework by collecting data from diverse devices. Moreover, Rong et al. in [39] and Berta et al. in [13] created an edge computing framework that can gather data and connect to the actuator as the system output, which is similar to our edge device framework. Our framework is a general framework for edge computing and has the ability to connect with several IoT networks and to offer multiple output components that utilize the acquired data.

8.2.2. Edge Devices

Multiple works have used personal computers for installing and operating the frameworks. Nevertheless, they do not support the GPIO connectivity that is commonly used in sensor devices. Chen et al. in [36] and Berta et al. in [13] have implemented framework systems using single-board computer devices, such as the *Raspberry Pi*, which has significant benefits. Hence, we chose to deploy the proposed framework on these devices. This approach enables sensors to connect directly to the single-board computer devices for data collections, making the development of IoT application systems more straightforward.

8.2.3. Framework Features

In terms of framework features, all the related works offer capabilities for gathering data from IoT devices and sending them to a cloud server. However, as in our proposal, the works by Chen et al. in [36] and Rong et al. in [39] included the feature to process sensor data by converting them based on user-defined configurations.

All the works examined provided the capability to dynamically set up and deploy the framework using the connected devices as the main requirement. Some works required direct access to the devices for operations. Notably, Banerjee et al. [35], Chen et al. [36],

Rong et al. [39], Berta et al. [13], and our proposed framework allow users to remotely update the configuration from the cloud server.

8.2.4. Scalability, Interoperability, and Cross-Vendor Capabilities

All the works that have been reviewed focus on incorporating the scalability and interoperability in the functionality. However, some of them have the limited methods of connectivity for linking IoT devices to the edge framework. For instance, Sajjad et al.'s work [34] only allows the connectivity via Wi-Fi communications, whereas Zamora et al. [37] and Sharif et al.'s works [40] only permit connections from control unit devices to receive sensor data. Some works consider the cross-vendor capabilities of edge computing frameworks, particularly regarding data output components. Banerjee et al. [35], Ullah et al. [38], and the proposed framework allow the user to access to sensor data from edge devices using the *REST API*. However, only the proposed framework provides the additional features that allow data transmissions to various cloud computing vendors through *MQTT* and *HTTP POST* communications.

8.3. Discussions

This sub-section outlines the performance evaluation outcomes of the proposed edge device framework in this paper. The framework was developed for the universal edge computing device with the primary objective of enhancing the effectiveness of building IoT application systems. The framework provides the flexibility to specify system configurations, such as time intervals for collecting and transmitting data to the server periodically. As the optimal time interval may vary depending on the purpose, we assessed the computing performance of the edge device framework across various time interval settings.

Figures 14 and 15 exhibit the mean CPU and memory usage during the execution of the main services across different time intervals. The results indicate that shorter time intervals require higher percentages of the CPU usage, where all the experimental results fall below 25%. Moreover, the amount of the memory usage remains relatively stable across time intervals, suggesting that the proposed system operates without demanding excessive computational resources.

The advent of *SIoT* has increased the complexity and universality of IoT application systems by enabling other services to access to sensor data beyond merely transmitting them to the server. As a result, it is crucial to include features that simplify data accessibility. To this end, we integrated web services that enable users to access sensor data via *HTTP* communications. Furthermore, we evaluated the communication and computing performance of the edge device framework when multiple users access it simultaneously.

Figures 16 and 17 illustrate the average response time and the CPU usage rate when the number of virtual users increases from 5 to 150. The average response time is 824 ms, and the CPU usage rate is 55% for 100 devices with the response message containing 500 data records. These results indicate that the proposed edge device framework can accommodate hundreds of users with the reasonable response time and the CPU usage rate.

To illustrate the latest developments in edge computing frameworks, we evaluated several comparable models and extracted relevant information from their published papers. Our analysis results, presented in Table 3, lead us to believe that the edge device framework offers advanced features and functionality that are highly valuable, especially given the growing trend towards developing more complex and general IoT application systems.

Furthermore, the evaluation results for the fingerprint-based indoor localization system and the data logging system demonstrate that the edge device framework can automatically retrieve the edge configuration file from the server. It can execute the service program by following the configuration file, gather data from the sensor, convert it to the standard format, and send it to the server within the pre-defined time frame using the communication protocol service. In addition, Figures 9 and 13 indicate that it allows users to remotely update the device configurations through the web interface of *SEMAR*. Hence, the implemented edge device framework in this paper can enhance the usage of device sensors and contribute to the efficient development of IoT application systems utilizing *SEMAR*.

8.4. Generalization

According to the results of this paper, the *edge device framework* can improve utilizations of IoT devices by enabling users to remotely configure the parameters, including the connectivity of sensor interfaces, data conversions, data models, local data storage, local visualizations, and data transmission intervals to the server. All of the configuration parameters will be stored in the database to be used as templates for future use of similar sensors.

This framework was built on Python and can be utilized on any single-board computer supporting Python. It includes *NVIDIA Jetson Nano* [41], *BeagleBone Black* [42], *UDOO X86* [43], and *Odroid XU4* [44].

The *SIoT* architecture [25] was considered to develop a general edge computing device. Each IoT object in the *SIoT* architecture provides the computation and communication capabilities. The framework functions can be classified as *data input*, *data processing*, and *data output*. In *data input*, the functions to handle various sensor connectivities and data converters are implemented, including GPIO, GPIB, serial, and wireless communications. It offers multiple *data output* components to utilize the sensor data obtained. *REST API* data access is included for data transmissions to cloud services.

In the *MIoT* architecture [26], the function to manage the data communication model between IoT objects in static scenarios was implemented. This framework architecture includes the function that allows data communications in dynamic scenarios. It is possible to add new interactions between sensors and edge devices using *data input* components, and to manage data communications between the edge layer and the cloud layer through *data output* components, allowing cross-vendor data communications in the standard JSON format.

9. Conclusions

This paper presented the design and implementation of the *edge device framework* in the *SEMAR* IoT application server platform. It can remotely optimize device utilizations by configuring it through the *SEMAR* interface. The framework defines the connectivity of sensor interfaces, the data processing, the transmitted sensor elements, the communication protocol, the local data storage, the local visualization, and the data transmission interval on the server. It enables connection to a variety of sensor interfaces, transforms the data into a standard format, and provides multiple output components for data utilization.

Our evaluation results through applications with two IoT application systems verified the adaptability and validity of the proposed framework. IoT edge systems were developed in dynamic scenarios by allowing users to add or remove sensor devices flexibly.

In future works, we will continue enhancing the proposed framework, including implementations of the edge configuration validation function and the remote debugging function in *SEMAR*. They are necessary to prevent errors and guarantee consistency and reliability, and to find and fix problems in the edge systems. Then, we will continue to evaluate it through applications to other IoT application systems.

Author Contributions: Conceptualization, Y.Y.F.P., N.F. and S.S.; Methodology, Y.Y.F.P. and R.H.; Software, Y.Y.F.P., S.I. and J.S.; Writing—Original Draft Preparation, Y.Y.F.P.; Writing—Review and Editing, N.F.; Validation, M.K. and M.O. All authors have read and agreed to the published version of the manuscript.

Funding: This research received no external funding.

Institutional Review Board Statement: Not applicable.

Informed Consent Statement: Not applicable.

Data Availability Statement: Not applicable.

Acknowledgments: The authors thank the reviewers for their thorough reading and helpful comments.

Conflicts of Interest: The authors declare no conflict of interest.

References

1. Gubbi, J.; Buyya, R.; Marusic, S.; Palaniswami, M. Internet of things (IoT): A Vision, Architectural Elements, and Future Directions. *Future Gener. Comput. Syst.* **2013**, *29*, 1645–1660. [CrossRef]
2. Stankovic, J.A. Research Directions for the Internet of Things. *IEEE Internet Things J.* **2014**, *1*, 3–9. [CrossRef]
3. Noura, M.; Atiquzzaman, M.; Gaedke, M. Interoperability in Internet of Things: Taxonomies and Open Challenges. *Mob. Netw. Appl.* **2018**, *24*, 796–809. [CrossRef]
4. Cubo, J.; Nieto, A.; Pimentel, E. A Cloud-Based Internet of Things Platform for Ambient Assisted Living. *Sensors* **2014**, *14*, 14070–14105. [CrossRef] [PubMed]
5. Yar, H.; Imran, A.S.; Khan, Z.A.; Sajjad, M.; Kastrati, Z. Towards Smart Home Automation Using IoT-enabled Edge-Computing Paradigm. *Sensors* **2021**, *21*, 4932. [CrossRef]
6. Shi, W.; Cao, J.; Zhang, Q.; Li, Y.; Xu, L. Edge Computing: Vision and Challenges. *IEEE Internet Things J.* **2016**, *3*, 637–646. [CrossRef]
7. Roman, R.; Lopez, J.; Mambo, M. Mobile Edge Computing, Fog et al.: A Survey and Analysis of Security Threats and Challenges. *Future Gener. Comput. Syst.* **2018**, *78*, 680–698. [CrossRef]
8. Salkic, S.; Ustundag, B.C.; Uzunovic, T.; Golubovic, E. Edge Computing Framework for Wearable Sensor-Based Human Activity Recognition. In Proceedings of the International Symposium on Innovative and Interdisciplinary Applications of Advanced Technologies (IAT 2019), Sarajevo, Bosnia-Herzegovina, 20–23 June 2019; pp. 376–387.
9. Chen, X.; Shi, Q.; Yang, L.; Xu, J. ThriftyEdge: Resource-efficient Edge Computing for Intelligent IoT Applications. *IEEE Netw.* **2018**, *32*, 61–65. [CrossRef]
10. Das, A.; Patterson, S.; Wittie, M. Edgebench: Benchmarking edge computing platforms. In Proceedings of the 2018 IEEE/ACM International Conference on Utility and Cloud Computing Companion (UCC Companion), Zurich, Switzerland, 17–20 December 2018.
11. Rodríguez, A.; Valverde, J.; Portilla, J.; Otero, A.; Riesgo, T.; de la Torre, E. FPGA-Based High-Performance Embedded Systems for Adaptive Edge Computing in Cyber-Physical Systems: The ARTICO3 Framework. *Sensors* **2018**, *18*, 1877. [CrossRef]
12. Mavromatis, A.; Colman-Meixner, C.; Silva, A.P.; Vasilakos, X.; Nejabati, R.; Simeonidou, D. A Software-defined IoT Device Management Framework for Edge and Cloud Computing. *IEEE Internet Things J.* **2020**, *7*, 1718–1735. [CrossRef]
13. Berta, R.; Bellotti, F.; De Gloria, A.; Lazzaroni, L. Assessing Versatility of a Generic End-to-End Platform for IoT Ecosystem Applications. *Sensors* **2022**, *22*, 713. [CrossRef]
14. Panduman, Y.Y.; Funabiki, N.; Puspitaningayu, P.; Sakagami, M.; Sukaridhoto, S. Implementations of Integration Functions in IoT Application Server Platform. In Proceedings of the Fifth International Conference on Vocational Education and Electrical Engineering (ICVEE) 2022, Surabaya, Indonesia, 10–11 September 2021.
15. MQTT Org. Message Queuing Telemetry Transport Protocol. Available online: http://mqtt.org/ (accessed on 19 May 2023).
16. Huo, Y.; Puspitaningayu, P.; Funabiki, N.; Hamazaki, K.; Kuribayashi, M.; Kojima, K.A. Proposal of the Fingerprint Optimization Method for the Fingerprint-Based Indoor Localization System with IEEE 802.15.4 Devices. *Information* **2022**, *13*, 211. [CrossRef]
17. Puspitaningayu, P.; Huo, Y.; Funabiki, N.; Hamazaki, K.; Kuribayashi, M.; Kao, W. Investigations of Detection Accuracy Improvements for Fingerprint-based Indoor Localization System Using IEEE 802.15.4. In Proceedings of the Fourth International Conference on Vocational Education and Electrical Engineering (ICVEE) 2021, Surabaya, Indonesia, 2–3 October 2021; pp. 1–5.
18. Mahmood, O.A.; Abdellah, A.R.; Muthanna, A.; Koucheryavy, A. Distributed Edge Computing for Resource Allocation in Smart Cities based on The IoT. *Information* **2022**, *13*, 328. [CrossRef]
19. Sarangi, S.; Naik, V.; Choudhury, S.B.; Jain, P.; Kosgi, V.; Sharma, R.; Bhatt, P.; Srinivasu, P. An Affordable IoT Edge Platform for Digital Farming in Developing Regions. In Proceedings of the 2019 11th International Conference on Communication Systems & Networks (COMSNETS), Bengaluru, India, 7–11 January 2019.
20. Oueida, S.; Kotb, Y.; Aloqaily, M.; Jararweh, Y.; Baker, T. An Edge Computing Based Smart Healthcare Framework for Resource Management. *Sensors* **2018**, *18*, 430. [CrossRef]
21. Mach, P.; Becvar, Z. Mobile Edge Computing: A Survey on Architecture and Computation Offloading. *IEEE Commun. Surv. Tutorials* **2017**, *19*, 1628–1656. [CrossRef]
22. Yousafzai, A.; Yaqoob, I.; Imran, M.; Gani, A.; Noor, R.M. Process Migration-Based Computational Offloading Framework for IoT-Supported Mobile Edge/Cloud Computing. *IEEE Internet Things J.* **2020**, *7*, 4171–4182. [CrossRef]
23. Yang, W.; Liu, W.; Wei, X.; Guo, Z.; Yang, K.; Huang, H.; Qi, L. EdgeKeeper: A Trusted Edge Computing Framework for Ubiquitous Power Internet of Things. *Front. Inf. Technol. Electron. Eng.* **2021**, *22*, 374–399. [CrossRef]
24. Kim, W.; Ko, H.; Yun, H.; Sung, J.; Kim, S.; Nam, J. A Generic Internet of Things (IoT) Platform Supporting Plug-and-Play Device Management based on The Semantic Web. *J. Ambient. Intell. Humaniz. Comput.* **2019**, 1–11. [CrossRef]
25. Atzori, L.; Iera, A.; Morabito, G.; Nitti, M. The Social Internet of Things (SIoT)–When social networks Meet the Internet of Things: Concept, Architecture and Network Characterization. *Comput. Netw.* **2012**, *56*, 3594–3608. [CrossRef]

26. Cauteruccio, F.; Cinelli, L.; Fortino, G.; Savaglio, C.; Terracina, G.; Ursino, D.; Virgili, L. An approach to Compute The Scope of A Social Object in A Multi-IoT Scenario. *Pervasive Mob. Comput.* **2020**, *67*, 101223m. [CrossRef]
27. Al-Fuqaha, A.I.; Guizani, M.; Mohammadi, M.; Aledhari, M.; Ayyash, M. Internet of Things: A Survey on Enabling Technologies, Protocols, and Applications. *IEEE Commun. Surv. Tutor.* **2015**, *17*, 2347–2376. [CrossRef]
28. Lombardi, M.; Pascale, F.; Santaniello, D. Internet of Things: A General Overview Between Architectures, Protocols and Applications. *Information* **2021**, *12*, 87. [CrossRef]
29. Chalapathi, G.S.S.; Chamola, V.; Vaish, A.; Buyya, R. *Industrial Internet of Things (IIoT) Applications of Edge and Fog Computing: A Review and Future Directions*; Springer: Cham, Switzerland, 2021; pp. 293–325. [CrossRef]
30. Panduman, Y.Y.; Funabiki, N.; Puspitaningayu, P.; Kuribayashi, M.; Sukaridhoto, S.; Kao, W.-C. Design and Implementation of SEMAR IoT Server Platform with Applications. *Sensors* **2022**, *22*, 6436. [CrossRef] [PubMed]
31. MongoDB, Mongodb: The Application Data Platform. Available online: https://www.mongodb.com/ (accessed on 19 May 2023).
32. Mono Wireless. Mono Wireless Product Information. Available online: https://mono-wireless.com/jp/products/index.html (accessed on 19 May 2023).
33. Graphtec, Wireless LAN—Midi Logger GL240: Graphtec. Available online: https://www.graphtec.co.jp/en/instruments/gl240/wireless.html (accessed on 19 May 2023).
34. Sajjad, H.P.; Danniswara, K.; Al-Shishtawy, A.; Vlassov, V. SpanEdge: Towards Unifying Stream Processing over Central and Near-the-Edge Data Centers. In Proceedings of the 2016 IEEE/ACM Symposium on Edge Computing (SEC), Washington, DC, USA, 27–28 October 2016; Institute of Electrical and Electronics Engineers (IEEE): Piscataway, NJ, USA, 2016; pp. 168–178.
35. Banerjee, S.; Liu, P.; Patro, A.; Willis, D. ParaDrop: An Edge Computing Platform in Home Gateways. In *Fog for 5G and IoT*; Wiley: Hoboken, NJ, USA, 2017; p. 13.
36. Chen, B.; Wan, J.; Celesti, A.; Li, D.; Abbas, H.; Zhang, Q. Edge Computing in IoT-Based Manufacturing. *IEEE Commun. Mag.* **2018**, *56*, 103–109. [CrossRef]
37. Zamora-Izquierdo, M.A.; Sant, J.; Martínez, J.A.; Martínez, V.; Skarmeta, A.F. Smart farming IoT platform based on edge and cloud computing. *Biosyst. Eng.* **2019**, *177*, 4–17. [CrossRef]
38. Ullah, R.; Rehman, M.A.U.; Kim, B.-S. Design and Implementation of an Open Source Framework and Prototype For Named Data Networking-Based Edge Cloud Computing System. *IEEE Access* **2019**, *7*, 57741–57759. [CrossRef]
39. Rong, G.; Xu, Y.; Tong, X.; Fan, H. An edge-cloud collaborative computing platform for building AIoT applications efficiently. *J. Cloud Comput.* **2021**, *10*, 36. [CrossRef]
40. Sharif, Z.; Jung, L. T.; Ayaz, M.; Yahya, M.; Khan, D. Smart Home Automation by Internet-of-Things Edge Computing Platform. *Int. J. Adv. Comput. Sci. Appl.* **2022**, *13*. [CrossRef]
41. NVIDIA Jetson Nano. Available online: https://www.nvidia.com/en-us/autonomous-machines/embedded-systems/jetson-nano (accessed on 19 May 2023).
42. BeagleBone—Black. Available online: https://beagleboard.org/black (accessed on 19 May 2023).
43. Discover UDOO x86 II: The Most Powerful Maker Board Ever. Available online: https://www.udoo.org/discover-udoo-x86-ii/ (accessed on 19 May 2023).
44. Odroid XU4—Octa Core Odroid Computer. Available online: https://www.odroid.co.uk/hardkernel-odroid-xu4/odroid-xu4 (accessed on 19 May 2023).

information

Article

IoT Device Identification Using Unsupervised Machine Learning

Carson Koball, Bhaskar P. Rimal, Yong Wang *, Tyler Salmen and Connor Ford

The Beacom College of Computer and Cyber Sciences, Dakota State University, Madison, SD 57042, USA
* Correspondence: yong.wang@dsu.edu

Abstract: Device identification is a fundamental issue in the Internet of Things (IoT). Many critical services, including access control and intrusion prevention, are built on correctly identifying each unique device in a network. However, device identification faces many challenges in the IoT. For example, a common technique to identify a device in a network is using the device's MAC address. However, MAC addresses can be easily spoofed. On the other hand, IoT devices also include dynamic characteristics such as traffic patterns which could be used for device identification. Machine-learning-assisted approaches are promising for device identification since they can capture dynamic device behaviors and have automation capabilities. Supervised machine-learning-assisted techniques demonstrate high accuracies for device identification. However, they require a large number of labeled datasets, which can be a challenge. On the other hand, unsupervised machine learning can also reach good accuracies without requiring labeled datasets. This paper presents an unsupervised machine-learning approach for IoT device identification.

Keywords: internet of things; device identification; machine learning; unsupervised machine learning

Citation: Koball, C.; Rimal, B.P.; Wang, Y.; Salmen, T.; Ford, C. IoT Device Identification Using Unsupervised Machine Learning. *Information* **2023**, *14*, 320. https://doi.org/10.3390/info14060320

Academic Editors: Spyros Panagiotakis and Evangelos K. Markakis

Received: 4 March 2023
Revised: 23 May 2023
Accepted: 25 May 2023
Published: 31 May 2023

1. Introduction

The Internet of Things (IoT) is a term used to describe the interconnection of computing devices embedded in everyday objects to the Internet through the home, business, or institutional networks [1]. IoT devices and applications present significant security challenges, including limited device capabilities, lack of standardization, insufficient trust and integrity, and software vulnerabilities [2]. As a result, device identification is challenging in the IoT. Many critical services, such as access control and intrusion detection, are built on correctly identifying each unique device [3].

As users are identified in a digital network by their unique identities, IoT devices also require their unique identities when connecting to a network. Identities of Things (IDoT), a general term, has been adopted to describe IoT entities (e.g., users and devices). Four primary authentication factors could be used to identify users: something you know (e.g., username and password), something you possess (e.g., a physical token or a smart card), something you are (e.g., fingerprint or face recognition), and something you do (e.g., voice or sign). IoT devices can only be identified by something they have. A common technique to identify a device in a network is using the device's MAC address. However, MAC addresses can be easily spoofed.

Identity in IoT devices consists of attributes and dynamic values along with the member in varying contexts [4]. It can be a collection of things, should have a purpose, and should be treated uniformly across platforms. There are many representations of identities, and they can rely on globally unique identifiers [5,6], a combination of user characteristics [7], a set of attributes of the users [8], or a set of claims [9]. These approaches all possess a commonality based on the fact that they link an identity unique to a particular entity [4].

Machine-learning (ML)-assisted approaches attract many research interests because they can capture dynamic characteristics from the devices for device identification [10–17].

ML-assisted approaches fall into two general categories, i.e., supervised ML-assisted approaches [10,11] and unsupervised ML-assisted approaches [12–15,17]. Supervised ML-assisted approaches demonstrate great accuracy when used for device identification. However, they often require a large number of labeled datasets, which could be a challenge. On the other hand, unsupervised ML can reach reasonable accuracies without labeled datasets [12–15,17]. Additionally, the unsupervised method allows for greater flexibility regarding dynamic IoT networks where devices may join or leave at any time. Therefore, this paper focuses on unsupervised ML approaches for device identification.

The unsupervised approach presented in the paper utilizes an ensemble-based approach for device identification. A K-Nearest Neighbors classifier is used to identify each IoT device. The dataset proceeds through four steps: preprocessing, outlier removal, feature selection, and clustering before a device prediction is made. The key contributions of this paper include, but are not limited to, (1) demonstrating how to use unsupervised ML for device identification; (2) evaluating the performance of supervised ML and unsupervised ML using the same dataset; (3) discussing possible benefits that unsupervised ML may bring to the field of device identification.

The remainder of this paper is structured as follows. Section 2 introduces related work. Section 3 presents the proposed unsupervised ML-assisted approach for IoT device identification. The evaluation results are presented in Section 4. Finally, the conclusion and discussion for future research are discussed in Section 5.

2. Related Work

The surveys in [18,19] show that supervised, unsupervised, semi-supervised, and deep-learning approaches could all be used for device identification. Supervised ML is used in [10,11,20–22]. The authors in [20] proposed a supervised ML-assisted approach for identifying known IoT devices. A proprietary tool developed in [23] was used in [20] to extract features from captured network traffic. Using the same feature extraction tool, a two-stage meta classifier for IoT device identification was studied in [10]. The first stage classifier differentiates IoT and non-IoT devices. The second stage classifier identifies a specific IoT device class. The classifier considered in [20] is Random Forest-based. Other classifiers considered in the supervised ML include Decision Trees, Logistic Regression models, Support Vector Machines (SVM), GBM, and XGBoost models [10,11]. These papers show the utmost accuracy in identifying devices in the IoT. However, supervised ML requires labeling to train the models, which may be expensive or impossible to acquire.

In [12], authors used unsupervised clustering to identify IoT device types in network flow traffic. Network traffic was broken up into time granularities of 1–8-min packet flows for each device on the network. Depending on the device, the final clustering used K-Means with 128 or 256 clusters. This research takes a heuristic approach to identify flows in a packet capture by taking the data in 1–8-min intervals of packets. In [13], authors used data captured directly from the devices to identify cycles in the flow data relating to how often and how predictable the transmission of data is. They then use K-Nearest Neighbors and arbitrary labeling to cluster devices. This approach is much slower than other algorithms.

Unsupervised deep learning is used in [14,15], where ML autoencoders are combined with clustering algorithms to identify arbitrary device types. Work in [14] focused on identifying compromised devices using packet statistics, whereas [15] considered how variational autoencoders arbitrarily identify devices on a network using a combination of periodic features such as those in [13] and flow statistics [15].

As discussed in [12–15,17], unsupervised learning can reach accuracies as good as or better than those in supervised approaches, while having much higher accuracy in both unseen and compromised devices.

3. Unsupervised Machine-Learning-Assisted Approach for IoT Device Identification

The unsupervised process detailed in this paper utilizes an ensemble-based approach to device identification, where each base model that comprises the ensemble network is a one-class classifier. Figure 1 details the various steps that comprise the one-class classifiers implemented in this paper. These steps will be discussed more thoroughly later in this section.

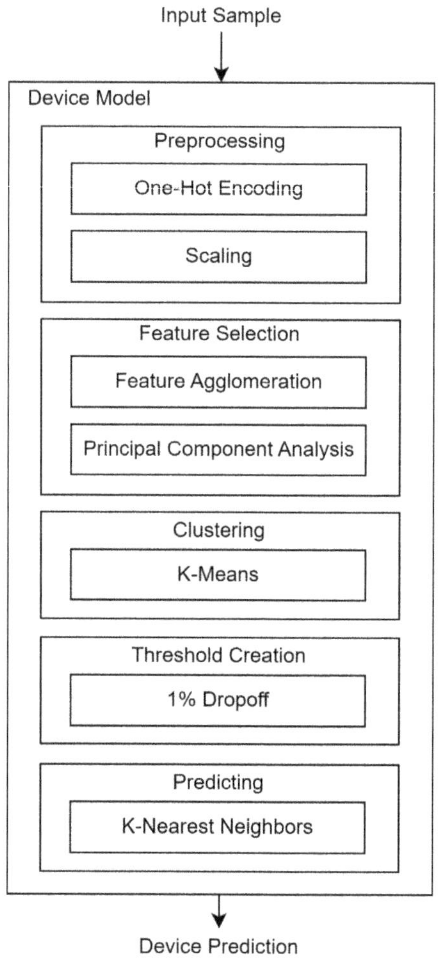

Figure 1. An overview of a one-class classifier.

The combined results from each base model will ultimately decide what a given sample will be classified as, as seen in Figure 2. The goal of a one-class classifier is to distinguish which samples do and do not belong to the class it represents. In the discussion of IoT device identification, the class would be an IoT device. This approach differs from a one-vs-rest scheme for two reasons. The first reason is that introducing a new device or removing an existing device from the network will require the entire ensemble network to be retrained in the one-vs-rest scheme. In the one-class scheme, however, introducing a new device will only require a new model for that device to be trained, whereas removing an existing device will only require that the existing model for the said device is discarded. The second reason this approach differs from a one-vs-rest scheme is that this approach allows a sample to be considered to be an unknown sample, such as the case where the sample belongs to a device not in the network. In many one-vs-rest schemes, this unknown sample would have been predicted as one of the devices in the trained model network.

As a result, the approach used in this paper allows for better scalability as the number of devices increases in a given network.

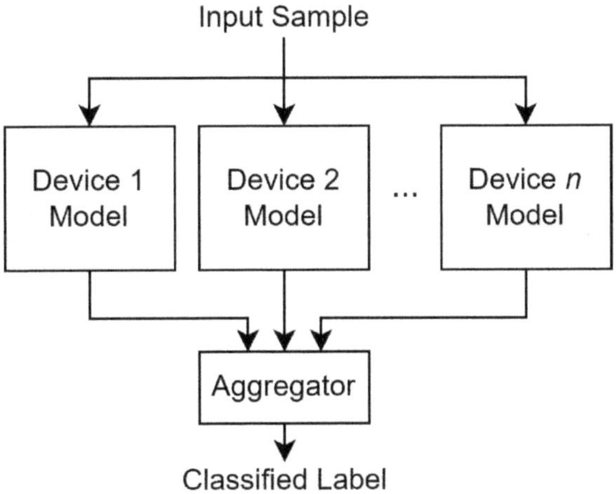

Figure 2. An overview of a one-class ensemble network.

3.1. Pre-Processing

The first step of the training stage is to process the data to augment the model's ability to capture information from the dataset. One such method of processing is through one-hot encoding. To include nominal features, such as the network protocol used in a flow, said features must be encoded to remove any inferred order between values. Additionally, since the proposed model utilizes distance-based algorithms in both the training and predicting stages of its lifespan, it is imperative to properly scale the raw data the model takes as input. Without scaling, features with inherently larger ranges will dominate features with inherently smaller ranges in Euclidean-based distance comparisons. Furthermore, the information found in these features with smaller ranges can be overlooked without scaling. In the implementation of the model in this paper, standardization is performed on each feature, where each feature is independently scaled to fit a normal distribution.

3.2. Feature Selection

The second step in the training stage is the selection of relevant features that will be used for analysis. The first portion of this step is using Feature Agglomeration (FA) [24]. A total of 100 clusters from the FA algorithm were used in the model. The second portion of this step is the use of Principal Component Analysis (PCA) [25]. All 100 features extracted from the PCA were also used in the model.

3.3. Clustering

The third step in the training stage is clustering the processed data. Clustering allows for more effective capture of each device's traffic patterns and reduces computation time in predicting. The model in this paper utilizes K-Means clustering [26]. A low K value would result in the K-Means algorithm overgeneralizing the data, creating centroids that are insignificant to the traffic patterns of each device. Alternatively, a K value too high would result in the K-Means algorithm overfitting on the data, creating centroids that are fitted extremely well to sampled noise in the training data, leaving little room for generalization in unseen data. As with [12], the K value for each model was determined by testing an incrementally larger K to a maximum of 1000. The sum of averaged squared distances of each sample to its closest centroid, or inertia, was recorded for each value of K tested. An elbow point was then identified where an increase in K would not substantially reduce the inertia of the data points, and this K value was chosen for the model. Additionally, each centroid will be assigned a corresponding training distribution value based on the total training samples assigned to the centroid. This will be used as a tie-breaking metric in the predicting stage of the model's lifespan, detailed in Section 3.5.

3.4. Threshold Creation

The final step in the training stage is to create the threshold that decides classification behavior. This threshold defines whether an input sample is or is not predicted as the device of a given model. As with [12], each centroid created with the K-Means algorithm has an assigned set of samples that belong to it. The distance between each point and its centroid is then calculated, and the threshold for the centroid is defined as the distance that includes no more than 99% of the samples that belong to it. As a result, each centroid will have a corresponding threshold distance value that will exclude 1% of the samples that belong to it from the training dataset. Another method of identifying a distance threshold was explored utilizing DBSCAN [27].

3.5. Predicting

From the training stage of the model, a series of centroids, as well as their respective training distribution values and distance thresholds, are stored. Additionally, the preprocessing models, FA, and PCA models are stored after being fitted from the training dataset. The model is given the sample as input to predict a new test sample, where the preprocessing models transform its feature values. Next, the sample's features are selected based on the FA and PCA models, which ultimately reduce the number of features that will be observed. To make the prediction, the model implements a K-Nearest Neighbors classifier fitted with the centroids defined by the K-Means algorithm in the training stage. The centroid closest to the transformed test sample is then considered the centroid to which the testing sample belongs. However, if the distance between the centroid and the testing sample is greater than the distance threshold set for the centroid, the testing sample will be considered a negative sample. Conversely, if the distance between the centroid and the testing sample is equal to or smaller than the distance threshold set for the centroid, the testing sample would be considered a positive sample, as seen in Figure 3.

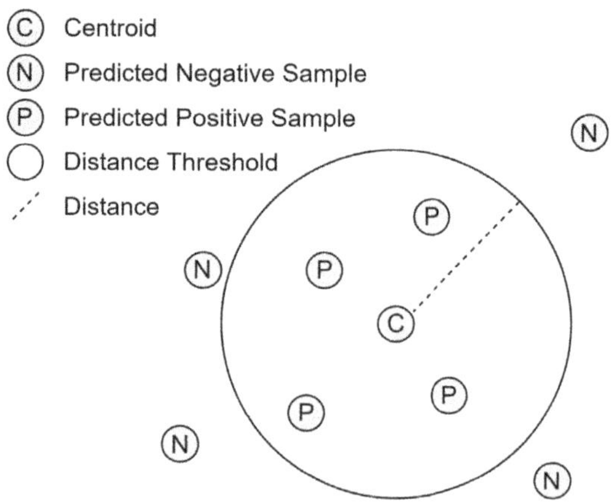

Figure 3. Example of a distance threshold and its effect on predicted samples in a 2-dimensional space.

This model could effectively act as a classifier for the device it represents in a single-device environment. However, in environments with more than one device, any given sample can be predicted positively for multiple device models. Consequently, the network must distinguish which positive sample will be assigned to which device. This final tie-breaking comparison utilizes the training distribution value assigned to each cluster in the clustering step. The training distribution value for each centroid of these device models is compared to all the devices considered sample positive. The device model whose centroid has the highest training distribution value will ultimately claim the sample as its own.

3.6. Testing Dataset

As shown in Figure 4, a testing network was established to generate a testing dataset. A total of 16 IoT devices are connected to the testing network. The network traffic is collected through nTAP and RaspAP [28]. nTAP is a passive, full-duplex monitoring device that provides visibility into the network regardless of traffic. nTAP collects network traffic before it reaches the firewall. RaspAP provides Internet access for IoT devices and is used to collect network traffic at the Wi-Fi access point. *tcpdump* is used to collect network traffic in nTap and RaspAP.

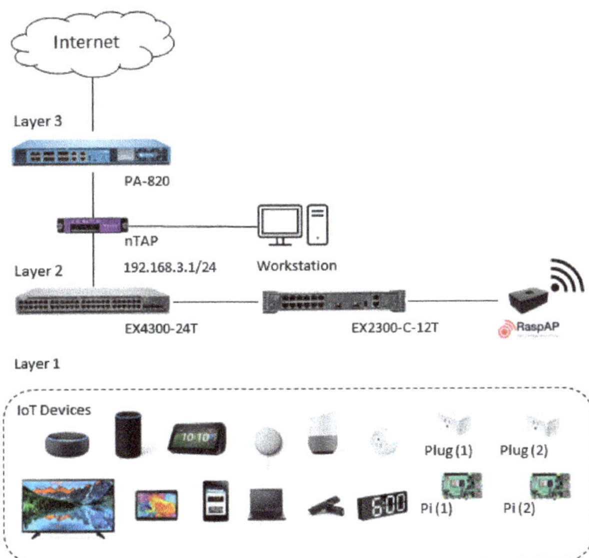

Figure 4. Experimental network testbed.

Data were collected between 31 March 2022 and 9 May 2022. Multiple datasets were collected during different periods during the experiment. tcpdump was used on both the RaspAP and the nTAP data to collect network traffic. Data collected through RaspAP is used for ML training. Furthermore, data collected through nTAP is used to validate the ML classifier on the network perimeter. Approximately 150 GB and 200 GB of data were collected from RaspAP and nTAP, respectively.

3.7. Feature Extraction

Network traffic from IoT devices is collected through tools such as tcpdump and Wireshark. The network traffic is saved in pcap files. An open-source tool, CICFlowMeter, extracts network features from pcap files [29,30]. CICFlowMeter can generate bidirectional flows and calculate time-related features in both the forward and backward directions. Originally, CICFlowMeter was created to identify malicious traffic that might contain malware. The features that could be extracted from each traffic flow include flow duration, total forward packets, and total backward packets. In addition to the flow features, flow ID, source IP, source port number, destination IP, destination port number, protocol, and timestamps are also recorded for each flow. After feature extraction, CICFlowMeter creates a CSV file with all the features for each pcap file. In this study, we have developed a tool, *CICFlowMeter++* by enhancing the original CICFlowMeter. CICFlowMeter++ can extract

233 features from a TCP flow and includes many new features from [10] for comparison. In addition to these 233 features, 9 more features, including source device, destination device, source medium, destination medium, source state, destination state, source manufacturer, destination manufacturer, and flow device, are also added to the CSV file. A total of 242 features are available in the CSV file.

Our preliminary testing and results indicated that ML was ineffective in identifying a device if the device did not generate sufficient data for training. Furthermore, similar devices, e.g., Amazon Echo, Amazon Echo Dot, and Amazon Echo Show, or the same type of devices, e.g., K Smart Plug 1 and 2, also present challenges for device identification. After removing three devices that did not meet the criteria and combining similar devices, eight devices remained in the dataset. Table 1 shows the number of TCP flows for each device in the dataset.

Table 1. Datasets generated by the devices in the experiments (* combined TCP flows from similar devices).

Device	Dataset
Amazon Echo Show	47,804
Lenovo Chromebook	9307
Google Nexus Tablet	9388
K Smart Plug *	79,160
Raspberry Pi *	19,481
ZMI Smart Clock	7185
Amazon Smart Plug	5227
Samsung Smart TV	94,976

4. Results and Discussions

Using the approach presented in Section 3, the performance of the unsupervised ML approach for device identification is studied. We also compare the supervised ML approaches for device identification using the same dataset. For the unsupervised approach, a maximum of 10,000 samples for each of the eight devices are selected for training from the training subset. Exactly 1000 samples for each of the eight devices are selected for testing from the testing subset.

4.1. Feature Selection

A total of 242 features are available after data processing. Using FA and PCA, the number of features used by the model decreased from 242 to 100. Since the PCA model has transformed these 100 features, these features can be seen as combinations of the original features. Thus, the 100 features used for unsupervised ML are not directly mapped to any original features.

4.2. Clustering

Based on the elbow point method for identifying K values for K-Means clustering, each device had varying values of K used. Table 2 shows the number of clusters used for each device in the clustering step.

Table 2. K values identified for each device's model for K-Means clustering.

Device	Number of Clusters
Amazon Echo Show	400
Lenovo Chromebook	100
Google Nexus Tablet	300
K Smart Plug	60
Raspberry Pi	300
ZMI Smart Clock	70
Amazon Smart Plug	70
Samsung Smart TV	100

4.3. Device Identification

Two approaches, DBSCAN, and 1% Dropoff, are used for threshold creation. Table 3 shows the accuracies from both approaches. As shown in Table 3, the testing accuracy from using 1% Dropoff for threshold creation is slightly higher than when using the DBSCAN method.

Table 3. Distance threshold methods and their results.

Metric	DBSCAN	1% Dropoff
Macro Precision	0.821	0.828
Macro Recall	0.777	0.799
Macro F1 Score	0.797	0.813

Figure 5 shows the confusion matrix for device identification. As shown in Figure 5, unsupervised ML shows excellent accuracy values when identifying the Smart TV, K Smart Plug, and Amazon Smart Plug devices. The accuracy for identifying the Nexus Tablet device is not ideal.

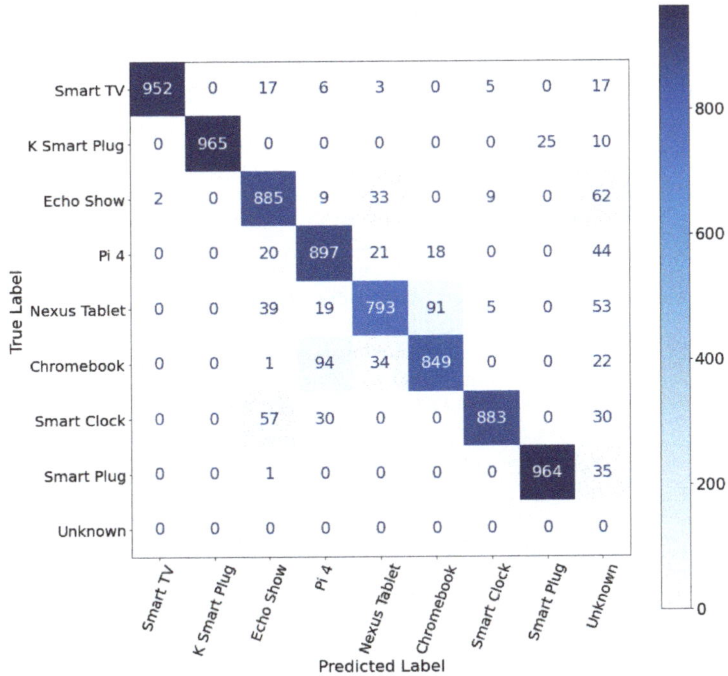

Figure 5. Confusion matrix utilizing 1% dropoff.

4.4. Unsupervised ML vs. Supervised ML

We evaluate the supervised ML approach for device identification using the same dataset, as shown in Table 1. The testing dataset is divided into two datasets, i.e., the training dataset and the testing dataset, using an 80/20 split. The Random Forest classifier is used to evaluate the importance of the features for device identification. We further evaluate six supervised ML classifiers for device identification, including AdaBoost, Decision Tree, K-Nearest Neighbor, Logistic Regression, Random Forest, and LinearSVC. Our evaluation shows that the AdaBoost with 200 features achieves the best testing accuracies for device identification. Table 4 shows the accuracy values for the eight IoT devices from the AdaBoost with 200 features. Furthermore, Table 5 shows the precision, recall, f1 score, and accuracy values from the proposed unsupervised approach.

Table 4. Supervised ML vs. Unsupervised ML.

Device	Supervised ML	Unsupervised ML
Amazon Echo Show	92.4%	88.5%
Lenovo Chromebook	87.5%	84.9%
Google Nexus Tablet	100.0%	79.3%
K Smart Plug	91.0%	96.5%
Raspberry Pi	97.5%	89.7%
ZMI Smart Clock	98.8%	88.3%
Amazon Smart Plug	90.4%	96.4%
Samsung Smart TV	99.1%	95.2%

Table 5. Unsupervised ML.

Device	Precision	Recall	F1 Score	Accuracy
Amazon Echo Show	0.87	0.89	0.88	88.5%
Lenovo Chromebook	0.89	0.85	0.87	84.9%
Google Nexus Tablet	0.90	0.79	0.84	79.3%
K Smart Plug	1.00	0.96	0.98	96.5%
Raspberry Pi	0.85	0.90	0.87	89.7%
ZMI Smart Clock	0.98	0.88	0.93	88.3%
Amazon Smart Plug	0.97	0.96	0.97	96.4%
Samsung Smart TV	1.00	0.95	0.97	95.2%

As shown in Table 4, supervised ML generally provides better accuracies in device identification than unsupervised ML. For certain devices, e.g., K Smart Plug and Amazon Smart Plug, the unsupervised ML performs better than the supervised ML approach.

Our results show that both supervised and unsupervised ML could be used for device identification. Although supervised ML methods provide better accuracy values for device identification in this environment, the supervised paradigm requires labeled datasets that may not be attainable. In scenarios where labeling is not possible, our results further indicate that an unsupervised approach to device identification may still be a viable option. Additionally, the one-class nature that is inherent to our method allows for separate device classifiers to be added and removed from an ensemble model without the need for retraining the entire model. This modular property suggests that an unsupervised approach may be preferred in dynamic networks where devices frequently join and leave a network. Consequently, a hybrid approach including both supervised ML and unsupervised ML can be considered.

5. Conclusions and Outlook

This paper studies unsupervised ML for device identification. Our unsupervised ML approach employs a series of one-class classifiers that each includes five steps, i.e., preprocessing, feature selection, clustering, threshold creation, and predicting. The presented approach was applied to a dataset that includes eight devices. The obtained results show reasonable accuracies for these eight devices during our testing. Our analysis indicates that unsupervised ML may have similar challenges as supervised ML in identifying similar devices or devices of the same type. However, an unsupervised approach may provide additional benefits, such as scalability in dynamic networks and the removal of labeling processes not found in supervised methods, to the challenge of IoT device classification. Potential avenues for future work include utilizing a hybrid approach, including both supervised ML and unsupervised ML approaches for device identification, and studying how ML-assisted approaches perform in untrusted environments and in real-time traffic environments.

Author Contributions: Conceptualization, B.P.R. and Y.W.; methodology, C.K.; software, C.K.; validation, C.K.; formal analysis, C.K.; investigation, C.K., T.S., and C.F.; resources, C.K.; data curation, C.K.; writing—original draft preparation, C.K., T.S., and C.F.; writing—review and editing, B.P.R. and Y.W.; visualization, C.K.; supervision, B.P.R. and Y.W.; project administration, B.P.R. and Y.W.; funding acquisition, B.P.R. and Y.W. All authors have read and agreed to the published version of the manuscript.

Funding: This research is partially supported by the National Centers of Academic Excellence in Cybersecurity (NCAE-C).

Data Availability Statement: The data presented in this study are not publicly available due to proprietary tools used in the research.

Conflicts of Interest: The funders had no role in the design of the study; in the collection, analyses, or interpretation of data; in the writing of the manuscript; or in the decision to publish the results.

References

1. Al-Fuqaha, A.; Guizani, M.; Mohammadi, M.; Aledhari, M.; Ayyash, M. Internet of things: A survey on enabling technologies, protocols, and applications. *IEEE Commun. Surv. Tutor.* **2015**, *17*, 2347–2376. [CrossRef]
2. Bhattarai, S.; Wang, Y. End-to-End Trust and Security for Internet of Things Applications. *Computer* **2018**, *51*, 20–27. [CrossRef]
3. Muthusamy Ragothaman, K.N.; Wang, Y. A Systematic Mapping Study of Access Control in the Internet of Things. In Proceedings of the 54th Hawaii International Conference on System Sciences, Kauai, HI, USA, 5–8 January 2021; pp. 7090–7099. Available online: http://hdl.handle.net/10125/71474 (accessed on 25 February 2023).
4. Pal, S.; Hitchens, M.; Varadharajan, V. Modeling Identity for the Internet of Things: Survey, Classification and Trends. In Proceedings of the 12th International Conference on Sensing Technology (ICST), Limerick, Ireland, 4–6 December 2018; pp. 45–51. [CrossRef]
5. Koo, J.; Kim, Y.G. Interoperability of device identification in heterogeneous IoT platforms. In Proceedings of the 2017 13th International Computer Engineering Conference (ICENCO), Cairo, Egypt, 27–28 December 2017; pp. 26–29. [CrossRef]
6. Ning, H.; Zhen, Z.; Shi, F.; Daneshmand, M. A Survey of Identity Modeling and Identity Addressing in Internet of Things. *IEEE Internet Things J.* **2020**, *7*, 4697–4710. [CrossRef]
7. Jøsang, A.; Fabre, J.; Hay, B.; Dalziel, J.; Pope, S. Trust requirements in identity management. In *Proceedings of the Australasian Workshop on Grid Computing and e-Research-Volume 44*; Australian Computer Society, Inc.: Darlinghurst, NSW, Australia, 2005; pp. 99–108.
8. Alpár, G.; Batina, L.; Batten, L.; Moonsamy, V.; Krasnova, A.; Guellier, A.; Natgunanathan, I. New directions in IoT privacy using attribute-based authentication. In Proceedings of the ACM International Conference on Computing Frontiers, Como, Italy, 16–19 May 2016; pp. 461–466.
9. Cameron, K. The laws of identity. *Microsoft Corp.* **2005**, *12*, 8–11.
10. Meidan, Y.; Bohadana, M.; Shabtai, A.; Guarnizo, J.D.; Ochoa, M.; Tippenhauer, N.O.; Elovici, Y. ProfilIoT: A Machine Learning Approach for IoT Device Identification Based on Network Traffic Analysis. In Proceedings of the Symposium on Applied Computing, Marrakech, Morocco, 4–6 April 2017; pp. 506–509.
11. Wang, Y.; Rimal, B.P.; Elder, M.; Maldonado, S.I.C.; Chen, H.; Koball, C.; Ragothaman, K. IoT Device Identification Using Supervised Machine Learning. In Proceedings of the 2022 IEEE International Conference on Consumer Electronics (ICCE), Las Vegas, NV, USA, 7–9 January 2022; pp. 1–6. [CrossRef]
12. Sivanathan, A.; Gharakheili, H.H.; Sivaraman, V. Inferring IoT Device Types from Network Behavior Using Unsupervised Clustering. In Proceedings of the 2019 IEEE 44th Conference on Local Computer Networks (LCN), Osnabrueck, Germany, 14–17 October 2019; pp. 230–233.
13. Marchal, S.; Miettinen, M.; Nguyen, T.D.; Sadeghi, A.-R.; Asokan, N. AuDI: Toward Autonomous IoT Device-Type Identification Using Periodic Communication. *IEEE J. Sel. Areas Commun.* **2019**, *37*, 1402–1412. [CrossRef]

14. Bhatia, R.; Benno, S.; Esteban, J.; Lakshman, T.V.; Grogan, J. Unsupervised Machine Learning for Network-Centric Anomaly Detection in IoT. In Proceedings of the 3rd ACM CoNEXT Workshop on Big Data, Machine Learning and Artificial Intelligence for Data Communication Networks, Orlando, FL, USA, 9 December 2019; pp. 42–48.
15. Zhang, S.; Wang, Z.; Yang, J.; Bai, D.; Li, F.; Li, Z.; Wu, J.; Liu, X. Unsupervised IoT Fingerprinting Method via Variational Auto-encoder and K-means. In Proceedings of the ICC 2021—IEEE International Conference on Communications, Montreal, QC, Canada, 14–23 June 2021; pp. 1–6. [CrossRef]
16. Sikeridis, D.; Rimal, B.P.; Papapanagiotou, I.; Devetsikiotis, M. Unsupervised Crowd-Assisted Learning Enabling Location-Aware Facilities. *IEEE Internet Things J.* **2018**, *5*, 4699–4713. [CrossRef]
17. Liu, X.; Abdelhakim, M.; Krishnamurthy, P.; Tipper, D. Identifying Malicious Nodes in Multihop IoT Networks Using Diversity and Unsupervised Learning. In Proceedings of the 2018 IEEE International Conference on Communications (ICC), Kansas City, MO, USA, 20–24 May 2018; pp. 1–6. [CrossRef]
18. Liu, Y.; Wang, J.; Li, J.; Niu, S.; Song, H. Machine Learning for the Detection and Identification of Internet of Things Devices: A Survey. *IEEE Internet Things J.* **2022**, *9*, 298–320. [CrossRef]
19. Safi, M.; Dadkhah, S.; Shoeleh, F.; Mahdikhani, H.; Molyneaux, H.; Ghorbani, A.A. A Survey on IoT Profiling, Fingerprinting, and Identification. *ACM Trans. Internet Things* **2022**, *3*, 1–39. [CrossRef]
20. Meidan, Y.; Bohadana, M.; Shabtai, A.; Ochoa, M.; Tippenhauer, N.O.; Guarnizo, J.D.; Elovici, Y. Detection of unauthorized IoT devices using machine learning techniques. *arXiv* **2017**, arXiv:1709.04647.
21. Aksoy, A.; Gunes, M.H. Automated IoT Device Identification using Network Traffic. In Proceedings of the IEEE ICC, Shanghai, China, 20–24 May 2019; pp. 1–7.
22. Hamad, S.A.; Zhang, W.E.; Sheng, Q.Z.; Nepal, S. IoT Device Identification via Network-Flow Based Fingerprinting and Learning. In Proceedings of the 18th IEEE International Conference on Trust, Security and Privacy in Computing and Communications/13th IEEE International Conference on Big Data Science and Engineering (TrustCom/BigDataSE), Rotorua, New Zealand, 5–8 August 2019; pp. 103–111.
23. Bekerman, D.; Shapira, B.; Rokach, L.; Bar, A. Unknown malware detection using network traffic classification. In Proceedings of the 2015 IEEE Conference on Communications and Network Security (CNS), Florence, Italy, 28–30 September 2015; pp. 134–142. [CrossRef]
24. Kassambara, A. Practical Guide to Cluster Analysis in R: Unsupervised Machine Learning. 2017; Volume 1. Available online: http://www.sthda.com/english/ (accessed on 25 February 2023).
25. Jolliffe, I.T. *Principal Component Analysis for Special Types of Data*; Springer: Berlin/Heidelberg, Germany, 2002.
26. Lloyd, S. Least squares quantization in PCM. *IEEE Trans. Inf. Theory* **1982**, *28*, 129–137. [CrossRef]
27. Schubert, E.; Sander, J.; Ester, M.; Kriegel, H.P.; Xu, X. DBSCAN revisited, revisited: Why and how you should (still) use DBSCAN. *ACM Trans. Database Syst.* **2017**, *42*, 1–21. [CrossRef]
28. RaspAP. RaspAP: Simple Wireless AP Setup & Management for Debian-Based Devices. Available online: https://github.com/RaspAP (accessed on 25 February 2023).
29. Lashkari, A.H.; Draper-Gil, G.; Mamun, M.S.I.; Ghorbani, A.A. Characterization of tor traffic using time based features. In Proceedings of the ICISSP, Porto, Portugal, 19–21 February 2017; pp. 253–262.
30. Draper-Gil, G.; Lashkari, A.H.; Mamun, M.S.I.; Ghorbani, A.A. Characterization of encrypted and vpn traffic using time-related. In Proceedings of the 2nd International Conference Information Systems Security and Privacy, Rome, Italy, 19–21 February 2016; pp. 407–414.

information

MDPI

Article

Multimodal Approaches for Indoor Localization for Ambient Assisted Living in Smart Homes

Nirmalya Thakur * and Chia Y. Han

Department of Electrical Engineering and Computer Science, University of Cincinnati, Cincinnati, OH 45221-0030, USA; han@ucmail.uc.edu
* Correspondence: thakurna@mail.uc.edu

Abstract: This work makes multiple scientific contributions to the field of Indoor Localization for Ambient Assisted Living in Smart Homes. First, it presents a Big-Data driven methodology that studies the multimodal components of user interactions and analyzes the data from Bluetooth Low Energy (BLE) beacons and BLE scanners to detect a user's indoor location in a specific 'activity-based zone' during Activities of Daily Living. Second, it introduces a context independent approach that can interpret the accelerometer and gyroscope data from diverse behavioral patterns to detect the 'zone-based' indoor location of a user in any Internet of Things (IoT)-based environment. These two approaches achieved performance accuracies of 81.36% and 81.13%, respectively, when tested on a dataset. Third, it presents a methodology to detect the spatial coordinates of a user's indoor position that outperforms all similar works in this field, as per the associated root mean squared error—one of the performance evaluation metrics in ISO/IEC18305:2016—an international standard for testing Localization and Tracking Systems. Finally, it presents a comprehensive comparative study that includes Random Forest, Artificial Neural Network, Decision Tree, Support Vector Machine, k-NN, Gradient Boosted Trees, Deep Learning, and Linear Regression, to address the challenge of identifying the optimal machine learning approach for Indoor Localization.

Keywords: big data; machine learning; indoor localization; ambient assisted living; internet of things; smart homes; elderly population; indoor location; human–computer interaction; assistive technology

Citation: Thakur, N.; Han, C.Y. Multimodal Approaches for Indoor Localization for Ambient Assisted Living in Smart Homes. *Information* **2021**, *12*, 114. https://doi.org/10.3390/info12030114

Academic Editor: Spyros Panagiotakis

Received: 24 January 2021
Accepted: 5 March 2021
Published: 7 March 2021

1. Introduction

Technologies like Global Positioning Systems (GPS) and Global Navigation Satellite Systems (GNSS) have revolutionized navigation research by being able to track people, objects, and assets in real-time. Despite the significant success of these technologies in outdoor environments, they are still ineffective in indoor settings [1]. This is primarily for two reasons, first, these technologies depend on line of sight communication between GPS satellites and receivers which is not possible in an indoor environment and second, GPS provides a maximum accuracy of up to five meters [2]. With Industry 4.0, there has been an increasing need for developing systems for indoor navigation and localization for the future of living and working environments, which would involve human–computer, human–machine, and human–robot interactions in a myriad of ways. These environments could involve Smart Homes, Smart Cities, Smart Workplaces, Smart Industries, and Smart Vehicles, just to name a few. There are multiple application domains that are in need for a standard methodology for Indoor Localization. A system for Indoor Localization may broadly be defined as a system of interconnected devices, networks, and technologies that help to detect, track, and locate the position of people and objects inside closed or semi-closed environments, where technologies such as GPS or GNSS do not work [3]. As per [3], the market opportunities of Indoor Localization related systems are expected to be in the order of USD 10 billion by 2024 due to the diverse societal needs that such systems can address. Some potential applications of such Indoor Localization related technologies could include (1) tracking the location of products during smart manufacturing in automated or semi-automated manufacturing sites; (2) tracking the location and operation of unmanned vehicles or robots in industrial settings; (3) detecting the precise location of an elderly fall for communicating the same to emergency responders; (4) helping older adults with various forms of Cognitive Impairments (CI) to perform their daily routine tasks by directing them to specific locations for performing these activities; (5) tracking the precise location of individuals with Dementia or Alzheimer's when they face freezing of gait to alert caregivers; (6) assisting the visually impaired to reach specific objects of interest in both living and working environments; (7) helping individuals suffering from delirium to navigate from one place to the other for performing different activities; (8) detection of the precise location of the elderly when they face cramps or other forms of motor impairments; (9) detecting the location of patients in hospitals to avoid the need for in-person monitoring; and (10) automated tracking of different kinds of physical assets in Internet of Things (IoT)-based functional and work-related environments.

Only one area of interest, Ambient Assisted Living (AAL) of Elderly People during Activities of Daily Living (ADLs) in the future of technology-laden living environments, for instance, Smart Homes and Smart Cities, will be addressed in this work. AAL may broadly be defined as a computing paradigm that uses information technology and its applications to enhance user abilities, performance, and quality of life through interconnected systems that can sense, anticipate, adapt, predict, and respond to human behavior and needs. Human behavior in the confines of their living and functional environments is characterized by activities that they perform in these environments. In a broad scope, an activity may be defined as an interaction between a subject and an object, for the subject to achieve a desired end goal or objective. Here, the subject is the user who performs the activity and the set of environment parameters that they interface with during this activity are known as the objects. Based on the variations in the environment in which the activity is performed, the same activity may involve different objects that a user interfaces with, to reach the end goal. Similarly, the diversities in the user can also lead to different interaction patterns with objects for performing the same activity in the same or in a different environment [4]. Activities can have various characteristic features. These include—sequential, concurrent, interleaved, false start, and social interactions. Those activities that are crucial for one's sustenance and which one performs on a daily routine basis are known as Activities of Daily Living (ADLs). There are five broad categories of ADLs—Personal Hygiene, Dressing, Eating, Maintaining Continence, and Mobility [5].

People live longer these days due to advanced healthcare facilities. The population of elderly people has been on a constant rise and there are around 962 million elderly people [6] across the world. According to [7], by 2050, the population of elderly people is expected to become around 1.6 billion and outnumber the population of younger people globally. To add, the population of older adults, aged 80 years or more, is expected to increase three times and reach around 425 million by 2050. Increasing age is associated with physical disabilities, cognitive impairments, memory issues, and disorganized behavior which limit a person's ability to carry out their daily routine tasks in an independent manner. The worldwide costs of looking after elderly people with various forms of cognitive impairments, such as Dementia, is estimated to be around USD 818 billion and is increasing at a very fast rate [8]. In the United States alone, approximately 5.8 million elderly people currently have Dementia and 1 in every 3 seniors dies from Dementia. In 2020, care of people with Dementia accounted for approximately USD 305 billion to the U.S. economy, out of which the caregiver costs are estimated to be around USD 244 billion. It is predicted that these costs are going to rise steeply over the next few years [8]. A major challenge in this field is to make the future of Internet of Things (IoT)-based ubiquitous living environments, such as Smart Homes and Smart Cities aware, adaptive, and personalized so that they can contribute towards independent living and healthy aging of the elderly while fostering their biological, psychological, behavioral, physical, mental, and emotional well-being. Indoor Localization has an immense role to play towards addressing these challenges—both in terms of independent living and healthy aging of the elderly as well as for addressing the huge burden of their caregiving costs.

Despite several advances [9–59] in this field, which have been reviewed in a detailed manner in Section 2, multiple research challenges remain to be addressed. These include—(1) inability of the activity recognition and the activity analysis-based AAL systems to track the indoor location of the user during ADLs; (2) dependency of Indoor Localization systems on context parameters local to specific IoT-based settings which limit the functionalities of such systems to those specific environments; (3) need for better precision and accuracy for detection of the indoor location of a user; and (4) need to deduce and identify the optimal machine learning-based approach in view of the wide variety of learning approaches that have been investigated by researchers for development of Indoor Localization systems. Thus, addressing these above-mentioned challenges by exploring the intersections of Big Data, Machine Learning, Indoor Localization, Ambient Assisted Living, Internet of Things, Activity Centric Computing, Human–Computer Interaction, Pattern Recognition, and Assisted Living Technologies to provide a long-term, robust, feasible, easily implementable, sustainable, and economic solution to these global research challenges serves as the main motivation for the work presented in this paper. To summarize, the scientific contributions of this paper are as follows:

1. Big-Data driven methodology that studies the multimodal components of user interactions and analyzes the data from BLE beacons and BLE scanners to track a user's indoor location in a specific 'activity-based zone' during Activities of Daily Living. This approach was developed by using a k-nearest neighbor (k-NN)-based learning approach. When tested on a dataset it achieved a performance accuracy of 81.36%.
2. A context independent approach that can interpret the accelerometer and gyroscope data from diverse behavioral patterns to detect the 'zone-based' indoor location of a user in any IoT-based environment. Here, the 'zone-based' mapping of a user's location refers to mapping the user in one of the multiple 'activity-based zones' that any given IoT-based environment can be classified into based on the associated context attributes. This methodology was developed by using a Random Forest-based learning approach. When tested on a dataset it achieved a performance accuracy of 81.13%.
3. A methodology to detect the spatial coordinates of a user's indoor position based on the associated user interactions with the context parameters and the user-centered local spatial context, by using a reference system. The performance characteristics of this system were evaluated as per three metrics stated in ISO/IEC18305:2016 [31], which is an international standard for testing Localization and Tracking Systems. These metrics included root mean squared error (RMSE) in X-direction, RMSE in Y-direction, and the Horizontal Error, which were found to be 5.85 cm, 5.36 cm, and 7.93 cm, respectively. A comparison of the performance characteristics of this approach with similar works in this field that used the RMSE evaluation method showed that our system outperformed all recent works that had a similar approach.
4. A comprehensive comparative study of different machine learning approaches that include—Random Forest, Artificial Neural Network, Decision Tree, Support Vector Machine, k-NN, Gradient Boosted Trees, Deep Learning, and Linear Regression, with an aim to address the research challenge of identifying the optimal machine learning-based approach for Indoor Localization. The performance characteristics of each of these learning methods were studied by evaluating the RMSE in X-direction, the RMSE in Y-direction, and the Horizontal Error as per ISO/IEC18305:2016 [31]. The results and findings of this study show that the Random Forest-based learning approach can be considered as the optimal learning method for development of Indoor Localization and tracking related technologies.

This paper is organized as follows. We present a comprehensive overview of the related works in this field in Section 2. In Section 3, a brief overview is given about RapidMiner [60], a data science and machine learning software development platform, that has been used for development of all the methodologies proposed in this paper. Section 4 presents the methods and the steps associated with the development of the three novel methodologies for Indoor Localization that have been proposed in this work. Section 5 discusses the results and findings associated with each of these methodologies. In Section 6, we present the comparative study of different machine learning approaches that include—Random Forest, Artificial Neural Network, Decision Tree, Support Vector Machine, k-NN, Gradient Boosted Trees, Deep Learning, and Linear Regression, with an aim to address the research challenge of identifying the optimal machine learning-based approach for Indoor Localization. Section 7 elaborates the research challenges in this field and discusses how the work presented in this paper outperforms all similar works while addressing the associated research challenges. It is followed by Section 8 where conclusion and scope for future work are outlined.

2. Literature Review

In this section, we have reviewed different kinds of AAL-based systems and technologies that have primarily focused on Indoor Localization, Activity Recognition, and Activity Analysis in Smart and Interconnected IoT-based environments, such as Smart Homes and Smart Cities.

Machine learning approaches have been widely used by researchers to track people and objects in indoor environments and settings. Musa et al. [9] developed a system that used a non-line of sight approach and multipath propagation in the context of using the ultra-wide band methodology. The system used a cross-fold validation method to train a decision tree that could detect the indoor location of a user. A similar decision-tree driven machine learning framework was developed by Yim et al. [10]. The framework was equipped with the functionality to build the decision tree in the off-line phase and it used the fingerprinting approach for Indoor Localization. Sjoberg et al. [11] developed a visual recognition approach using the support vector machine (SVM) classifier. The system consisted of a visual bag-of-words model and other visual features of the environment that were used to train this classifier for Indoor Localization. A method of 2.5D indoor positioning was proposed by Zhang et al. [12]. Here, the SVM classifier was trained to detect the specific floor where a user is located based on the WiFi signal strength and thereafter it obtained the user's position information by analysis of other characteristics of the associated altitude data. Zhang et al. [13] developed a k-NN classification approach for Indoor Localization that used the signal strength fingerprint technology. The system assigned weights to the samples based on their associated signal strengths to divide them into clusters, where each cluster represented a specific location.

In [14], Ge et al. proposed an algorithm for indoor location tracking that was developed by using the k-NN approach. The algorithm used signal processing principles to detect and analyze the data coming from access points in the user's location to train the k-NN classifier. In [15], Hu et al. proposed another k-NN based learning approach that detected the location of the user based on the nearest access points of the user. One of the key findings of the work was that the condition of k = 1 led to the best positioning performance accuracy. The artificial neural network approach (ANN) was used by Khan et al. [16] for developing an indoor position detection system. The architecture of the approach involved studying and interpreting the data from Wireless Local Area Network (WLAN) access points and Wireless Sensor Networks (WSN) to train the artificial neural network (ANN) that could perform virtual tessellation of the available indoor space. Another neural network driven system was proposed by Labinghisa et al. [17]. This approach was based on the concept of virtual access points with an aim to increase the number of access points without the requirements of any additional hardware. These additional access points helped to track more user movement related data for training of the neural network.

A Wi-Fi fingerprint-based indoor positioning system was proposed by Qin et al. [18] that was neural network driven. The system used a convolutional denoising autoencoder to analyze and extract key features from the RSSI data, which were then used to train the neural network. The authors evaluated their system on two datasets to discuss its performance characteristics. In [19], Varma et al. used the Random Forest learning approach to perform Indoor Localization in an Internet of Things (IoT)-based environment during real-time experiments. The authors set up an IoT-based space with 13 beacons. The signals coming from these beacons, based on the user's varying position, were used to train the Random Forest model. A Wi-Fi signal analysis based indoor position detection system was proposed by Gao et al. [20] that also used the Random Forest classifier. It was developed and implemented by using the region-based division of location grids method to minimize the maximum error. The system adopted the method of adjusted cosine similarity to match the user's position with the exact grid by analyzing the fingerprint information. Linear regression methods have also been used by researchers to develop Indoor Localization systems and technologies. For instance, in [21], the authors developed a learning model where each anchor node had its own linear ranging method. Their linear regression model studied the distance between different anchors to perform anchor distance-based location detection.

The work proposed by Barsocchi et al. [22] involved development of a linear regression-based learning approach that calculated the user's distance from a reference point based on RSSI values. The approach consisted of the methodology to map these values into distances to detect the position of the user in the given environment. Zhang et al. [23] developed a deep learning-based 3D positioning framework for a hospital environment. The system used the data from the cell phone network as well as the Wi-Fi access points to determine the exact position of the user in terms of the latitude, longitude, and the level of the building at which they were present. Another deep learning-based Indoor Localization system was proposed by Poulose et al. [24]. The system developed heat maps from the RSSI signals obtained from the access points, to train the deep learning model. By conducting experiments, the authors evaluated the effectiveness of their system for its deployment in an autonomous environment. A Gradient Boosted Decision Tree approach was proposed by Wang et al. [25] that used the fingerprint methodology to detect the location of a user in an indoor setting. The authors used the concept of wavelet transform to filter the noises in the channel state information data, which was then used by the system to update the associated fingerprint information for the machine learning model. As can be seen from [9–25] a range of machine learning approaches have been used for development of various types of Indoor Localization systems for IoT-based environments. However, none of these works implemented multiple machine learning models to evaluate and compare their working to deduce the best machine learning approach in terms of the associated performance accuracy. Due to the differences in the datasets used or the real-time data that was collected, the associated data preprocessing steps that varied from system to system, the differences in train to test ratio of the data, and several other dissimilar steps that were associated with the developments of each of these machine learning models, their final performance accuracies cannot be directly compared to deduce the best approach. The challenge is therefore to identify the optimal machine learning model that can be used to develop the future of Indoor Localization systems, Indoor Positioning systems, and Location-Based Services.

In addition to machine learning-based approaches, context reasoning-based approaches have also been investigated by researchers to develop Indoor Localization systems. Such approaches are limited and functional only in the confines of the specific environments for which they were developed. In [26], Lin et al. developed an indoor positioning system specific for the factory environments of the Hon-Hai Precision Industry. An intelligent indoor parking system was developed by Liu et al. [27] that could help with indoor parking. The system was implemented in a shopping mall environment to test its performance characteristics. Jiang et al. [28] proposed an Indoor Localization system for hospital environments that used concepts from GPS and UWB technologies. Barral et al. [29] developed a methodology that could track the location of forklift trucks in an industry-based environment. Zadeh et al. [30] proposed an Indoor Localization framework for an academic environment to help with taking attendance of students. As can be seen from [26–30], these Indoor Localization systems are specific to certain environments, as they are dependent on the associated features and characteristics of the environment for which they were designed. For instance, the system proposed in [29] cannot be deployed in any of the environments described in [26–28,30]. The environments described in [26–30] represent just a few of the IoT-based environments associated with the living and functional areas of humans in the future of interconnected Smart Cities. As such context reasoning-based systems are not functional in other IoT-based settings, the challenge is thus to develop a means for Indoor Localization that is not environment dependent and can be seamlessly deployed in any IoT-based setting irrespective of the associated context parameters and their attributes.

Here, we also review ISO/IEC 18305:2016, which is an international standard for evaluating localization and tracking systems [31]. This standard was jointly prepared and developed by the Joint Technical Committee ISO/IEC JTC 1, Information technology, Subcommittee SC 31. ISO/IEC JTC 1/SC 31 is a standardization subcommittee of the joint committee ISO/IEC JTC 1 of the International Organization for Standardization (ISO) and the International Electrotechnical Commission (IEC), which develops and facilitates international standards, technical reports, and technical specifications in the field of automatic identification and data capture techniques. This standard is one of the outcomes of the EU FP7 project EVARILOS—Evaluation of RF-based Indoor Localization Solutions for the Future Internet [32]. The main objective of this standard is to define a standard set of testing and evaluation measures or methods that can be used to evaluate the performance metrics of different types of Indoor Localization systems, Indoor Positioning systems, and Location Based Services in different scenarios. It provides a comprehensive list of 14 such scenarios and 5 types of buildings where this standard can be implemented. It is worth mentioning here that the standard discusses these settings from the viewpoint of localization of a person, object as well as a robot in such scenarios. Such definitions of scenarios include the characterization of the associated motion as well as the definition of the number of entities (human, objects, or robots) that are required to be tracked in that given scenario. It also lists 30 metrics for evaluating the performance characteristics in each case. The metrics related to calculation of different kinds of errors are introduced in Chapter 8 of this standard. Some of these metrics that have been widely used by researchers since the inception of this standard include—RMSE, Mean of Error, Covariance Matrix of Error, Mean of Absolute Error, and Mean and Standard Deviation of Vertical Error.

While several of these metrics have been used by researchers to evaluate their Indoor Localization systems, we focus on one specific performance metric—the RMSE. The standard presents the formulae for determination of the RMSE in the X-direction, RMSE in the Y-direction, and RMSE in the X-Y plane. When the RMSE is determined in the X-Y plane, it is referred to as Horizontal Error as per the definitions of the standard [31]. Next, we review some of the works related to Indoor Localization systems that have used the RMSE method for evaluating their performance characteristics. In [33], the authors analyzed the RSSI data coming from multiple anchor nodes set up in a Wireless Sensor Network system. They used Kalman filter to determine the direction and speed of the user and their system had a RMSE of 1.4 m. In [34], the authors developed a new RFID-based device that could sense proximity tags in the environment to detect the indoor location of a user with a RMSE of 0.32 m. Angermann et al. [35] developed a Bayesian estimation-based framework for pedestrian localization and mapping that had a RMSE of 1 to 2 m. Evennou et al. [36] used signal processing-based methods to develop an Indoor Localization system that had a RMSE of 1.53 m. Wang et al. [37] used particle filters and extended the traditional WLAN methods to develop a pedestrian tracking system that had a RMSE of 4.3 m. A Monte Carlo-based Indoor Localization algorithm was proposed in [38]. The RMSE of this system was 1.2 m. A SVM classifier was proposed in [39] that used smartphone data for performing Indoor Localization. The performance characteristics presented in the paper show that the value of the RMSE of this system was 4.55 m. Another smart phone-based system, known as HIVE [40], that was Hidden Markov Model driven was proposed by Liu et al.

The system had a RMSE of 3.1 m. The work by Chen et al. [41] involved fusion of data coming from smartphone sensors, WiFi, and Landmarks, which were analyzed by a Kalman Filter for Indoor Localization. This method had a RMSE of 1 m. Li et al. [42] proposed a sensor technology-driven smartphone-based pedestrian location detection system that had a RMSE of 2.9 m. In [43], the authors analyzed iBeacon measurements and a calibration range, which was thereafter used to develop a Kalman filter for pedestrian dead reckoning. The system had a RMSE of 1.28 m. As per [44] and [45], (1) the average dimensions of newly built one-bedroom apartments and two-bedroom apartments in United States in 2018 were 757 square feet (70.3276 square meters) and 1138 square feet (105.7236 square meters), respectively. Considering such dimensions of living spaces in the context of AAL in Smart Homes, it is the need of the hour that the Indoor Localization systems and tracking related technologies become more precise in terms of detecting the exact indoor location of the user. The challenge is this context is thus to develop Indoor Localization systems that have lesser RMSE as compared to [33–43].

AAL in Smart Homes is not only about tracking the indoor location of the user, it also involves analyzing their behaviors and activity patterns to enhance their quality of life and user experience in the context of diverse user interactions. Next, we review some of the recent AAL-based technologies that have focused on activity recognition and analysis in Smart Homes. An activity recognition framework was proposed by Ranieri et al. [46] for AAL of elderly in Smart Homes. The framework focused on studying various parameters of user interaction data from videos, wearable sensors, and ambient sensors to analyze activities. An activity analysis approach for monitoring elderly behavior during daily activities was proposed by Fahad et al. [47]. The objective of the work was to track elderly behavior and detect any possible anomalies in the same that could have resulted from cognitive or physical impairments or decline in abilities. A smart phone accelerometer-based activity recognition framework was proposed by Suriani et al. [48]. The authors developed the system by using spiking neural networks and used datasets to evaluate its performance characteristics. A similar smartphone-based application was proposed by Mousavi et al. in [49] which could detect falls in elderly. The system was trained using SVM and had a performance accuracy of 96.33%. A wearable sensor driven fall detection system was proposed by Alarifi [50]. The wearables were placed on six different locations on the user's body to track multimodal components of motion and behavior data, which were studied and analyzed by a convolution neural network. In [51], Al-Okby proposed a smart wearable device for detection of elderly falls. The device analyzed multimodal components of the user's motion and had the functionality to alert caregivers in the event of a fall. As can be seen from these works [46–51], multiple components of human postures, pose, motion, and behavior can be tracked and analyzed for development of AAL-based activity recognition and fall detection systems. However, the main limitation of these systems is their inability to track the location of the user. For instance, consider the example of an elderly staying alone in an apartment located in a multistoried building. When this elderly person experiences a fall, a fall detection system such as [51] could alert caregivers but the lack of the precise location information could cause delay of medical attention or assistive care. This is because the location of the elderly can be tracked in terms of the building information from GPS, but the specific floor, apartment, or room-related information is not available to the caregivers or emergency responders. Such delay of care can have both short-term and long-term health-related impacts to the elderly. Thus, it is the need of the hour that AAL-based systems not only track, monitor, and analyze elderly behavior but they should also be equipped with the functionality to detect the indoor location of the elderly.

Cloud computing-based approaches have also been used in recent studies [52–55] for development of AAL-based systems and applications that can monitor human behavior and trigger alarms as well as track the location of the user. While this concept of cloud computing applied to AAL technologies holds potential, but these existing systems also have several limitations. For instance—the system proposed by Navarro et al. [53] is environment specific and was designed, adapted, and built specifically for Fundació Ave Maria [56], which is a non-profit organization in Spain, so, the same design cannot not be seamlessly applied to any other environment consisting of varying environment parameters that would be associated with diverse range of human behavior and user interactions; the system proposed by Nikoloudakis et al. [52] uses an outdoor positioning mechanism that can only detect whether the user leaves the premises of their location; the work proposed by Facchinetti et al. [55] is a mobile app and that brings into context these challenges—(i) elderly people are less likely to download a mobile app as compared to the other age groups [57], (ii) elderly people are naturally resistant to using different kinds of technology-based apps on their phones, tablets, and other interactive devices [58], and (iii) older adults face multiple usability issues with such apps [59]; even though another system proposed by Nikoloudakis et al. [54] presents approaches for both indoor and outdoor positioning, it cannot model and analyze the fine grain levels of human activity such as atomic activity, context attributes, core atomic activity, and core context attributes along with their associated weights, in the context of dynamic user interactions during ADLs. To add to the above, for all these systems [52–55], the RMSE for detection of the indoor location of the user is also not less than 1 m. Thus, to summarize, the main research challenges in this field are as follows:

1. The AAL-based activity recognition, activity analysis, and fall detection systems currently lack the ability to track the indoor location of the user. It is highly essential that in addition to being able to track, analyze, and interpret human behavior, such systems are also able to detect the associated indoor location information, so that the same can be communicated to caregivers or emergency responders, to facilitate a timely care in the event of a fall or any similar health related emergencies. Delay in care from a health-related emergency, such as a fall, can have both short-term and long-term health related impacts.
2. Several Indoor Localization systems are context-based and are functional only in the specific environments in which they were developed. For instance, [26] was developed for factory environments, [27] was developed for indoor parking, [28] was developed for hospital settings, [29] was developed for tracking forklift trucks in industry-based settings, and [30] was developed for performing Indoor Localization in academic environments for taking attendance of students. The future of interconnected Smart Cities would consist of a host of indoor environments in the living and functional spaces of humans, which would be far more diverse, different, and complicated as compared to the environments described in [26–30]. The challenge is thus to develop a means for Indoor Localization that is not environment dependent and can be seamlessly deployed in any IoT-based setting irrespective of the associated context parameters and their attributes.
3. In view of the average dimensions of the living spaces in Smart Homes, the RMSE of the existing Indoor Localization systems are still high and greater precision and accuracy for detection of indoor location in the need of the hour.
4. A range of machine learning-based approaches—Random Forest, Artificial Neural Network, Decision Tree, Support Vector Machine, k-NN, Gradient Boosted Trees, Deep Learning, and Linear Regression, have been used by several researchers [9–25] for development of various types of Indoor Localization systems for IoT-based environments. Identification of the optimal machine learning model that can be used to develop the future of Indoor Localization systems, Indoor Positioning Systems, and Location-Based Services is highly necessary.

Addressing these above-mentioned challenges by exploring the intersections of Big Data, Machine Learning, Indoor Localization, Ambient Assisted Living, Internet of Things, Activity Centric Computing, Human–Computer Interaction, Pattern Recognition, and Assisted Living Technologies to contribute towards AAL in Smart Homes serves as the main motivation for this work.

3. Technology Review

This section briefly reviews RapidMiner, formerly known as Yet Another Learning Environment (YALE) [60], which we have used for the work presented in this paper. RapidMiner is a software tool that allows development and implementation of a wide range of Machine Learning, Data Science, Artificial Intelligence, and Big Data related algorithms and models. The initial version of this tool was developed back in 2001 at the Technical University of Dortmund. From 2006, a company called Rapid-I started implementing additional functionalities and features in the tool. A year later, the name of the software was changed from YALE to RapidMiner and six years from then, the name of the company was changed from Rapid-I to RapidMiner. As of current day, RapidMiner is used both for educational research and for development of commercial applications and products.

RapidMiner is available as an integrated development environment that consists of—(1) RapidMiner Studio, (2) RapidMiner Auto Model, (3) RapidMiner Turbo Prep, (4) RapidMiner Go, (5) RapidMiner Server, and (6) RapidMiner Radoop. For all the work related to the methodologies proposed in this paper, we used RapidMiner Studio. For the remainder of this paper, wherever we have mentioned "RapidMiner", we have referred to "RapidMiner Studio" and not any of the other development environments associated with this software tool.

RapidMiner is developed as an open core model that provides a rich Graphical User Interface (GUI) to allow users to develop different kinds of applications, generate workflows, and implement various algorithms. These applications, workflows, or algorithms are known as "processes" and they consist of multiple "operators". In a RapidMiner "process", each of its "operators" are associated with a specific functionality which is required for working of the "process". RapidMiner provides a range of built-in "operators" that can be directly used with or without any modifications for development of a specific "process". There is also a certain category of "operators" that can be used to modify the characteristic features of other "operators". The tool also allows developers to create their own "operators" and these can be shared and made available to all other users of RapidMiner via the RapidMiner Marketplace.

For development of any RapidMiner "process", the associated "operators" are always connected either in a linear fashion or in a hierarchical manner as shown in Figures 1 and 2, respectively. In Figure 1, 'Learner', 'Model Applier', and 'Evaluator' refer to different "operators" in RapidMiner. The inputs to these "operators" are shown by arrows pointing towards these respective "operators" and the outputs of these "operators" are shown by arrows pointing away from these respective "operators". For instance, the input to the 'Evaluator' "operator" is Example-Set and the output produced by this "operator" is the Performance Vector. Here, only three "operators" have been shown for representation of the linear arrangement amongst "operators", however, in an actual RapidMiner "process", the number of "operators" can vary as well as the specific "operators" could be different from the three "operators" shown in Figure 1. Similarly, a typical RapidMiner "process" is shown in Figure 2 which shows hierarchical arrangement amongst Methods 1, 2, and 3, each of which are "operators" Here, the "operators" 'Learner', 'Model Applier', and 'Evaluator' are also connected in a hierarchical manner. In an actual RapidMiner "process" the number and types of "operators" connected hierarchically could be different as compared to the number and types of "operators" shown in Figure 2.

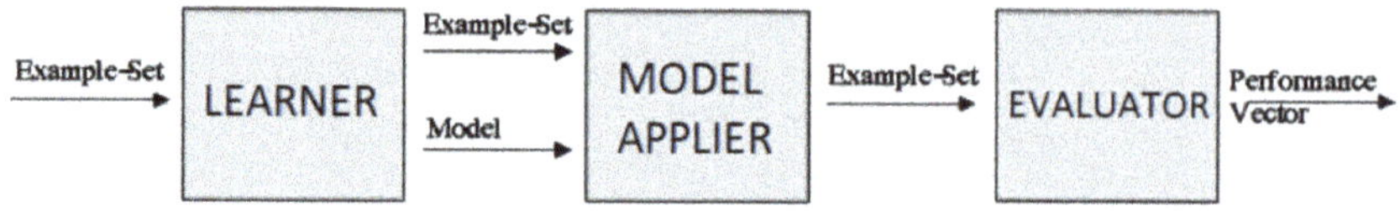

Figure 1. Layout of a typical linear "process" in RapidMiner Studio. This "process" shows linear arrangement amongst three RapidMiner "operators"—the 'Learner' operator, the 'Model Applier', and the 'Evaluator' [4].

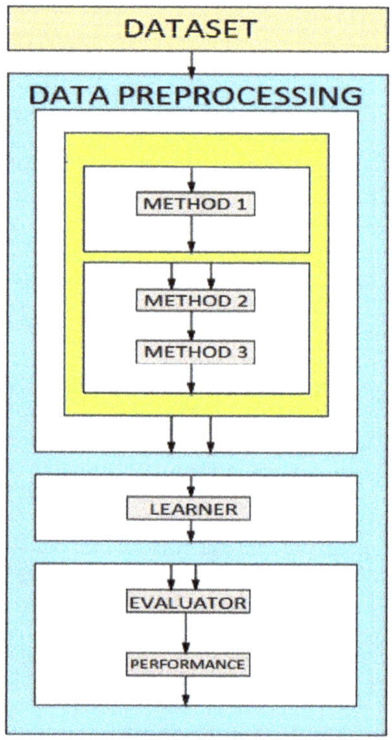

Figure 2. Layout of a typical hierarchical "process" in RapidMiner Studio. This "process" shows hierarchical arrangement amongst Methods 1, 2, and 3 each of which are "operators". The "operators"—'Learner', 'Model Applier', and 'Evaluator' are also connected in a hierarchical manner [4].

The following are some of the salient features of RapidMiner Studio:

1. It provides built-in "operators" with distinct functionalities that can be directly used or modified for development and implementation of Machine Learning, Data Science, Artificial Intelligence, and Big Data related algorithms and applications.
2. RapidMiner is developed using Java. This makes RapidMiner "processes" platform independent and Write Once Run Anywhere (WORA), which is a characteristic feature of Java.
3. The tool allows downloading multiple extensions for seamless communication and integration of RapidMiner "processes" with other software and hardware platforms.
4. Scripts written in any programming language, such as Python and R can also be integrated in a RapidMiner "process" to add additional functionalities to the same.
5. The tool allows development of new "operators" and seamless sharing of the same via the RapidMiner community.

6. It also consists of "operators" that allow this software tool to connect with social media profiles of the user, such as Twitter and Facebook, to extract tweets, comments, posts, reactions, and related social media activity.

RapidMiner is developed as a client-server model. The server is made available as on-premise, in public or as a private cloud infrastructure. There are two different versions of RapidMiner available—the free version and the commercial version. The primary difference between these two versions is that the free version has a data processing limit of 10,000 rows for any "process". For all the work presented in this paper, the free version of RapidMiner 9.8.001 was downloaded and installed on a Microsoft Windows 10 Computer with Intel (R) Core (TM) i7-7600U CPU @ 2.80GHz, 2904 MHz, 2 Core(s) and 4 Logical Processor(s). The datasets that were studied and analyzed in this paper did not have more than 10,000 rows of data so this limitation of the free version of RapidMiner did not have any effect on the methodologies or the associated results and findings.

There are a few similar software tools that allow seamless development and implementation of Machine Learning, Data Science, Artificial Intelligence, and Big Data related algorithms and applications. Two such software tools which are very popular are—(1) Waikato Environment for Knowledge Analysis (WEKA) [61] and (2) MLC++ [62]. WEKA, developed in Java, allows development and implementation of various kinds of machine learning methods—such as regression, classification, feature selection, cross-validation, and bootstrapping. MLC++ is a C++ library that allows development and implementation of only supervised machine learning algorithms. The primary limitation of both WEKA and MLC++ is that they do not allow nesting of "operators" as supported by RapidMiner. In RapidMiner, "operator" nesting can be done either in a linear form or in a hierarchical form as shown in Figures 1 and 2, respectively. The only means for implementing such a feature in WEKA or MLC++ is by creating duplicate copies of the original dataset. However, this process is time consuming and requires a lot of memory space. To add to the above limitation of WEKA, it needs the current dataset to be available in the main memory of the system in which it is being executed. This also contributes towards consumption of computer memory. Neither WEKA nor MLC++ allow inclusion of programming scripts, such as scripts written in Python or R in their respective applications. To add to the above, MLC++ is not platform independent and neither does it have the WORA feature. In view of the above limitations of WEKA and MLC++ and the contrasting salient features of RapidMiner, RapidMiner was used for development of all the methodologies outlined in this paper.

4. Development of the Proposed Methodologies for Indoor Localization

For development of the proposed methodologies for Indoor Localization for AAL during ADLs performed in an IoT-based environment, such as a Smart Home, we posit the following:

(a) The Received Signal Strength indicator (RSSI) data coming from BLE scanners and BLE beacons can be studied and analyzed to detect the changes in a user's instantaneous location during different activities, which are a result of the varying user interactions with dynamic context parameters.

(b) The dynamic changes in the spatial configurations of a user during different activities can be interpreted by the analysis of the behavioral patterns that are localized and distinct for different activities.

(c) Tracking and analyzing the user interactions with the context parameters along with the associated spatial information, by using a reference system, helps to detect the dynamic spatial configurations of the user.

Here, we use the concept of complex activity analysis for analysis of ADLs at a macro and micro level. A complex activity may be broadly defined as a collection of atomic activities (Ati), context attributes (Cti), core atomic activities (γAt), and core context attributes (ρCt) along with their characteristics [63]. The atomic activities refer to the small actions and tasks associated with an activity. The environment variables or parameters on which these tasks are performed are known as the context attributes. The core atomic activities refer to the atomic activities that are crucial for completion of the given complex activity and the context attributes on which these core atomic activities occur are known as the core context attributes. The atomic activities that are performed at the beginning and end of a given complex activity are known as start atomic activities (AtS) and end atomic activities (AtE), respectively. The associated context parameters are known as start context attributes (CtS) and end context attributes (CtE), respectively. This is further elaborated in Figure 3. Figure 4 describes a few atomic activities and complex activities in a typical environment [4]. Tables 1 and 2 show the complex activity analysis of two typical ADLs, Preparing Breakfast and Eating Lunch [4], studied in terms of the associated atomic activities, context attributes, core atomic activities, and core context attributes along with their associated weights, which can be determined by probabilistic reasoning principles [63]. As per [63], a greater weight of an Ati or Cti signifies greater relevance of the same towards the given complex activity. Therefore, weights of all γAt and ρCt are higher as compared to the rest of the Ati or Cti. The weights associated with the Ati, Cti, γAt, ρCt, AtS, AtE, CtS, and CtE are used to determine the threshold function of the given complex activity. The threshold function underlines the condition for completion of the complex activity [63].

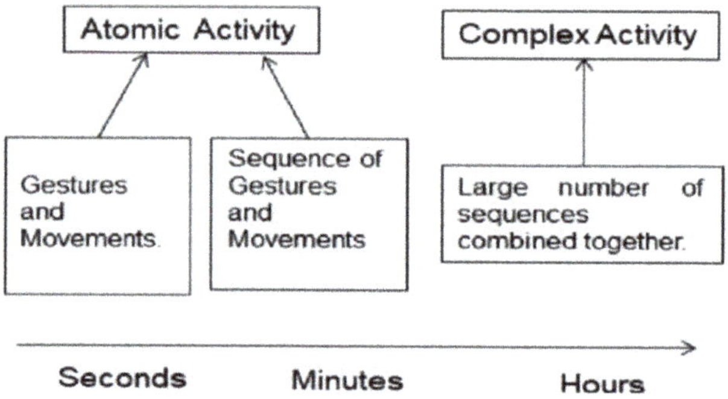

Figure 3. Representation of Atomic Activities and Complex Activities [4].

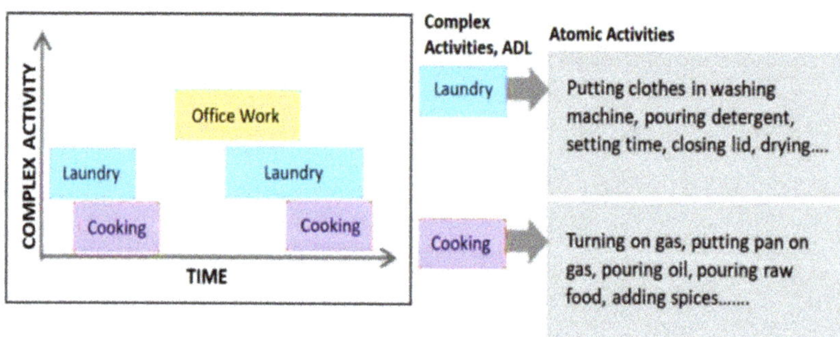

Figure 4. Description of Atomic Activities associated with two different Complex Activities in a typical environment [4].

Table 1. Analysis of a typical ADL, Preparing Breakfast (PB) from [4], studied in terms of the associated atomic activities, context attributes, core atomic activities, core context attributes and their threshold values.

Complex Activity WCAtk (PB Atk)—PB (0.73)	
Ati	At1: Standing (0.10) At2: Walking Towards Toaster (0.12) At3: Putting bread into Toaster (0.15) At4: Setting the Time (0.15) At5: Turning off toaster (0.25) At6: Taking out bread (0.18) At7: Sitting Back (0.05)
Cti	Ct1: Lights on (0.10) Ct2: Kitchen Area (0.12) Ct3: Bread Present (0.15) Ct4: Time settings working (0.15) Ct5: Toaster Present (0.25) Ct6: Bread cool (0.18) Ct7: Sitting Area (0.05)
AtS and CtS	At1, At2, and Ct1, Ct2
AtE and CtE	At6, At7, and Ct6, Ct7
γAt and ρCt	At3, At4, At5, At6 and Ct3, Ct4, Ct5, Ct6

Table 2. Analysis of a typical ADL, Eating Lunch (EL) from [4], studied in terms of the associated atomic activities, context attributes, core atomic activities, core context attributes, and their threshold values.

Complex Activity WCAtk (EL Atk)—EL (0.72)	
Ati	At1: Standing (0.08) At2: Walking Towards Dining Table (0.20) At3: Serving Food on a Plate (0.25) At4: Washing Hand/Using Hand Sanitizer (0.20) At5: Sitting Down (0.08) At6: Starting to Eat (0.19)
Cti	Ct1: Lights on (0.08) Ct2: Dining Area (0.20) Ct3: Food Present (0.25) Ct4: Plate Present (0.20) Ct5: Sitting Options Available (0.08) Ct6: Food Quality and Taste (0.19)
AtS and CtS	At1, At2, and Ct1, Ct2
AtE and CtE	At5, At6, and Ct5, Ct6
γAt and ρCt	At2, At3, At4 and Ct2, Ct3, Ct4

As can be seen from Table 1, for the complex activity of Preparing Breakfast in the environment described in [4], the atomic activities (Ati) are—Standing, Walking Towards Toaster, Putting bread into Toaster, Setting the Time, Turning off toaster, Taking out bread, and Sitting Back. The weights associated with these Ati are—0.10, 0.12, 0.15, 0.15, 0.25, 0.18, and 0.05, respectively. The associated context attributes (Cti) are—Lights on, Kitchen Area, Bread Present, Time settings working, Toaster Present, Bread cool, and Sitting Area. The weights associated with these Cti are—0.10, 0.12, 0.15, 0.15, 0.25, 0.18, and 0.05, respectively. These weights were assigned as per the probabilistic reasoning principles outlined in [63]. The Ati with the highest weights were identified as the Core Atomic Activities (γAt) as per the definition of γAt in [63]. The corresponding context attributes were identified as Core

Context Attributes (ρCt). Therefore, the list of γAt for this complex activity are—At3, At4, At5, and At6. Their corresponding context attributes—Ct3, Ct4, Ct5, and Ct6 were therefore considered as Core Context Attributes (ρCt). The Ati associated with the start and end of this complex activity, or in other words the Start Atomic Activities (AtS) and End Atomic Activities (AtE) are—At1, At2 and At6, At7, respectively. The corresponding context attributes—Ct1, Ct2 and Ct6, Ct7 were therefore considered as Start Context Attributes (CtS) and End Context Attributes (CtE), respectively. It is worth mentioning here that this complex activity analysis was performed based on the specific environment parameters described in [4] and this analysis can change if the same complex activity is performed in an environment which has a different set of environment parameters as compared to the environment described in [4].

As can be seen from Table 2, for the complex activity of Eating Lunch in the environment described in [4], the atomic activities (Ati) are—Standing, Walking Towards Dining Table, Serving Food on a Plate, Washing Hand/Using Hand Sanitizer, Sitting Down, and Starting to Eat. The weights associated with these Ati are—0.08, 0.20, 0.25, 0.20, 0.08, and 0.19, respectively. The associated context attributes (Cti) are—Lights on, Dining Area, Food Present, Plate Present, Sitting Options Available, and Food Quality and Taste. The weights associated with these Cti are—0.08, 0.20, 0.25, 0.20, 0.08, and 0.19, respectively. These weights were assigned as per the probabilistic reasoning principles outlined in [63]. The Ati with the highest weights were identified as the Core Atomic Activities (γAt) as per the definition of γAt in [63]. The corresponding context attributes were identified as Core Context Attributes (ρCt). Therefore, the list of γAt for this complex activity are—At2, At3, and At4. Their corresponding context attributes—Ct2, Ct3, and Ct4 were therefore considered as Core Context Attributes (ρCt). The Ati associated with the start and end of this complex activity or in other words the Start Atomic Activities (AtS) and End Atomic Activities (AtE) are—At1, At2 and At5, At6, respectively. The corresponding context attributes—Ct1, Ct2 and Ct5, Ct6 were therefore considered as Start Context Attributes (CtS) and End Context Attributes (CtE), respectively. It is worth mentioning here that this complex activity analysis was performed based on the specific environment parameters described in [4] and this analysis can change if the same complex activity is performed in an environment which has a different set of environment parameters as compared to the environment described in [4].

In the following sub sections, we outline the methodologies for development of the multimodal approaches for Indoor Localization for AAL for testing and evaluation of the above three hypotheses.

4.1. Indoor Localization from BLE Beacons and BLE Scanners Data during ADLs

The following are the steps for development of this proposed functionality:

i. Set up an IoT-based environment, within a spatial context such as indoor layout of rooms with furniture's and appliances, using wearables and wireless sensors to collect the Big Data related to different ADLs. The associated representation scheme involves mapping the entire spatial location into non-overlapping 'activity-based zones', distinct to different complex activities, by performing complex activity analysis [63].

ii. Analyze the ADLs in terms of the associated atomic activities, context attributes, core atomic activities, and core context attributes and their associated threshold values by probabilistic reasoning principles [63].

iii. Infer the semantic relationships between the changing dynamics of atomic activities, context attributes, core atomic activities, and core context attributes associated to different ADLs to study and interpret the spatial and temporal features of these ADLs.

iv. Study the characteristics of the data coming from the wireless sensors to analyze the associated RSSI data from BLE beacons and BLE scanners, recorded during different ADLs, based on the user's proximity to the context attributes in each 'activity-based zone'. This information helps to infer the user's presence or absence in each of these 'activity-based zones'. For instance, when the user performs a typical complex activity—cooking using microwave, based on the user's proximity to the microwave, the user's presence can be deduced in a 'zone' where the microwave is present.

v. Associate the relationships from (iii) with the characteristics of the RSSI data from BLE beacons and BLE scanners and map the entire IoT-based environment into non-overlapping 'activity-based zones', that are distinct to each ADL, by taking into consideration all possible complex activities that may be performed in the confines of the given IoT-based space. For instance, in a typical IoT-based environment [4], if the complex activities performed include—Watching TV, Using Laptop, Listening to Subwoofer, Using Washing Machine, Cooking Food, and Taking Shower; the associated 'activity-based zones' could be TV zone, Laptop zone, Subwoofer zone, Washing Machine zone, Cooking zone, and Bathroom zone. This inference of the respective 'zones' is based on the complex activity analysis [63] of all these activities as presented in [4].

vi. Split the data into training set and test set and train a learning model to study these relationships and patterns in the data to detect the indoor location associated with the given ADL, based on detecting the user's presence or absence in a specific 'activity-based zone' at a specific point of time.

vii. Analyze the performance characteristics of the learning model by using a confusion matrix.

Step (i) above involves setting up a Big Data collection methodology to study, track, and interpret the multimodal components of user interactions associated with different complex activities performed in each of these 'activity-based zones'. The data collection could be performed by using the Context-Driven Human Activity Recognition Framework that has already been developed, implemented, and tested at the Multimedia and Augmented Reality Lab, in the Department of Electrical Engineering and Computer Science at the University of Cincinnati. The results of the same were published in [64]. This framework was developed based on the work by Ma [65]. This framework, developed in the form of a software package, has multiple functionalities related to Big Data collection and we briefly review the same here.

First, it uses Microsoft Kinect Sensors to track the varying changes in the postures, gestures, and user behavior by performing skeletal tracking. In this process of skeletal tracking, the human body is represented in the form of a skeletal with 20 joint points and their associated characteristics. These joint points include—hip center, spine, shoulder center, head, left shoulder, left elbow, left wrist, left hand, right shoulder, right elbow, right wrist, right hand, left hip, left knee, left ankle, left foot, right hip, right knee, right ankle, and right foot. Characteristics of these joint points—such as joint point distance, joint point rotation, and joint point speed are tracked by this framework to detect and reason various movements and behavioral patterns for performing activity recognition. For instance, when a person is clapping, the distance between the joint point pairs (7,11) and (8,12) decreases and then increases in a periodic manner, where 7, 11, 8, and 12 represent the left wrist, right wrist, left hand, and right hand joint points, respectively. Based on these joint point characteristics, the framework can classify behaviors as Type-1 and Type-2. Behaviors associated with the lower limb are classified as Type-1 and behaviors associated with the upper limb are classified as Type-2, respectively. The Type-1 behaviors that can be recognized by this framework include—standing, walking, and sitting. The Type-2 behaviors that can be recognized by this framework include—waving, talking over cell phone, reading book or magazine, sleeping while seated, seated relaxed, and making hand gestures while talking. The third layer of the framework analyzes the relationships between changing behaviors associated with different activities by interpreting the user interactions with context parameters as well as the object behaviors in the user's spatial environment. It uses context-driven reasoning principles to perform complex activity recognition and analysis. To add, this layer can also track the sequence in which different complex activities are performed. The fourth layer of this framework allows capturing of social interactions and can perform all the functionalities of the above three layers while considering two users in the given IoT-based space at any point of time. This framework was developed as a Microsoft Windows-based application [64] that can seamlessly communicate and connect with both wireless and wearable sensors used to collect Big Data related to different complex activities. It consists of an intuitive user interface that shows the real-time Big Data coming from the IoT-based sensors as well as it shows the analysis of the same to deduce the associated activity being performed as per the methodologies outlined above.

For development of the proposed Big-Data driven methodology that studies the multimodal components of user interactions and analyzes the data from BLE beacons and BLE scanners to track a user's indoor location in a specific 'activity-based zone' during ADLs, we used an open-source dataset by Tabbakha et al. [66]. This dataset was chosen because its attributes were same as the real-time data that could be collected and analyzed by the Context-Driven Human Activity Recognition Framework [64] as outlined above. This dataset contains activity and human behavior related data collected from both wireless sensors and wearables in an IoT-based environment. The data attributes present in this dataset include the accelerometer data, gyroscope data, and the RSSI data obtained from BLE beacons and BLE scanners. The simulated smart home environment in which this data was collected consisted of four rooms or 'zones'—kitchen, bedroom, office, and toilet. For collecting the data as presented in this dataset [66], the authors developed a wearable device by using the Linkit 7697 and the MPU6050 sensors. This device was placed on the user's waist during each experimental trial. This wearable device tracked the behavior related information of the user as well as collected position related data with respect to the user's location in each of the four rooms or 'zones'. A BLE beacon was incorporated in this wearable and Raspberry Pi-based BLE scanners were installed at different locations of the IoT-based space. These scanners tracked the position of the user by sensing the BLE beacon and interpreting the associated RSSI data. Each of these rooms or 'zones' had one BLE scanner placed on or near a context attribute associated with the distinct complex activity that would be performed in that 'zone'. The scanners were placed on the kitchen table in the kitchen 'zone', on the bed in the bedroom 'zone', next to the working table in the 'office' zone, and next to the toilet door in the toilet 'zone'. The authors set the advertising interval of the BLE beacon to 100 ms with the transmitting power of up to −30 dBm. The scanner was programmed to report to the data collection server if the beacon was missing or in other words if the user was not present in that 'zone'. For such scenarios, the associated RSSI value was updated to −120 in the dataset to indicate that the BLE beacon was out of range. The behavior related data was collected by this wearable by tracking the acceleration data (along the X, Y, and Z axes) and gyroscope data (along the X, Y, and Z axes) associated with the user's movements. The data of the accelerometer and gyroscope from the wearable were sampled at the rate of 20 Hz. A total of 20 volunteers (10 males and 10 females) had participated in the experimental trials. The attributes that we used for development of this functionality included the RSSI data from BLE beacons and BLE scanners in the simulated environment and the location information that was associated with the different ADLs performed in this environment.

To study, analyze, and interpret these relationships that exist in the dataset, to detect the associated spatial context, i.e., to infer the location of the user within a specific 'activity-based zone' during ADLs, in the indoor room layout at a specific point of time, we used RapidMiner [60], because of its salient features and characteristics that make it highly suitable for development of such an application. These features and characteristics of RapidMiner as well as additional details about the relevance for selection of RapidMiner for this work were outlined in Section 3.

4.1.1. System Architecture of the Methodology for Indoor Localization from BLE Beacons and BLE Scanners Data during ADLs

The flowchart of this proposed methodology is shown in Figure 5. This figure outlines the operation of this methodology, as discussed in the previous section, at a broad level. Here, by "Split Data", we refer to splitting the data into training set and test set, with

80% data being selected for the training and the remaining 20% being selected for the testing. The "performance" in Figure 5 refers to evaluation of the performance of the k-NN approach when tested on the test dataset. This was evaluated by using a confusion matrix. Next, we outline how this flowchart was used for development of this methodology. We used an open-source dataset by Tabbakha et al. [66] for development of this methodology. This dataset was chosen because its attributes were same as the real-time data that could be collected and analyzed by the Context-Driven Human Activity Recognition Framework [64] as outlined above (Step i). The functionalities of this proposed approach, i.e., Steps (ii) to (vii), were developed and implemented in RapidMiner as a "process" as shown in Figure 6.

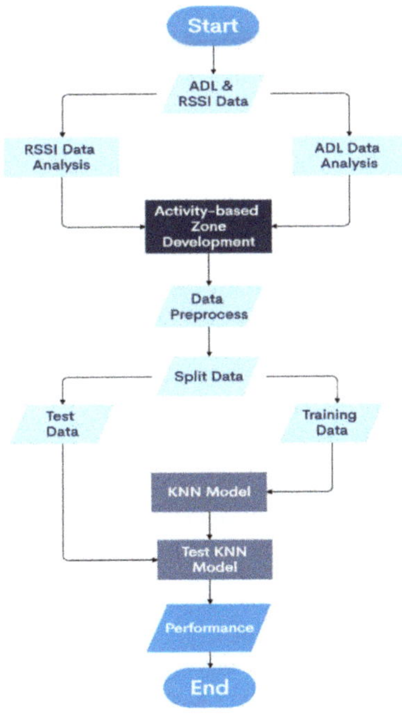

Figure 5. The flowchart for the proposed methodology for detection of indoor location in a specific 'activity-based zone' by analysis of the RSSI data coming from BLE beacons and BLE scanners during different ADLs.

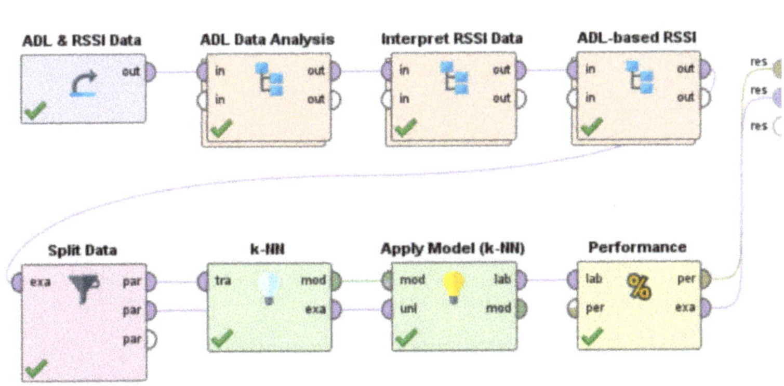

Figure 6. The RapidMiner "process" for detection of indoor location in a specific 'activity-based zone' by analysis of the RSSI data coming from BLE beacons and BLE scanners during different ADLs.

This "process" was developed as a combination of built-in "operators" and user defined "operators" in RapidMiner. An overview of built-in "operators" and user-defined "operators" in RapidMiner was presented in Section 3. We used the 'Dataset' "operator" to import this dataset into the RapidMiner "process". This "operator" was then renamed to 'ADL & RSSI Data' as for development of this "process" we needed to use only the activity-based data and the associated RSSI signals coming from BLE Beacons and BLE scanners during these ADLs. The semantic relationships between the changing dynamics of atomic activities, context attributes, core atomic activities, and core context attributes associated to different ADLs were then studied to interpret the spatial and temporal features of these ADLs (Step iii). This was done as per the methodology used to analyze complex activities (examples shown in Tables 1 and 2) and the associated "operator" that was developed was named as 'ADL Data Analysis'. The characteristics of the data coming from the sensors to analyze the associated RSSI data from BLE beacons and BLE scanners, recorded during different ADLs, were then studied and analyzed (Step iv) and the corresponding "operator" that was developed was named as 'Interpret RSSI Data'. Thereafter, we associated these relationships obtained from the 'ADL Data Analysis' "operator" with the characteristic features obtained from the 'Interpret RSSI Data' (Step v) to develop the 'activity-based zones'. We performed the same by developing an "operator"—'ADL-based RSSI'. Then, we used the built-in 'Split Data' "operator" to split the data into training set and test set with 80% of the data for training and 20% of the data for testing. A k-NN learning approach was used to develop the machine learning model which was tested on the test set by using the 'Apply Model (k-NN)' "operator". K-NN and 'Apply Model' are built-in "operators" in RapidMiner that can be directly used in any "process". Thereafter, we used the built-in 'Performance' "operator" in RapidMiner to evaluate the performance characteristics of the model in the form of a confusion matrix.

Figure 7 further clarifies the architecture of the proposed methodology as developed in RapidMiner [60]. This figure shows the flow of control depicting the sequence of operation of the different "operators" in this RapidMiner "process". As can be seen from Figure 7, the "operator" 'ADL & RSSI Data' is executed first, which is followed by the executions of the 'ADL Data Analysis', 'Interpret RSSI Data', 'ADL-based RSSI', 'Split Data', 'k-NN', 'Apply Model (k-NN)', and 'Performance' "operators", respectively. This RapidMiner "process", that studies the multimodal components of user interactions during ADLs and analyzes the data from BLE beacons and BLE scanners to track a user's indoor location in a specific 'activity-based zone'—which could be either the kitchen or the bedroom or the office or the toilet 'zone', achieved an overall performance accuracy of 81.36%. Further discussion of these results, the associated performance characteristics, and the rationale behind using confusion matrix for evaluation of this methodology are presented in Section 5.1.

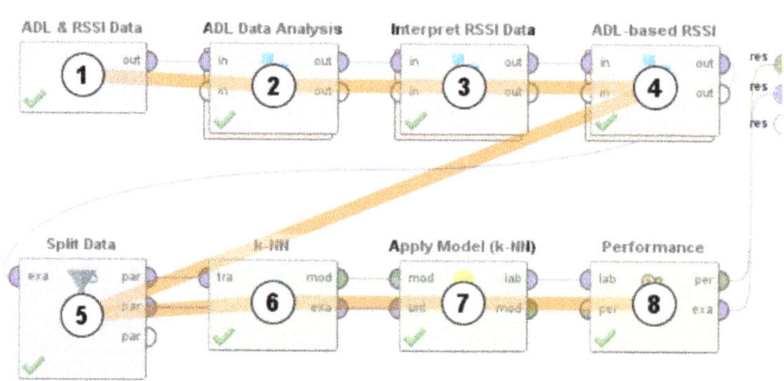

Figure 7. The flow of control showing the sequence of operation of the different "operators" in the RapidMiner "process" for detection of indoor location in a specific 'activity-based zone' by analysis of the RSSI data coming from BLE beacons and BLE scanners during different ADLs.

4.2. Context Independent Indoor Localization from Accelerometer and Gyroscope Data

The following are the steps for development of this functionality:

i. Set up an IoT-based environment, within a spatial context such as indoor layout of rooms with furniture's and appliances, using wearables and wireless sensors to collect the Big Data related to different ADLs. The associated representation scheme involves mapping the entire spatial location into non-overlapping 'activity-based zones', distinct to different complex activities, by performing complex activity analysis [63] as outlined in Section 4.1.

ii. Analyze the ADLs in terms of the associated atomic activities, context attributes, core atomic activities, and core context attributes, and their associated threshold values based on probabilistic reasoning principles [63].

iii. Infer the semantic relationships between the changing dynamics of atomic activities, context attributes, core atomic activities, and core context attributes along with the associated spatial and temporal information.

iv. Study and analyze the semantic relationships between the accelerometer data (in X, Y, and Z directions), gyroscope data (in X, Y, and Z directions) and the associated atomic activities, context attributes, core atomic activities, and core context attributes within each 'activity-based zone'.

v. Study and analyze the semantic relationships between the accelerometer data (in X, Y, and Z directions), gyroscope data (in X, Y, and Z directions) and the associated atomic activities, context attributes, core atomic activities, and core context attributes across different 'activity-based zones' based on the sequence in which the different ADLs took place and the related temporal information.

vi. Integrate the findings from (iv) and (v) to interpret the interrelated and semantic relationships between the accelerometer data and the gyroscope data with respect to different ADLs performed in all the 'activity-based zones' in the given IoT-based space.

vii. Split the data into training set and test set and develop a machine learning-based model to detect the location of a user, in terms of these spatial 'zones' based on the associated accelerometer data (in X, Y, and Z directions) and gyroscope data (in X, Y, and Z directions).

viii. Evaluate the performance characteristics of the model by using a confusion matrix.

4.2.1. System Architecture of the Methodology for Context Independent Indoor Localization from Accelerometer and Gyroscope Data

The flowchart of the proposed methodology is shown in Figure 8. For convenient representation in this flowchart, accelerometer has been represented as "Acc" and gyroscope has been represented as "Gyro". This figure outlines the operation of this methodology, as discussed in the previous section, at a broad level. Here, by "Split Data", we refer to splitting the data into training set and test set, with 70% data being selected for the training and the remaining 30% being selected for the testing. The "performance" in Figure 8 refers to evaluation of the performance of the Random Forest approach when tested on the test dataset. This was performed by using the confusion matrix. Next, we outline the steps that we followed for implementation of this methodology as a RapidMiner process. As outlined in Section 4.1, for implementation of Step (i) above and for collection of Big Data related to ADLs, the Context-Driven Human Activity Recognition Framework could be used, that has already been developed, implemented, and tested at the Multimedia and Augmented Reality Lab, in the Department of Electrical Engineering and Computer Science at the University of Cincinnati. The results of the same were published in [64]. As this dataset by Tabbakha et al. [66] already had the data that we could have collected by setting up this data collection framework, so we used this dataset for development of the remaining functionalities of this methodology from Step (ii) in the form of a RapidMiner "process" as shown in Figure 9 by following the flowchart shown in Figure 8.

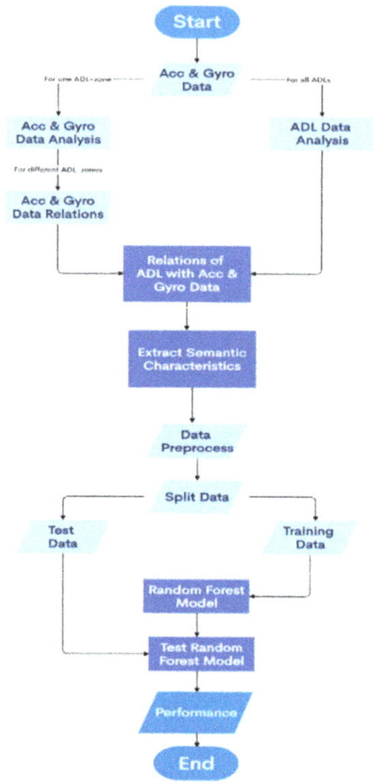

Figure 8. Flowchart of the proposed methodology for detection of indoor location based on the varying accelerometer and gyroscope data associated with distinct behavioral patterns related to distinct ADLs performed in distinct 'activity-based zones'.

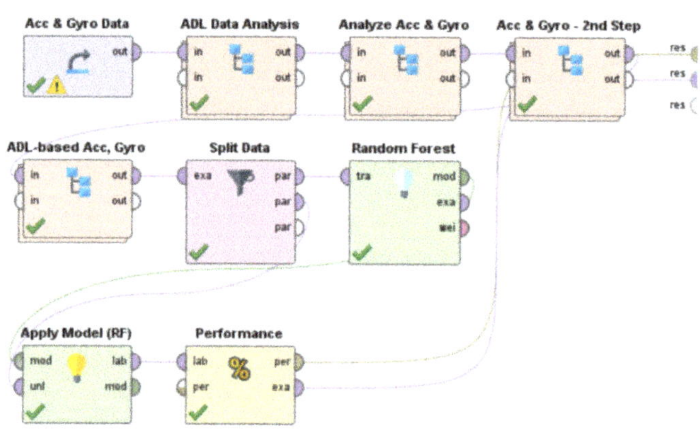

Figure 9. The RapidMiner "process" for detection of indoor location based on the varying accelerometer and gyroscope data associated with distinct behavioral patterns related to distinct ADLs performed in distinct 'activity-based zones'.

This "process" was developed as a combination of built-in "operators" and user defined "operators" in RapidMiner. An overview of built-in "operators" and user-defined "operators" in RapidMiner was presented in Section 3. We used the 'Dataset' "operator" to import this dataset into the RapidMiner "process". This "operator" was then renamed to 'Acc & Gyro Data' as for development of this "process" as we needed to use only the activity-based data and the associated accelerometer and gyroscope data. The semantic relationships between the changing dynamics of atomic activities, context attributes, core atomic activities, and core context attributes associated to different ADLs were then studied to interpret the spatial and temporal features of these ADLs (Step iii). This was done as per as per the methodology used to analyze complex activities (examples shown in Tables 1 and 2) and the associated "operator" that was developed was named as 'ADL Data Analysis'. The functionality to study the semantic relationships between the accelerometer data (in X, Y, and Z directions), gyroscope data (in X, Y, and Z directions) and the associated atomic activities, context attributes, core atomic activities, and core context attributes within each 'activity-based zone' was then developed (Step iv) in the form of an "operator" which we named as 'Analyze Acc & Gyro'. Thereafter, we developed the functionality to study the semantic relationships between the accelerometer data (in X, Y, and Z directions), gyroscope data (in X, Y, and Z directions), and the associated atomic activities, context attributes, core atomic activities, and core context attributes across different 'activity-based zones' based on the sequence in which the ADLs took place as well as the associated temporal information, in the form of an "operator" which we named as 'Acc & Gyro 2nd Step' (Step v). The characteristic features of these "operators" were then merged to develop the functionality to interpret the interrelated and semantic relationships between the accelerometer data and the gyroscope data with respect to different ADLs performed in all the 'activity-based zones' (Step vi). This was done by developing an "operator" which we named as 'ADL-based Acc, Gyro'. Then, we used the built-in 'Split Data' "operator" to split the data into training set and test set with 70% of the data for training and 30% of the data for testing. A Random Forest-based learning approach was used to develop the machine learning model which was tested on the test set by using the 'Apply Model' "operator". 'Random Forest' and 'Apply Model' are built-in "operators" in RapidMiner that can be directly used in any "process". Thereafter, we used the built-in 'Performance' "operator" in RapidMiner to evaluate the performance characteristics of the model in the form of a confusion matrix.

Figure 10 further clarifies the architecture of the proposed methodology as developed in RapidMiner [60]. This figure shows the flow of control depicting the sequence of operation of the different "operators" in this RapidMiner "process". As can be seen from Figure 10, the "operator" 'Acc & Gyro Data' is executed first, which is followed by the executions of the 'ADL Data Analysis', 'Analyze Acc & Gyro', 'Analyze Acc & Gyro—2nd Step', 'ADL-based Acc, Gyro', 'Split Data', 'Random Forest', 'Apply Model (RF)', and 'Performance' "operators", respectively.

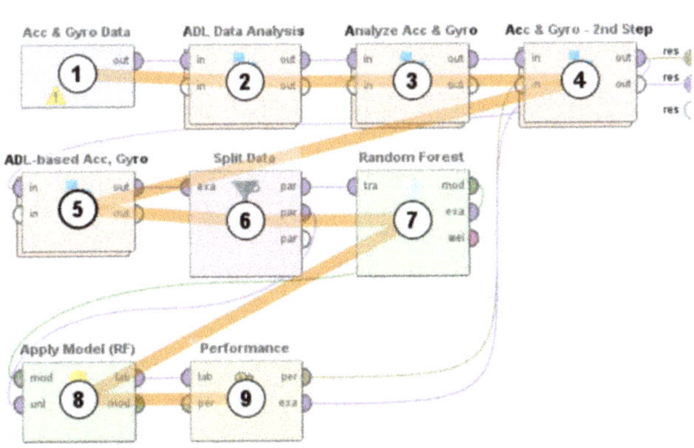

Figure 10. The flow of control showing the sequence of operation of the different "operators" in the RapidMiner "process" for detection of indoor location based on the varying accelerometer and gyroscope data associated with distinct behavioral patterns related to distinct ADLs performed in distinct 'activity-based zones'.

This methodology is based on interpretation of the accelerometer and gyroscope data from diverse behavioral patterns to detect the 'zone-based' indoor location of a user in any IoT-based environment. Here, the 'zone-based' mapping of a user's location refers to mapping the user in one of the multiple 'activity-based zones' that any given IoT-based environment can be classified into based on the specific activity being performed by the user. This classification of any given environment can be performed by the 'ADL Data Analysis' "operator" by using the complex activity recognition and analysis principles. The functionality of this "operator", as described above, is neither environment specific nor context parameter specific, thus its methodology can be applied for spatial mapping of any given IoT-based space. The accelerometer and gyroscope data that are analyzed and interpreted by this approach are a result or function of human behavior—that can be studied in the form of associated postures, gestures, movements, and motions, found in any IoT-based environment. Analysis of such behavior by the 'Analyze Acc & Gyro', 'Acc & Gyro 2nd Step', and 'ADL-based Acc, Gyro' is thus not dependent on a specific set of context parameters local to any specific IoT-based setting. All built-in data analysis related "operators" in RapidMiner are developed in a way so that they can be applied to any kind of data and they do not need any specific features in the environment to be present for their operation or function. The other "operators" that are a part of this "process"—'Split Data', 'Random Forest', 'Apply Model', and 'Performance' are built-in "operators" in RapidMiner and are therefore context independent. To summarize, all the operators that were used to develop this RapidMiner process, shown in Figure 9, are associated with distinct functionalities and characteristics that are not a function of any specific context-based or environment-based features local to any specific environment. In other words, these "operators" and thus the entire RapidMiner "process" as shown in Figure 9, would function for analysis and interpretation of any kind of user interaction data for performing the Indoor Localization of the user in that environment, based on the associated behavioral patterns distinct to different 'zones' local to that environment. This upholds the context independent nature of the entire RapidMiner "process" and thus the proposed methodology in Section 4.2. When tested on a dataset, this methodology, as shown in Figure 9, achieved a performance accuracy of 81.13%. Further discussion of these results, the associated performance characteristics, and the rationale behind using confusion matrix for evaluation of this methodology are presented in Section 5.2.

4.3. *Detection of the Spatial Coordinates of the User in any 'Activity-Based Zone'*

The following are the steps for development of this functionality in the proposed framework:

i. Set up an IoT-based environment, within a spatial context such as indoor layout of rooms with furniture's and appliances, using wearables, and wireless sensors to collect the Big Data related to different ADLs.

ii. The associated representation scheme involves setting up a context-based reference system in the given IoT-based environment. This system would track the instantaneous X and Y coordinates of the user's position information with respect to the origin of this reference system.

iii. Study each ADL performed in a specific 'activity-based zone' in terms of the multimodal user interactions performed on the context parameters local to that 'zone'. This involves studying the atomic activities, context attributes, core atomic activities, core context attributes, start atomic activities, start context attributes, end atomic activities, and end context attributes.

iv. For each of these user interactions with the context parameters, track the spatial configurations and changes in the user's position information, by using this reference system.

v. Study the changes in the instantaneous spatial configurations of the user as per this reference system with respect to the dynamic temporal information associated with each user interaction performed in the given 'activity-based zone'.

vi. Study and record all the user interactions as per (v), specific to the given ADL, in a given 'activity-based zone'.
vii. Split the data into training set and test set and use the training set to train a machine learning-based model for detection of the varying X and Y coordinates of the user's position information in any 'activity-based' zone, as per the dynamic user interactions with context parameters.
viii. Evaluate the performance characteristics of the model by using the root mean squared error method.

4.3.1. System Architecture of the Methodology for Detection of the Spatial Coordinates of the User in any 'Activity-Based Zone'

The flowchart for the proposed methodology is shown in Figure 11. This figure outlines the operation of this methodology, as discussed in the previous section, at a broad level. The distances from the three Bluetooth beacons that were used to develop the context-based reference system are represented as Distance A, Distance B, and Distance C, respectively. These distances were measured in meters. The actual X and Y coordinates of the user were measured in centimeters with an accuracy of +/−1 cm in the dataset [67] by analyzing the user's relative position with respect to these three beacons at a given point of time by using this reference system. In this figure, by "Split Data", we refer to splitting the data into training set and test set, with 70% data being selected for the training and the remaining 30% being selected for the testing. The "performance" in Figure 11 refers to evaluation of the performance of the machine learning approach when tested on the test dataset. This was evaluated by using the RMSE method, where RMSE errors were calculated separately in the X-direction and Y-direction as per ISO/IEC 18305:2016 [31]. This learning approach has been developed as a Random Forest model in this section. However, instead of a Random Forest model, other learning approaches such as Artificial Neural Network, Decision Tree, Support Vector Machine, k-NN, Gradient Boosted Trees, Deep Learning, and Linear Regression can also be seamlessly used for development of this methodology by following the flowchart shown in Figure 11. In Section 6, we have presented a comparative study where we implemented all these machine learning models—Random Forest, Artificial Neural Network, Decision Tree, Support Vector Machine, k-NN, Gradient Boosted Trees, Deep Learning, and Linear Regression for development of this methodology and compared their performance characteristics to deduce the optimal learning model for development of such an Indoor Localization system for detection of the spatial coordinates of the user in any 'activity-based zone'.

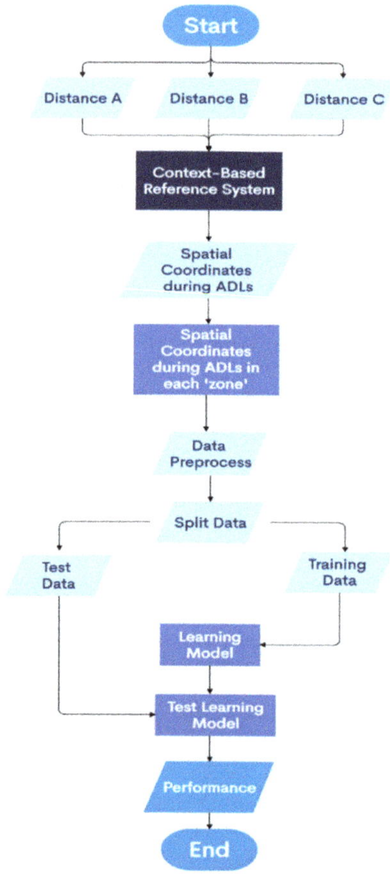

Figure 11. Flowchart for development of the methodology for detection of the varying X and Y coordinates of the user's position in any 'activity-based' zone, as per the dynamic user interactions with context parameters related to different activities.

Next, we outline the steps for development of the methodology as a RapidMiner process. As outlined in Section 4.1, for implementation of the Steps (i) and (ii) above and for collection of Big Data, the Context-Driven Human Activity Recognition Framework could be used, that has already been developed, implemented, and tested at the Multimedia and Augmented Reality Lab, in the Department of Electrical Engineering and Computer Science at the University of Cincinnati. The results associated with the same were published in [64]. As this dataset [67] already had the data and the corresponding data attributes, that we could have collected by setting up this data collection framework, so we used this dataset for development of the remaining functionalities of this methodology from Step (iii) in the form of a RapidMiner "process", as shown in Figure 12. This "process" was developed as a combination of built-in "operators" and user defined "operators" in RapidMiner. An overview of built-in "operators" and user-defined "operators" in RapidMiner was presented in Section 3.

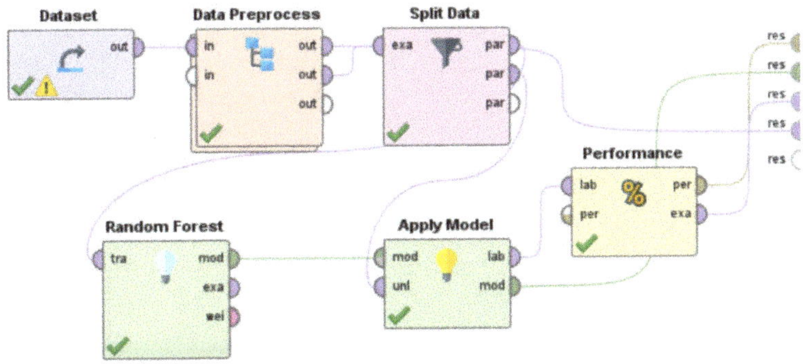

Figure 12. The RapidMiner "process" for detection of the varying X and Y coordinates of the user's position in any 'activity-based' zone, as per the dynamic user interactions with context parameters related to different activities.

This dataset used a reference system that was based out of comparing the user's spatial configuration, in terms of the actual distances—Distance A, Distance B, and Distance C, with respect to 3 Bluetooth beacons while analyzing the associated temporal information. This helped to detect the actual X and Y coordinates of the user in the given IoT-based space. The data consisted of 250 rows. We used RapidMiner [60] to develop a "process" to implement this functionality and evaluated its performance by using this dataset. The same version of RapidMiner and the same computer, as outlined in Section 3, were used for development of this "process". We used the 'Dataset' "operator" to import this dataset into the RapidMiner "process". The 'Data-Preprocess' "operator" was developed and then used to perform multiple preprocessing steps (Steps iii to vi) prior to splitting the data for training and testing. We used 70% of the data for training and the rest was used for testing. This data splitting was performed by the built in 'Split Data' "operator". A Random Forest learning approach was used to train the model by using the built-in 'Random Forest' "operator". The 'Apply Model' "operator", another built-in "operator", was used to apply the learning model on the test data and its performance characteristics were evaluated by using the 'Performance' "operator" in RapidMiner. We used the root mean square error (RMSE) method of evaluating the performance characteristics as per ISO/IEC18305:2016 [31]. This RapidMiner "process" is shown in Figure 12.

Figure 13 further clarifies the architecture of this proposed methodology as developed in RapidMiner [60]. This figure shows the flow of control depicting the sequence of operation of the different "operators" in this RapidMiner "process". As can be seen from Figure 13, the "operator" 'Dataset' is executed first, which is followed by the executions of the 'Data Preprocess', 'Split Data', 'Random Forest', 'Apply Model', and 'Performance' "operators", respectively. The RMSE for detection of X and Y coordinates of the user's position were found to be 5.85 cm and 5.36 cm, respectively. The Horizontal Error, as defined in ISO/IEC18305:2016 [31], was found to be 7.93 cm. Further discussion of these results, the associated performance characteristics, and the rationale behind using RMSE for evaluation of this methodology are presented in Section 5.3.

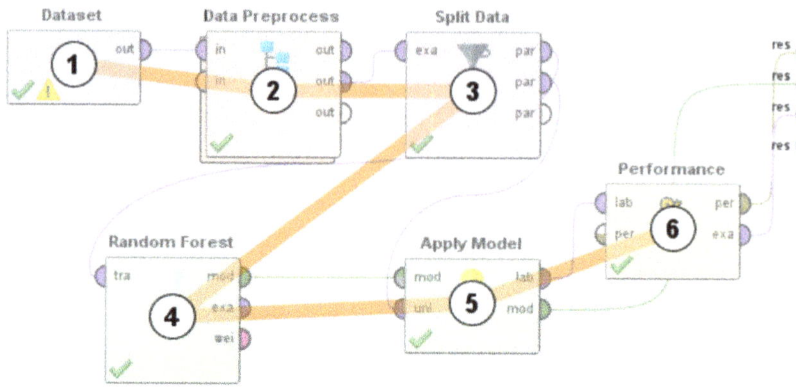

Figure 13. The flow of control showing the sequence of operation of the different "operators" in the RapidMiner "process" for detection of the varying X and Y coordinates of the user's position in any 'activity-based' zone, as per the dynamic user interactions with context parameters related to different activities.

5. Results and Findings

5.1. Indoor Localization from BLE Beacons and BLE Scanners Data during ADLs

In this section we present and discuss the results associated with the development of the proposed Big-Data driven methodology that studies the multimodal components of user interactions and analyzes the data from BLE beacons and BLE scanners to track a user's indoor location in a specific 'activity-based zone' during Activities of Daily Living, to test our first hypothesis—"*The RSSI data coming from BLE scanners and BLE beacons can be studied and analyzed to detect the changes in a user's instantaneous location during different activities, which are a result of the varying user interactions with dynamic context parameters*", as outlined in Section 4.1.

Upon development of the RapidMiner "process" as shown in Figure 6 by following the methods for development of this functionality (Section 4.1), we first studied the RSSI data from BLE beacons and BLE scanners associated with the varying atomic activities, context attributes, core atomic activities, and core context attributes associated with each of these ADLs performed in the different 'activity-based zones'. This is shown in Figures 14–16. In each of these figures, the *X*-axis represents the specific rooms or 'zones' modelled in the simulated Smart Home. The *Y*-axis represents the BLE scanner readings in different rooms or 'activity-based zones'. For instance, in Figure 14, the *X*-axis represents the different locations and the *Y*-axis represents the RSSI data recorded by the different BLE scanners present in the environment. From the dataset, it was observed that a BLE scanner provided an RSSI value of −120 when the BLE beacon was far away from the scanner or was out of its range. Therefore, for any value greater than −120, it could be concluded that the person was in that room or 'activity-based zone'. This is represented on the *Y*-axis. For instance, in Figure 14, the RSSI data is greater than −120 for the kitchen sensors but equal to −120 for sensors in all other rooms. This infers the presence of the person in the kitchen area.

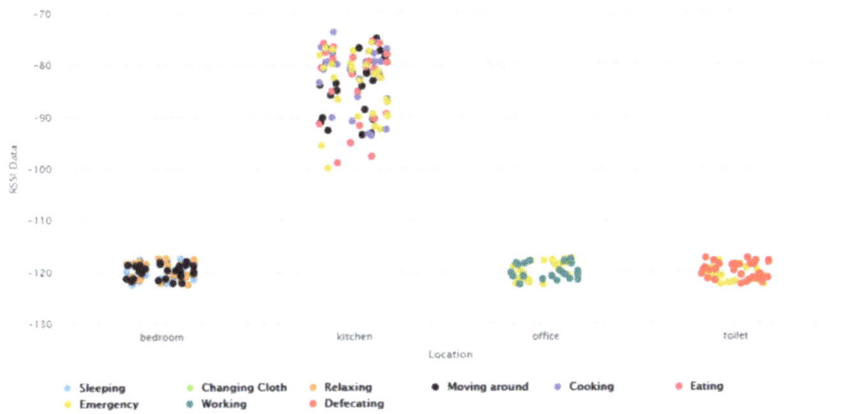

Figure 14. Analysis of RSSI data coming from different BLE scanners and BLE beacons when different activities were performed in the kitchen.

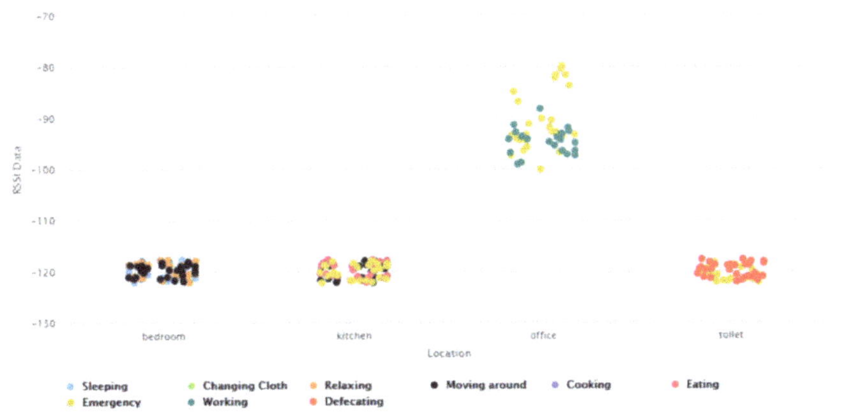

Figure 15. Analysis of RSSI data coming from different BLE scanners and BLE beacons when different activities were performed in the office area.

Figure 16. Analysis of RSSI data coming from different BLE scanners and BLE beacons when different activities were performed in the toilet.

The plots in each of these figures are color coded based on different complex activities that were performed by the user in each of these rooms or 'activity-based zones'—sleeping, changing clothes, relaxing, moving around, cooking, eating, working, defecating, and an emergency. An emergency constituted detecting the user in a lying position (either from a fall or unconsciousness) in an environment where a user is not supposed to lie down, for instance in the toilet. Figure 17 shows the output of the RapidMiner "process" (for the first 13 rows) which involved predicting the user's location in a specific 'activity-based zone' during different ADLs based on the associated RSSI data coming from the different BLE scanners and BLE beacons.

Row No.	Location	prediction(Location)	confidence(bedroom)	confidence(kitchen)	confidence(office)	confidence(toilet)
1	bedroom	bedroom	0.800	0.200	0	0
2	bedroom	kitchen	0	1	0	0
3	bedroom	bedroom	0.607	0.393	0	0
4	bedroom	bedroom	0.801	0.199	0	0
5	bedroom	bedroom	0.804	0.196	0	0
6	bedroom	bedroom	0.614	0.386	0	0
7	bedroom	kitchen	0.195	0.805	0	0
8	bedroom	kitchen	0.423	0.577	0	0
9	bedroom	bedroom	0.605	0.395	0	0
10	bedroom	bedroom	0.613	0.387	0	0
11	bedroom	bedroom	0.814	0	0.186	0
12	bedroom	bedroom	0.802	0.198	0	0
13	bedroom	bedroom	0.824	0.176	0	0

Figure 17. Output of the RapidMiner "process" (first 13 rows) shown in Figure 6 to detect a user's location during different ADLs based on the RSSI data coming from BLE scanners and BLE beacons.

The output of the RapidMiner "process" assigned a confidence value for predicting the user's location in each of these 'activity-based zones'. The 'activity-based zone' with the highest confidence value was the final prediction of the model—in terms of predicting the user's location. For example, in Row number 3 of Figure 17, the confidence values of the model for the user's presence in the bedroom, kitchen, office and toilet 'zones' were, respectively, 0.607, 0.393, 0, and 0. As the confidence value of the model was highest for the bedroom 'zone', so, the final prediction of the model (third attribute from the left in Figure 17) was that the user was present in the bedroom. Table 3 consists of the description of all the attributes represented in the output shown in Figure 17.

Table 3. Description of the attributes of the output of the RapidMiner "process" shown in Figure 17.

Attribute Name	Description
Row No	The row number in the output table
Location	The actual instantaneous zone-based location of the user
Prediction (Location)	The predicted instantaneous zone-based location of the user
Confidence (bedroom)	The degree of certainty that the user was in the bedroom
Confidence (kitchen)	The degree of certainty that the user was in the kitchen
Confidence (office)	The degree of certainty that the user was in the office area
Confidence (toilet)	The degree of certainty that the user was in the toilet

The performance characteristics of this model, in terms of predicting whether the user was in the bedroom or kitchen or office area or toilet were evaluated by using a confusion matrix. Here, the data attribute being predicted by this methodology was the 'activity-based zone' and the associated values of the same were bedroom, kitchen, office, and toilet, as per the characteristics of the dataset used [64] and the associated functionalities of this approach (Section 4.1). As can be seen from Figure 17 and Table 3, none of these values of the 'activity-based zone' were numerical values. Even though ISO/IEC18305:2016 [31] recommends evaluating Indoor Localization systems either by using RMSE, or Mean of

Error, or Covariance Matrix of Error, or Mean of Absolute Error, or Mean, or Standard Deviation of Vertical Error, etc.; such performance metrics can only be used when the predicted attribute is of numerical type—such as the numerical value of the X-coordinate, the numerical value of the Y-coordinate, the distance of the user from a specific reference point, etc. For non-numeric data types such performance metrics do not work. This is because one of the steps towards using the RMSE method of performance evaluation involves calculation of the arithmetic mean of the squares of a set of numbers [68] and similar mathematical operations are performed on the data when the other performance metrics as stated in ISO/IEC18305:2016 [31] are used. For non-numerical data neither can an arithmetic mean be computed nor can any mathematical operation be performed on the data. Evaluating the performance characteristics of an approach that involves prediction of non-numeric data by using a confusion matrix is a well-known practice in the field of machine learning, pattern recognition, data science, and their interrelated fields [69]. Therefore, we used a confusion matrix to study the performance characteristics of this methodology as proposed in Section 4.1.

The tabular representation and plot view of the performance characteristics, as obtained from RapidMiner, are shown in Figures 18 and 19, respectively. As can be observed from Figures 18 and 19, the model achieved an overall performance accuracy of 81.36%. The respective class recall values were 85.00%, 70.00%, 88.89%, and 90.00% for predicting the location of a user in bedroom, kitchen, office, and toilet, respectively. Further discussion about how this approach and the associated results and findings address multiple research challenges in this field is presented in Section 7.

accuracy: 81.36%

	true bedroom	true kitchen	true office	true toilet	class precision
pred. bedroom	17	1	0	0	94.44%
pred. kitchen	3	14	1	0	77.78%
pred. office	0	0	8	1	88.89%
pred. toilet	0	5	0	9	64.29%
class recall	85.00%	70.00%	88.89%	90.00%	

Figure 18. A confusion matrix (tabular view) representing the performance accuracy of the RapidMiner "process" shown in Figure 6 to detect a user's location during different ADLs based on the RSSI data coming from BLE scanners and BLE beacons.

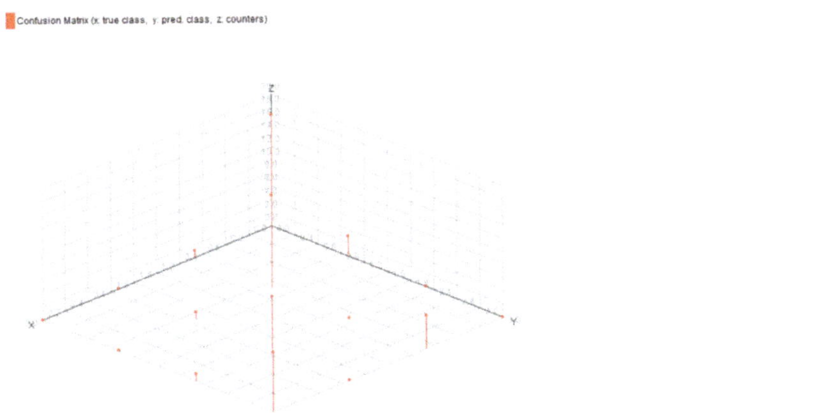

Figure 19. A confusion matrix (plot view) representing the performance accuracy of the RapidMiner "process" shown in Figure 6 to detect a user's location during different ADLs based on the RSSI data coming from BLE scanners and BLE beacons.

5.2. Context Independent Indoor Localization from Accelerometer and Gyroscope Data

In this section we present and discuss the results associated with the development of the proposed context independent approach that can interpret the accelerometer and gyroscope data from diverse behavioral patterns to detect the 'zone-based' indoor location of a user in any IoT-based environment, to test our second hypothesis—"*The dynamic changes in the spatial configurations of a user during different activities can be interpreted by the analysis of the behavioral patterns that are localized and distinct for different activities*", as outlined in Section 4.2. By using the RapidMiner "process" as shown in Figure 9 and by following the proposed functionalities of our framework (Section 4.2), we studied the variations of the accelerometer data (in X, Y, and Z directions) and gyroscope data (in X, Y, and Z directions) as per the variations in behavioral and user interaction patterns in the distinct 'activity-based zones' in the confines of the given IoT-based space. These variations in behavioral patterns were a result of the user performing different ADLs, characterized by distinct user interactions with the varying context parameters, in each of these 'activity-based zones'. This study is represented in Figures 20–25.

Figure 20. Analysis of the accelerometer data (in the X-direction) for different behavioral patterns associated with different ADLs performed in the given simulated smart home environment.

Figure 21. Analysis of the accelerometer data (in the Y-direction) for different behavioral patterns associated with different ADLs performed in the given simulated smart home environment.

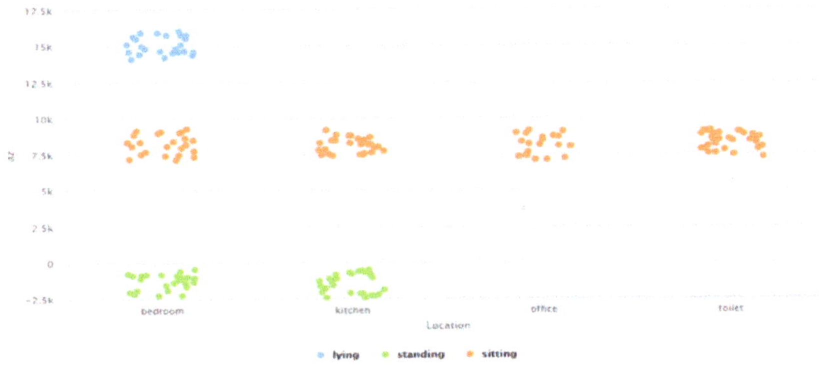

Figure 22. Analysis of the accelerometer data (in the Z-direction) for different behavioral patterns associated with different ADLs performed in the given simulated smart home environment.

Figure 23. Analysis of the gyroscope data (in the X-direction) for different behavioral patterns associated with different ADLs performed in the given simulated smart home environment.

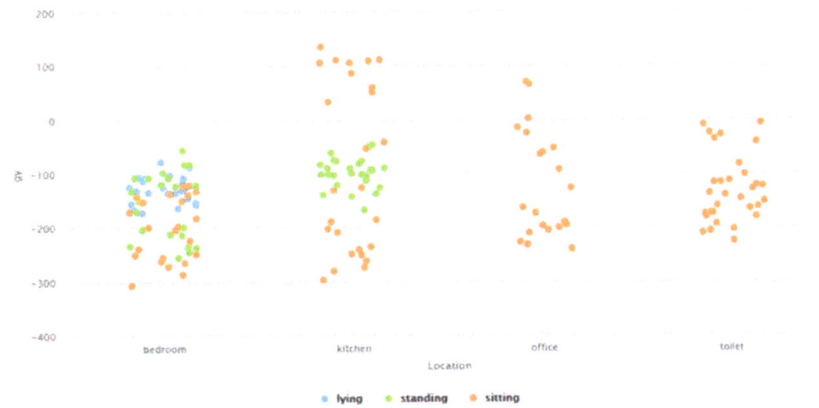

Figure 24. Analysis of the gyroscope data (in the Y-direction) for different behavioral patterns associated with different ADLs performed in the given simulated smart home environment.

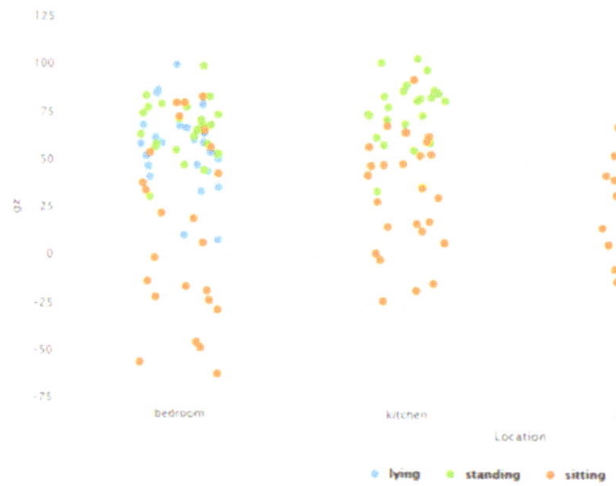

Figure 25. Analysis of the gyroscope data (in the Z-direction) for different behavioral patterns associated with different ADLs performed in the given smart home environment.

Figure 26 shows the output of the RapidMiner "process" (for the first 13 rows) which involved predicting the user's location in a specific 'activity-based zone' based on the associated accelerometer and gyroscope data. The output of the RapidMiner "process" assigned a confidence value for predicting the user's location in each of these 'activity-based zones'. The 'activity-based zone' with the highest confidence value was the final prediction of the model—in terms of predicting the user's location. For example, in Row number 7 of Figure 26, the confidence values of the model for the user's presence in the bedroom, kitchen, office, and toilet were, respectively, 0.985, 0.003, 0.010, and 0.002. As the confidence value of the model was highest for the bedroom so the final prediction of the model (third attribute from the left in Figure 26) was that the user was present in the bedroom. Table 4 consists of the description of all the attributes represented in the output shown in Figure 26.

Row No. ↑	Location	prediction(Location)	confidence(bedroom)	confidence(kitchen)	confidence(office)	confidence(toilet)
1	bedroom	bedroom	0.988	0	0	0.012
2	bedroom	bedroom	0.995	0.001	0	0.004
3	bedroom	bedroom	0.987	0.001	0	0.013
4	bedroom	bedroom	0.995	0.001	0	0.004
5	bedroom	bedroom	0.995	0.001	0	0.004
6	bedroom	bedroom	0.995	0.001	0	0.004
7	bedroom	bedroom	0.985	0.003	0.010	0.002
8	bedroom	bedroom	0.971	0.013	0	0.016
9	bedroom	bedroom	0.993	0.004	0	0.004
10	bedroom	bedroom	0.945	0.043	0.010	0.002
11	bedroom	bedroom	0.834	0.139	0.013	0.014
12	bedroom	bedroom	0.777	0.197	0.013	0.014
13	bedroom	bedroom	0.662	0.297	0.018	0.023

Figure 26. Output of the RapidMiner "process" (first 13 rows) shown in Figure 9 to detect a user's location during different ADLs based on the associated accelerometer and gyroscope data.

Table 4. Description of the attributes of the output of the RapidMiner "process" shown in Figure 26.

Attribute Name	Description
Row No	The row number in the output table
Location	The actual instantaneous zone-based location of the user
Prediction (Location)	The predicted instantaneous zone-based location of the user
Confidence (bedroom)	The degree of certainty that the user was in the bedroom
Confidence (kitchen)	The degree of certainty that the user was in the kitchen
Confidence (office)	The degree of certainty that the user was in the office area
Confidence (toilet)	The degree of certainty that the user was in the toilet

The performance characteristics of this model in terms of predicting whether the user was in the bedroom or kitchen or office area or toilet were evaluated by using a confusion matrix.

Here, the data attribute being predicted by this methodology was the 'activity-based zone' and the associated values of the same were bedroom, kitchen, office, and toilet, as per the characteristics of the dataset used [64] and the associated functionalities of this approach (Section 4.2). As can be seen from Figure 26 and Table 4, none of these values of the 'activity-based zone' were numerical values. Even though ISO/IEC18305:2016 [31] recommends evaluating Indoor Localization systems either by using RMSE, or Mean of Error, or Covariance Matrix of Error, or Mean of Absolute Error, or Mean, or Standard Deviation of Vertical Error, etc.; such performance metrics can only be used when the predicted attribute is of numerical type—such as the numerical value of the X-coordinate, the numerical value of the Y-coordinate, the distance of the user from a specific reference point, etc. For non-numeric data types such performance metrics do not work. This is because one of the steps towards using the RMSE method of performance evaluation involves calculation of the arithmetic mean of the squares of a set of numbers [68] and similar mathematical operations are performed on the data when the other performance metrics as stated in ISO/IEC18305:2016 [31] are used. For non-numerical data neither can an arithmetic mean be computed nor can any mathematical operation be performed on the data. Evaluating the performance characteristics of an approach that involves prediction of non-numeric data by using a confusion matrix is a well-known practice in the field of machine learning, pattern recognition, data science, and their interrelated fields [69]. Therefore, we used a confusion matrix to study the performance characteristics of this methodology as proposed in Section 4.2.

The tabular representation and plot view of the performance characteristics as obtained from RapidMiner are shown in Figures 27 and 28, respectively. As can be observed from Figures 27 and 28, the model achieved an overall performance accuracy of 81.13%. The class recall values were 86.36%, 68.75%, 83.33%, and 88.89% for predicting the location of a user in bedroom, kitchen, office, and toilet, respectively. Further discussion about how this approach and the associated results and findings address multiple research challenges in this field is presented in Section 7.

accuracy: 81.13%

	true bedroom	true kitchen	true office	true toilet	class precision
pred. bedroom	19	4	0	0	82.61%
pred. kitchen	1	11	0	1	84.62%
pred. office	0	0	5	0	100.00%
pred. toilet	2	1	1	8	66.67%
class recall	86.36%	68.75%	83.33%	88.89%	

Figure 27. A confusion matrix (tabular view) representing the performance accuracy of the Rapid-Miner "process" shown in Figure 9 to detect a user's indoor location based on the associated accelerometer and gyroscope data.

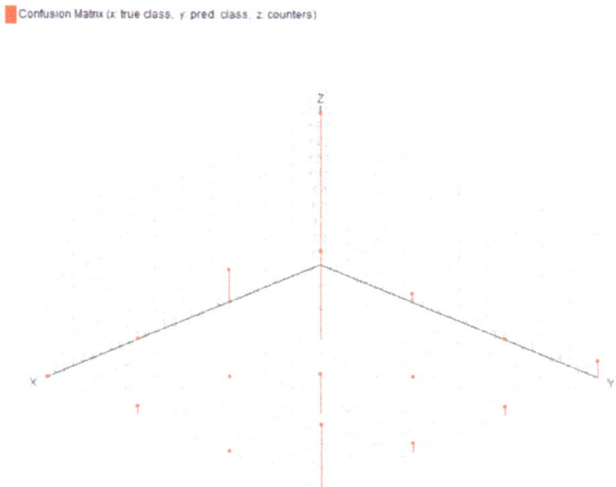

Figure 28. A confusion matrix (plot view) representing the performance accuracy of the RapidMiner "process" shown in Figure 9 to detect a user's indoor location based on the associated accelerometer and gyroscope data.

5.3. Detection of the Spatial Coordinates of the User in Any 'Activity-Based Zone'

In this section we present and discuss the results associated with the development of the proposed methodology to detect the spatial coordinates of a user's indoor position based on the associated user interactions with the context parameters and the user-centered local spatial context, by using a reference system, to test our third hypothesis—"*Tracking and analyzing the user interactions with the context parameters along with the associated spatial information, by using a reference system, helps to detect the dynamic spatial configurations of the user*", as outlined in Section 4.3. The process involved setting up a context-based reference system to track the user's location in the confines of the given IoT-based environment during different ADLs and consisted of multiple steps. The first step was to study each ADL performed in a specific 'activity-based zone' in terms of the multimodal user interactions performed on the context parameters local to that 'activity-based zone'. This involved studying the atomic activities, context attributes, core atomic activities, core context attributes, start atomic activities, start context attributes, end atomic activities, and end context attributes associated with all the complex activities. The second step involved tracking the spatial configurations and changes in the user's position information, by using the reference system (Section 4.3), during all the varying interactions with the context parameters. The methodology then studied the changes in the instantaneous spatial configurations of the user as per this reference system with respect to the dynamic temporal information associated with each user interaction performed in any given 'activity-based zone' to train a machine learning model. Upon development of the RapidMiner "process" as shown in Figure 12, we first studied the dynamic changes in the user's distance from the three beacons that were used to develop the reference system of this dataset [67]. This is shown in Figures 29–32. The distances from these three beacons, measured in meters, are represented as Distance A, Distance B, and Distance C, respectively, in these figures. These distances were measured in meters and the actual X and Y coordinates of the user were measured in centimeters with an accuracy of +/−1 cm in the dataset [67].

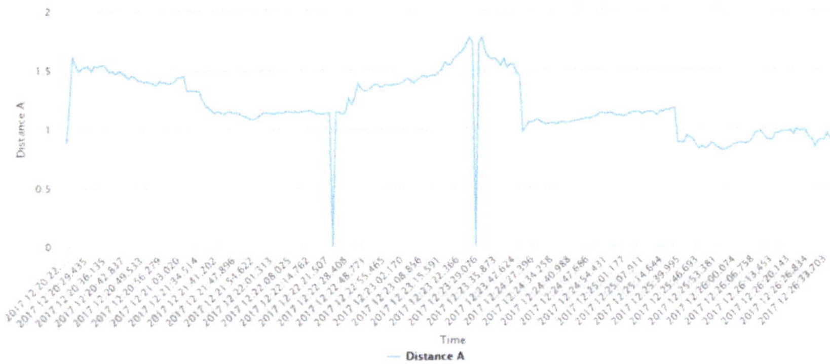

Figure 29. Analysis of the variation of Distance A (user's actual distance from one of the beacons) at different timestamps, based on the associated changes in behavioral patterns. Due to paucity of space the data associated with a few timestamps are shown here.

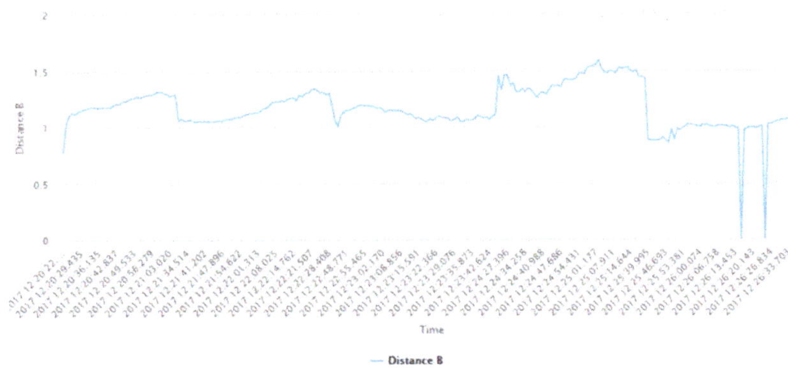

Figure 30. Analysis of the variation of Distance B (user's actual distance from one of the beacons) at different timestamps, based on the associated changes in behavioral patterns. Due to paucity of space the data associated with a few timestamps are shown here.

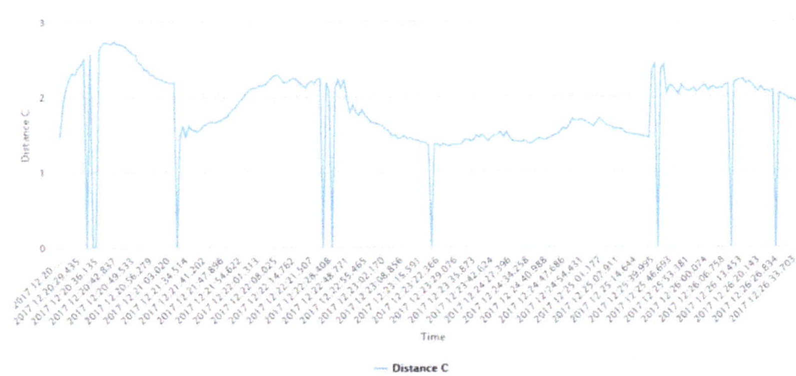

Figure 31. Analysis of the variation of Distance C (user's actual distance from one of the beacons) at different timestamps, based on the associated changes in behavioral patterns. Due to paucity of space the data associated with a few timestamps are shown here.

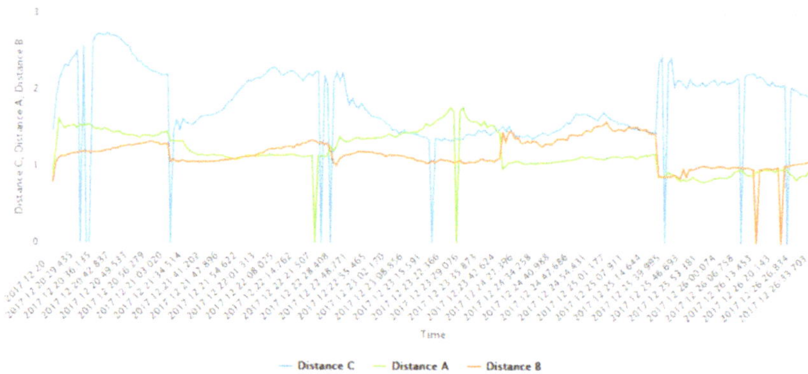

Figure 32. Analysis of the variation of Distances A, B, and C plotted together at different timestamps, based on the associated changes in behavioral patterns. Due to paucity of space the data associated with a few timestamps are shown here.

The Random Forest model that we developed in RapidMiner (Figure 12) consisted of 100 random trees, and used the least square criterion for splitting at each node. The maximum depth of a tree was 10. This learning model assigned weights to each of these distances—Distance A, Distance B, and Distance C to accurately track the user based on the provided reference system. The weights that the model associated with Distance A, Distance B, and Distance C were 0.531, 0.287, and 0.183, respectively, for determination of the X-coordinate of the user's location and these respective weights were 0.639, 0.170, and 0.191 for determination of the Y-coordinate of the user's location.

Figure 33 shows one of these random trees that was developed by the Random Forest model, as shown in Figure 12, and the reasoning-based description of this tree is shown in Figure 34. This random tree was associated with detecting different values of the X-coordinate of the user based on the associated rules at each node. We explain the working of the tree for one such detection here, when the user was located at the X-coordinate—79.000 as per this reference system. This is marked in blue in Figure 33. The comparison started at the topmost node—which for this tree was Distance A. This distance was lesser than or equal to 1.331 m, so the control moved to the right half of the tree. Then, it checked the value of Distance B, which was greater than 1.038 m, so it went to the left half of this node, where it checked the value of Distance A again. This value was greater than 1.078 m so it went to the right half and checked Distance B at the next node, which was less than or equal to 1.373 m, so the control moved to the right half of this node for checking Distance C. At this node, after performing the condition check, it moved to the left side of the node as Distance C was greater than 0.729 m. Then, the control compared the values of Distance A and Distance B with respect to a couple of more conditions at the respective child nodes to finally deduce the X-coordinate of the user as 79.000.

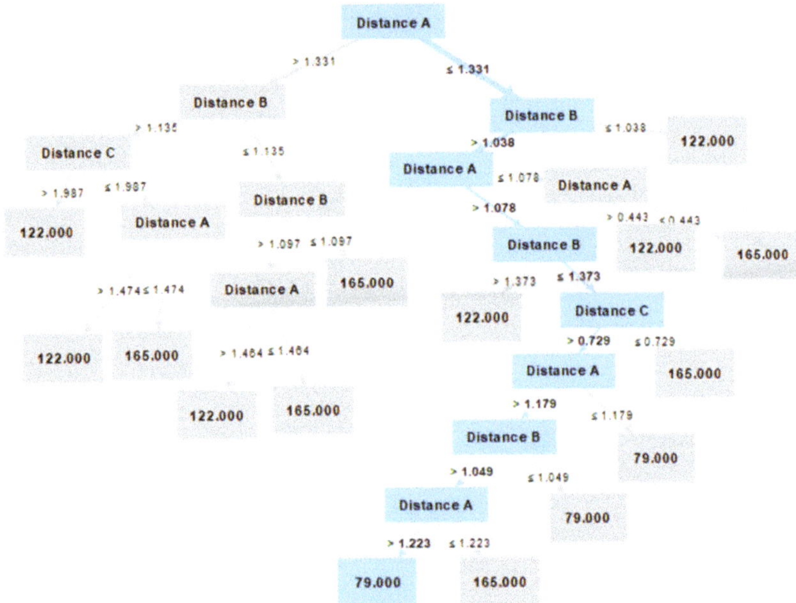

Figure 33. Representation of one of the Random Trees developed by the Random Forest-based RapidMiner "process" shown in Figure 12. This random tree was associated with detecting different values of the X-coordinate of the user based on the associated rules at each node.

```
Distance A > 1.344
|   Distance C > 2.147: 122.000 {count=30}
|   Distance C ≤ 2.147
|   |   Distance C > 0.674: 165.000 {count=28}
|   |   Distance C ≤ 0.674: 122.000 {count=3}
Distance A ≤ 1.344
|   Distance B > 1.335: 122.000 {count=30}
|   Distance B ≤ 1.335
|   |   Distance A > 1.076
|   |   |   Distance C > 1.798: 79.000 {count=24}
|   |   |   Distance C ≤ 1.798
|   |   |   |   Distance B > 1.112: 165.000 {count=1}
|   |   |   |   Distance B ≤ 1.112: 79.000 {count=14}
|   |   Distance A ≤ 1.076
|   |   |   Distance A > 0.408: 122.000 {count=43}
|   |   |   Distance A ≤ 0.408: 79.000 {count=2}
```

Figure 34. Reasoning-based description of the Random Tree shown in Figure 33, that was associated with detecting different values of the X-coordinate of the user based on the associated rules at each node.

Figure 35 shows another of these random trees that was developed by the Random Forest model, as shown in Figure 12, and the reasoning-based description of this tree is shown in Figure 36. This random tree was associated with detecting different values of the Y-coordinate of the user based on the associated rules at each node. We explain the working of the tree for one such detection here, when the user was located at the Y-coordinate—137.000 as per this reference system. This is marked in blue in Figure 35. The comparison starts at the topmost node—Distance A and it is greater than 1.0008 m, so the control moves to the left side of the tree to check Distance B. Here, the value of this distance was greater than 1.334 m, so the control moved to the right side of this node to check for another condition associated with Distance B. Here, it checked if the value of Distance B was greater than 1.286 m or not. For this specific condition as Distance B was greater than 1.286 m so the control traversed to the left side of the node to its child node which is associated with checking another condition at Distance A. This value was less than or equal to 1.095 m, so the Y-coordinate of the user was deduced to be 137.000.

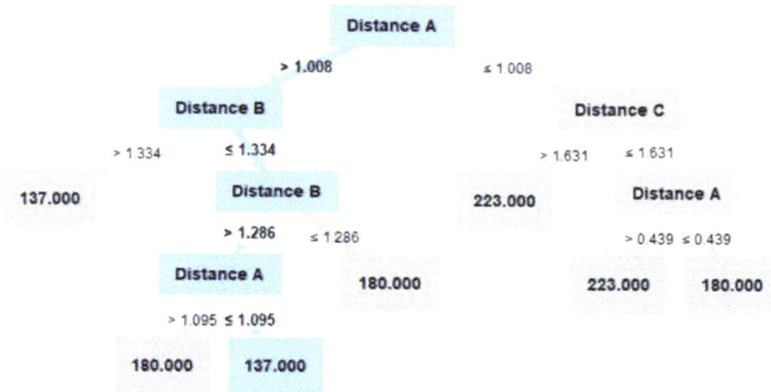

Figure 35. Representation of one of the Random Trees developed by the Random Forest-based RapidMiner "process" shown in Figure 12. This random tree was associated with detecting different values of the Y-coordinate of the user based on the associated rules at each node.

```
Distance A > 1.008
|   Distance B > 1.334: 137.000 {count=19}
|   Distance B ≤ 1.334
|   |   Distance B > 1.286
|   |   |   Distance A > 1.095: 180.000 {count=16}
|   |   |   Distance A ≤ 1.095: 137.000 {count=7}
|   |   Distance B ≤ 1.286: 180.000 {count=86}
Distance A ≤ 1.008
|   Distance C > 1.631: 223.000 {count=44}
|   Distance C ≤ 1.631
|   |   Distance A > 0.439: 223.000 {count=2}
|   |   Distance A ≤ 0.439: 180.000 {count=1}
```

Figure 36. Reasoning-based description of the Random Tree shown in Figure 35, that was associated with detecting different values of the Y-coordinate of the user based on the associated rules at each node.

Figures 37 and 38 show the output of the RapidMiner "process" (for the first 12 rows), shown in Figure 12, which detected the X-coordinate and Y-coordinate of the user's location based on this methodology, as outlined in Section 4.3. This output was shown by RapidMiner after taking into consideration all the predictions done by each of these 100 Random Trees which were a part of the developed Random Forest-based learning model (Figure 12). The maximum depth of all these random trees was 10. Tables 5 and 6 consist of the description of all the attributes represented in Figures 37 and 38, respectively.

Row No.	Position X	prediction(Position X)	Distance A	Distance B	Distance C	Time
1	122	119.850	0.877	0.769	1.457	2017 12:20:22.583
2	122	115.980	1.202	1.031	1.893	2017 12:20:23.683
3	122	128.364	1.533	1.123	2.233	2017 12:20:25.915
4	122	122.430	1.533	1.157	2.429	2017 12:20:30.560
5	122	129.238	1.533	1.174	0	2017 12:20:35.018
6	122	122.860	1.473	1.206	2.735	2017 12:20:42.837
7	122	122.860	1.452	1.202	2.699	2017 12:20:43.958
8	122	122	1.406	1.259	2.576	2017 12:20:49.533
9	122	122	1.401	1.259	2.460	2017 12:20:51.772
10	122	122	1.391	1.273	2.429	2017 12:20:52.880
11	122	126.730	1.382	1.280	2.362	2017 12:20:54.020
12	122	134.470	1.366	1.285	2.362	2017 12:20:55.151

Figure 37. Output (first 12 rows) of the RapidMiner "process" shown in Figure 12 for detection of the spatial coordinates (X-coordinate) of the user in each 'activity-based zone'.

Row No.	Position Y	prediction(Position Y)	Distance A	Distance B	Distance C	Time
1	180	219.990	0.877	0.769	1.457	2017 12:20:22.583
2	180	191.180	1.202	1.031	1.893	2017 12:20:23.683
3	180	180	1.533	1.123	2.233	2017 12:20:25.915
4	180	180	1.533	1.157	2.429	2017 12:20:30.560
5	180	180	1.533	1.174	0	2017 12:20:35.018
6	180	180	1.473	1.206	2.735	2017 12:20:42.837
7	180	180	1.452	1.202	2.699	2017 12:20:43.958
8	180	179.570	1.406	1.259	2.576	2017 12:20:49.533
9	180	179.570	1.401	1.259	2.460	2017 12:20:51.772
10	180	180	1.391	1.273	2.429	2017 12:20:52.880
11	180	180	1.382	1.280	2.362	2017 12:20:54.020
12	180	180	1.366	1.285	2.362	2017 12:20:55.151

Figure 38. Output (first 12 rows) of the RapidMiner "process" shown in Figure 12 for detection of the spatial coordinates (Y-coordinate) of the user in each 'activity-based zone'.

Table 5. Description of the attributes of the output of the RapidMiner "process" shown in Figure 37.

Attribute Name	Description
Row No	The row number in the output table
Position X	The actual X coordinate of the user's position
Prediction (Position X)	The predicted X coordinate of the user's position
Distance A	The actual distance of the user from the first Bluetooth beacon
Distance B	The actual distance of the user from the second Bluetooth beacon
Distance C	The actual distance of the user from the third Bluetooth beacon
Time	The associated timestamp information

Table 6. Description of the attributes of the output of the RapidMiner "process" shown in Figure 38.

Attribute Name	Description
Row No	The row number in the output table
Position Y	The actual Y coordinate of the user's position
Prediction (Position Y)	The predicted Y coordinate of the user's position
Distance A	The actual distance of the user from the first Bluetooth beacon
Distance B	The actual distance of the user from the second Bluetooth beacon
Distance C	The actual distance of the user from the third Bluetooth beacon
Time	The associated timestamp information

The performance characteristics of this RapidMiner "process", shown in Figure 12, were evaluated by using the RMSE in RapidMiner and the findings are outlined in Table 7. Here, as the predicted attributes were X-coordinate and the Y-coordinate values, both of which were of numerical type, so we were able to use the RMSE method for performance evaluation [68] as recommended by ISO/IEC18305:2016—an international standard for evaluating localization and tracking systems [31]. While RMSE is sometimes calculated by using vector analysis where a single value of RMSE is calculated instead of RMSE along X and Y directions, but as ISO/IEC18305:2016 [31] provides 3 different formulae in Chapter 8, for calculation of RMSE in X-direction, Y-direction, and the associated Horizontal Error, so we calculated these performance metrics separately. The formulae for calculation of these three performance characteristics, as mentioned in ISO/IEC18305:2016, are represented in Equations (1)–(3).

$$\varepsilon_{x,rms} = \sqrt{\frac{1}{\mathrm{N}}\sum_{\mathrm{i}=1}^{\mathrm{N}} \varepsilon_{x,i}^{2}} \tag{1}$$

$$\varepsilon_{y,rms} = \sqrt{\frac{1}{\mathrm{N}}\sum_{\mathrm{i}=1}^{\mathrm{N}} \varepsilon_{y,i}^{2}} \tag{2}$$

$$\varepsilon_{h,rms} = \sqrt{\varepsilon_{x,rms}^{2} + \varepsilon_{y,rms}^{2}} \tag{3}$$

where:

$\varepsilon_{x,rms}$ stands for RMSE in the X-direction
$\varepsilon_{y,rms}$ stands for RMSE in the Y-direction
$\varepsilon_{h,rms}$ stands for Horizontal Error that considers RMSE in the X-direction and RMSE in the Y-direction
$\varepsilon_{x,i}^{2}$ stands for squared errors in the X-direction
$\varepsilon_{y,i}^{2}$ stands for squared errors in the Y-direction
N stands for sample size

Table 7. Description of the performance characteristics of the Random Forest-based RapidMiner "process" shown in Figure 12.

Description of Performance Characteristic	Value
Root Mean Squared Error for detection of X-coordinate	5.85 cm
Root Mean Squared Error for detection of Y-coordinate	5.36 cm
Horizontal Error	7.93 cm

As can be observed from Table 7, the root mean squared error for detection of the instantaneous X-coordinate and Y-coordinate values of the user's position were found to be 5.85 cm and 5.36 cm, respectively. To add to the above, the Horizontal Error was found to be 7.93 cm. Further discussion about how this approach and the associated results and findings address multiple research challenges in this field is presented in Section 7.

6. Deducing the Optimal Machine Learning Model for Indoor Localization

As outlined in Section 2, one the research challenges in this field of Indoor Localization is the need to develop an optimal machine learning model for Indoor Localization systems, Indoor Positioning Systems, and Location-Based Services. In [9–25], researchers have used multiple machine learning approaches—Random Forest, Artificial Neural Network, Decision Tree, Support Vector Machine, k-NN, Gradient Boosted Trees, Deep Learning, and Linear Regression. However, none of these works implemented multiple machine learning models to evaluate and compare the associated performance characteristics to deduce the optimal machine learning approach. Due to the differences in the datasets used or the real-time data that was collected, the associated data preprocessing steps that were different, variations in train and test ratio of the data, and several other dissimilar steps that were associated with the developments of each of these machine learning models as presented in [9–25], their final performance accuracies cannot be directly compared to deduce the best approach. Thus, analyzing the performance characteristics of multiple machine learning models, developed, implemented, and tested as per the same methodology, to deduce the optimal approach for development of such Indoor Localization systems serves as the main motivation for the work presented in the section.

In Section 4.3, we outlined the steps associated with the proposed methodology to detect the spatial coordinates of a user's indoor position based on the associated user interactions with the context parameters and the user-centered local spatial context, by using a reference system. Upon development of the same, as a RapidMiner "process", as shown in Figure 12, by using the Random Forest-based learning approach, we evaluated its performance characteristics by calculating the RMSE for X coordinate, the RMSE for Y coordinate, and the Horizontal Error—these are performance evaluation metrics mentioned in ISO/IEC18305:2016 [31]. The RMSE for detection of the X-coordinate and Y-coordinate were found to be 5.85 cm and 5.36 cm as per Equations (1) and (2), respectively. The associated Horizontal Error as per Equation (3) was found to be 7.93 cm. These results are shown in Table 7. In this section, we followed the same steps as outlined in Section 4.3 and as per the flowchart shown in Figure 11, to develop, implement, and test this methodology by using all the machine learning methods that have been used by researchers [9–25] in this field. These machine learning methods included—Random Forest, Artificial Neural Network, Decision Tree, Support Vector Machine, k-NN, Gradient Boosted Trees, Deep Learning, and Linear Regression. As we had already developed this approach by using the Random Forest approach (Figure 12 and Section 5.3), so, we did not repeat the same in this section and we performed this study on all the other machine learning methods. For each of these methods we developed a RapidMiner "process" by using the same version of RapidMiner on the same computer as outlined in Section 3. Each of these processes were developed as a combination of built-in "operators" and user defined "operators" in RapidMiner. An overview of built-in "operators" and user-defined "operators" in RapidMiner was presented in Section 3. We used the same system architecture as outlined in Figure 11 for development of all the machine learning-based "processes" in RapidMiner as discussed here, so, a separate system architecture is not provided in this section. The specific steps that we followed for development of each of these RapidMiner processes corresponding to these different machine learning methods are as follows:

i. Use the 'Dataset' "operator" to import the dataset [67] into the RapidMiner "process".
ii. Utilize the 'Data-Preprocess' "operator" to perform multiple preprocessing steps (Steps iii to vi in Section 4.3) prior to splitting the data for training and testing. We developed this 'Data-Preprocess' "operator".
iii. Use the built-in "operator" called 'Split Data' to divide the dataset into training set and test set. The dataset [67] consisted of 250 rows. We used 70% of the data for training and the remaining 30% for testing.
iv. Use the specific machine learning model to train the system. By specific machine learning model, we mean either the usage of the Artificial Neural Network or Decision Tree or Support Vector Machine or k-NN or Gradient Boosted Trees or Deep Learning or Linear Regression. These machine learning models are present in RapidMiner as built-in "operators" that can be directly used. However, a few of these learning models in RapidMiner such as Artificial Neural Network, Support Vector Machines, and Linear Regression sometimes need the 'nominal to numerical' "operator" for training and testing of the model, based on the characteristics and nature of the dataset being used.
v. Utilize the built-in 'Apply Model' "operator" to apply the learning model on the test data. This "operator" was renamed in each of these "processes", as per the specific learning model that was being developed and evaluated, to indicate the differences

in the associated functionalities of this "operator" for each of these RapidMiner "processes". For the RapidMiner "process" that used the Artificial Neural Network, the 'Apply Model' "operator" was renamed to 'Apply Model—ANN'. Similarly, for the machine learning models—Decision Tree, Support Vector Machine, k-NN, Gradient Boosted Trees, Deep Learning, and Linear Regression, this "operator" was renamed to 'Apply Model—DT', 'Apply Model—SVM', 'Apply Model—kNN', 'Apply Model—GBT', 'Apply Model—DL', and 'Apply Model—LR', respectively.

vi. Use the built-in 'Performance' "operator" to evaluate the performance characteristics of the "process" by calculating the RMSE in X-direction, the RMSE in Y-direction, and the Horizontal Error as per Equations (1)–(3), respectively. This "operator" was renamed in each of these "processes", as per the specific learning model that was being developed and evaluated, to indicate the differences in the associated functionalities of this "operator" for each of these RapidMiner "processes". For the RapidMiner "process" that used the Artificial Neural Network, the 'Performance' "operator" was renamed to 'ANN-Performance'. Similarly, for the machine learning models—Decision Tree, Support Vector Machine, k-NN, Gradient Boosted Trees, Deep Learning, and Linear Regression, this "operator" was renamed to 'DT-Performance', 'SVM-Performance', 'kNN-Performance', 'GBT-Performance', 'DL-Performance', and 'LR-Performance', respectively.

These RapidMiner "processes", that were developed by using the learning approaches—Artificial Neural Network, Decision Tree, Support Vector Machine, k-NN, Gradient Boosted Trees, Deep Learning, and Linear Regression, are shown in Figures 39–45, respectively. The corresponding performance metrics in terms of the RMSE in X-direction, the RMSE in Y-direction, and the Horizontal Error are shown in Tables 8–14, respectively. We did not develop the RapidMiner "process" by using the Random Forest approach in this section as we had already developed the same in Figure 12 and discussed its performance characteristics in terms of the RMSE in X-direction, the RMSE in Y-direction, and the Horizontal Error in Table 7.

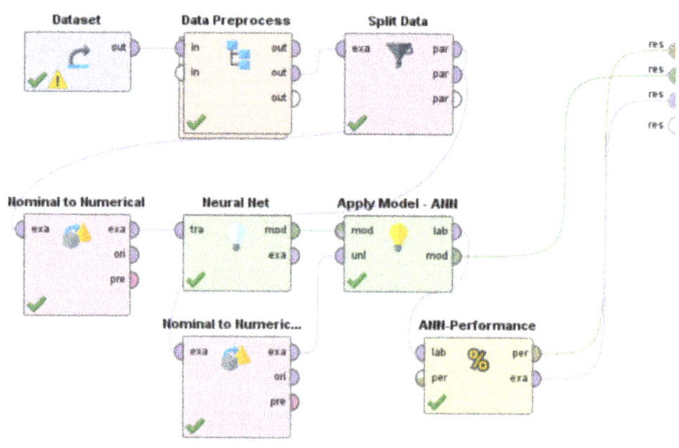

Figure 39. "Process" developed in RapidMiner that used an Artificial Neural Network (ANN)-based learning approach and followed the steps outlined in Section 4.3 for detection of the spatial coordinates of the user in each 'activity-based zone'.

Table 8. Description of the performance characteristics of the Artificial Neural Network (ANN)-based RapidMiner "process" shown in Figure 39.

Description of Performance Characteristic	Value
Root Mean Squared Error for detection of X-coordinate	28.00 cm
Root Mean Squared Error for detection of Y-coordinate	16.16 cm
Horizontal Error	32.33 cm

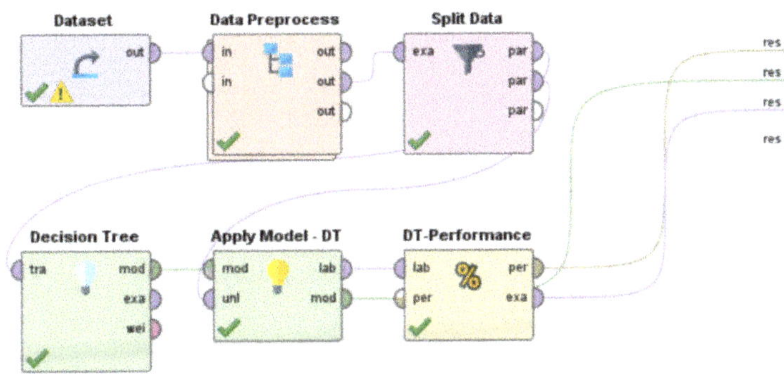

Figure 40. "Process" developed in RapidMiner that used a Decision Tree (DT)-based learning approach and followed the steps outlined in Section 4.3 for detection of the spatial coordinates of the user in each 'activity-based zone'.

Table 9. Description of the performance characteristics of the Decision Tree (DT)-based RapidMiner "process" shown in Figure 40.

Description of Performance Characteristic	Value
Root Mean Squared Error for detection of X-coordinate	12.52 cm
Root Mean Squared Error for detection of Y-coordinate	6.19 cm
Horizontal Error	13.97 cm

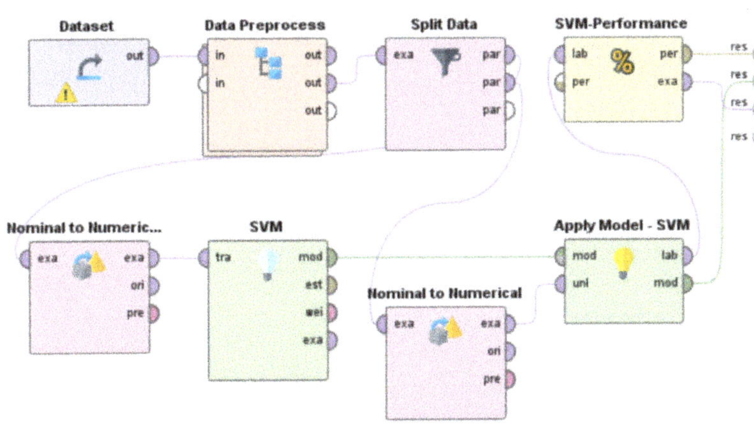

Figure 41. "Process" developed in RapidMiner that used a Support Vector Machine (SVM)-based learning approach and followed the steps outlined in Section 4.3 for detection of the spatial coordinates of the user in each 'activity-based zone'.

Table 10. Description of the performance characteristics of the Support Vector Machine (SVM)-based RapidMiner "process" shown in Figure 41.

Description of Performance Characteristic	Value
Root Mean Squared Error for detection of X-coordinate	27.92 cm
Root Mean Squared Error for detection of Y-coordinate	27.17 cm
Horizontal Error	38.96 cm

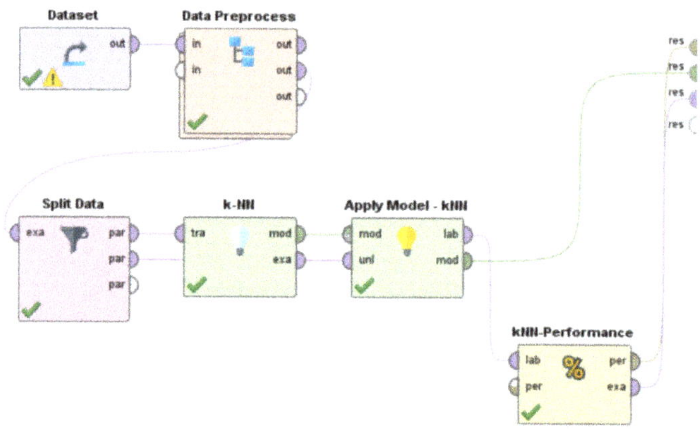

Figure 42. "Process" developed in RapidMiner that used a kNN-based learning approach and followed the steps outlined in Section 4.3 for detection of the spatial coordinates of the user in each 'activity-based zone'.

Table 11. Description of the performance characteristics of the kNN-based RapidMiner "process" shown in Figure 42.

Description of Performance Characteristic	Value
Root Mean Squared Error for detection of X-coordinate	10.11 cm
Root Mean Squared Error for detection of Y-coordinate	2.96 cm
Horizontal Error	10.54 cm

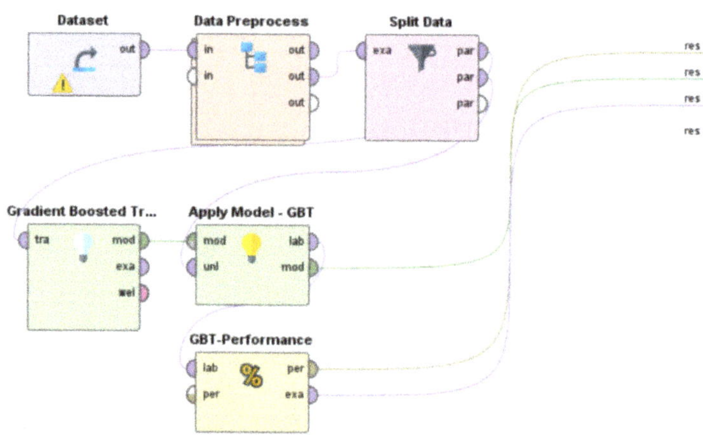

Figure 43. "Process" developed in RapidMiner that used a Gradient Boosted Trees (GBT)-based learning approach and followed the steps outlined in Section 4.3 for detection of the spatial coordinates of the user in each 'activity-based zone'.

Table 12. Description of the performance characteristics of the Gradient Boosted Trees (GBT)-based RapidMiner "process" shown in Figure 43.

Description of Performance Characteristic	Value
Root Mean Squared Error for detection of X-coordinate	28.12 cm
Root Mean Squared Error for detection of Y-coordinate	27.65 cm
Horizontal Error	39.44 cm

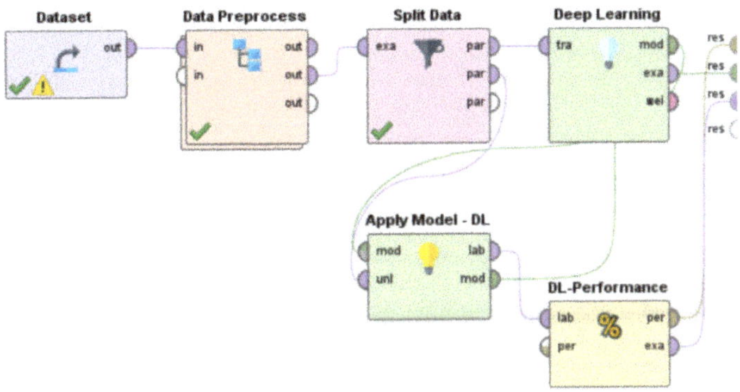

Figure 44. "Process" developed in RapidMiner that used a Deep Learning (DL)-based learning approach and followed the steps outlined in Section 4.3 for detection of the spatial coordinates of the user in each 'activity-based zone'.

Table 13. Description of the performance characteristics of the Deep Learning (DL)-based RapidMiner "process" shown in Figure 44.

Description of Performance Characteristic	Value
Root Mean Squared Error for detection of X-coordinate	29.67 cm
Root Mean Squared Error for detection of Y-coordinate	12.04 cm
Horizontal Error	32.02 cm

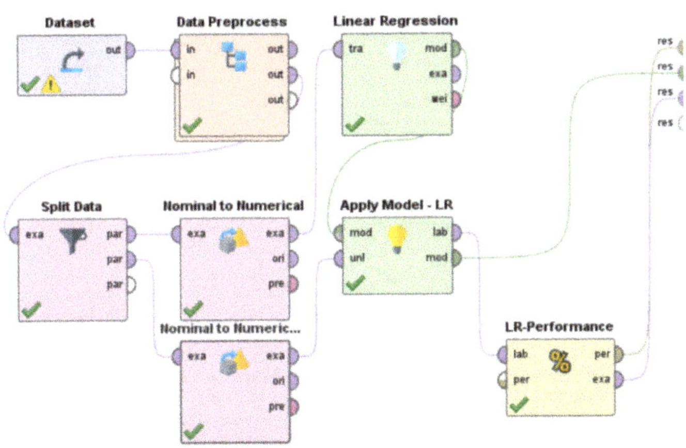

Figure 45. "Process" developed in RapidMiner that used a Linear Regression (LR)-based learning approach and followed the steps outlined in Section 4.3 for detection of the spatial coordinates of the user in each 'activity-based zone'.

Table 14. Description of the performance characteristics of the Linear Regression (LR)-based RapidMiner "process" shown in Figure 45.

Description of Performance Characteristic	Value
Root Mean Squared Error for detection of X-coordinate	28.064 cm
Root Mean Squared Error for detection of Y-coordinate	27.630 cm
Horizontal Error	39.382 cm

The performance metrics—RMSE in X-direction, RMSE in Y-direction, and Horizontal Error for all these machine learning models—Random Forest, Artificial Neural Network, Decision Tree, Support Vector Machine, k-NN, Gradient Boosted Trees, Deep Learning, and Linear Regression are summarized in Table 15. The analysis of these metrics is shown in Figure 46.

Table 15. Comparison of the performance metrics of the different learning approaches—Random Forest, Artificial Neural Network, Decision Tree, Support Vector Machine, k-NN, Gradient Boosted Trees, Deep Learning, and Linear Regression.

Learning Approach	Performance Metrics		
	RMSE in X-Direction	RMSE in Y-Direction	Horizontal Error
Random Forest	5.85 cm	5.36 cm	7.93 cm
Artificial Neural Network	28.00 cm	16.16 cm	32.33 cm
Decision Tree	12.52 cm	6.19 cm	13.97 cm
Support Vector Machine	27.92 cm	27.17 cm	38.96 cm
k-NN	10.11 cm	2.96 cm	10.54 cm
Gradient Boosted Trees	28.12 cm	27.65 cm	39.44 cm
Deep Learning	29.67 cm	12.04 cm	32.02 cm
Linear Regression	28.06 cm	27.63 cm	39.38 cm

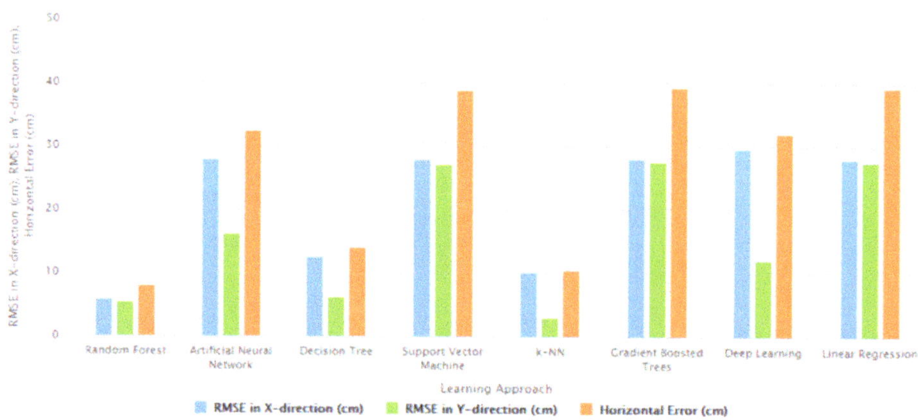

Figure 46. Comparison of the performance metrics of the different learning approaches—Random Forest, Artificial Neural Network, Decision Tree, Support Vector Machine, k-NN, Gradient Boosted Trees, Deep Learning, and Linear Regression, shown in the form a Bar (Column) Style Plot.

From Table 15 and Figure 46 the following can be observed and deduced:

i. The Random Forest-based learning approach has the least Horizontal Error of 7.93 cm and the Gradient Boosted Trees-based learning approach has the highest Horizontal Error of 39.44 cm. Considering Horizontal Error as a function, where Horizontal Error (x) gives the Horizontal Error of 'x', where 'x' is a machine learning model; the Horizontal Errors of these machine learning models can be arranged in an increasing to decreasing order as: Horizontal Error (Random Forest) < Horizontal Error (k-NN) < Horizontal Error (Decision Tree) < Horizontal Error (Deep Learning) < Horizontal Error (Artificial Neural Network) < Horizontal Error (Support Vector Machine) < Horizontal Error (Linear Regression) < Horizontal Error (Gradient Boosted Trees).

ii. The RMSE in X-direction is least for the Random Forest-based learning approach and is highest for the Deep Learning-based learning approach with the respective values being 5.85 cm and 29.67 cm, respectively. Considering RMSE in X-direction as a function, where RMSE in X-direction (p) gives the RMSE in X-direction of 'p', where 'p' is a machine learning model; the RMSE in X-directions of these machine learning models can be arranged in an increasing to decreasing order as: RMSE in X-direction (Random Forest) < RMSE in X-direction (k-NN) < RMSE in X-direction (Decision Tree) < RMSE in X-direction (Support Vector Machine) < RMSE in X-direction (Artificial Neural Network) < RMSE in X-direction (Linear Regression) < RMSE in X-direction (Gradient Boosted Trees) < RMSE in X-direction (Deep Learning).

iii. The RMSE in Y-direction is least for the k-NN-based learning approach with a value of 2.96 cm and this metric is highest for the Gradient Boosted Trees-based learning approach with a value of 27.65 cm. Considering RMSE in Y-direction as a function, where RMSE in Y-direction (q) gives the RMSE in Y-direction of 'q', where 'q' is a machine learning model; the RMSE in Y-directions of these machine learning models can be arranged in an increasing to decreasing order as: RMSE in Y-direction (k-NN) < RMSE in Y-direction (Random Forest) < RMSE in Y-direction (Decision Tree) < RMSE in Y-direction (Deep Learning) < RMSE in Y-direction (Artificial Neural Network) < RMSE in Y-direction (Support Vector Machine) < RMSE in Y-direction (Linear Regression) RMSE in Y-direction (Gradient Boosted Trees)

As can be seen from (i) and (ii) above, the Random Forest-based learning approach has the least RMSE in X-direction as well as the least Horizontal Error. The respective values being 5.85 cm and 7.93 cm, respectively. Even though the k-NN based learning approach has a lesser RMSE in Y-direction (2.96 cm) as compared to RMSE of Random Forest in Y-direction (5.36 cm), the overall Horizontal Error for the k-NN based learning approach is much higher as compared to the Horizontal Error of the Random Forest-based learning approach with the respective values being 10.54 cm and 7.93 cm. Thus, for all practical purposes it may be concluded that the Random Forest-based learning approach is the optimal machine learning model for development of Indoor Localization systems, Indoor Positioning Systems, and Location-Based Services. Further discussion about how this comparative study and the associated results and findings address multiple research challenges in this field is presented in Section 7.

7. Comparative Discussion

Despite several advances in the fields of Indoor Localization, Indoor Positioning Systems, Human Activity Recognition, Activity Analysis, and Ambient Assisted Living, there exist several research challenges in this field. The work presented in this paper at the intersection of Big Data, Machine Learning, Indoor Localization, Ambient Assisted Living, Internet of Things, Activity Centric Computing, Human–Computer Interaction, Pattern Recognition, and Assisted Living Technologies, and their related application domains aims to take a comprehensive approach to address these challenges. We introduced these research challenges in Section 2. In this section, we further discuss the same and outline how the work presented in this paper and the associated results and findings addresses these challenges and outperform similar works in this field. This is discussed as follows:

1. Need for AAL-based activity recognition and activity analysis-based systems to be able to track the indoor location of the user: The AAL-based systems currently lack the ability to track the indoor location of the user. There have been several works done [46–51] in these interrelated fields of activity recognition, activity analysis, and fall detection, but none of these works have focused on Indoor Localization. Being able to track the indoor location of a user is of prime importance and of crucial need for AAL-based systems to be able to contribute towards improving the quality of life of individuals in the future of living environments, such as, Smart Homes. For instance, an elderly person could be staying in an apartment which is a part of a multistoried building such as Taipei 101 [70] or Burj Khalifa [71]—both of which are amongst the tallest buildings in the world. When this elderly person experiences a fall, a fall detection system such as [51], could detect a fall and alert caregivers but the current GPS-based technologies would only provide the building level information.

The lack of the precise location information in terms of the specific floor, apartment, and room, could cause delay of medical attention or assistive care. Such delay of care can have both short-term and long-term health-related impacts to the elderly such as long lie [72], that can cause dehydration, rhabdomyolysis, pressure injuries, carpet burns, hypothermia, pneumonia, and fear of falling, which could lead to decreased independence and willingness in carrying out daily routine activities. Long lie can even lead to death in some cases. Thus, it is the need of the hour that AAL-based systems should not only be able to track, monitor, and analyze human behavior but they should also be equipped with the functionality to detect the indoor location of the users. The work presented in this paper addresses this challenge by proposing a novel Big-Data driven methodology that can study the multimodal components of user interactions during Activities of Daily Living (ADLs) (Tables 1 and 2) and analyze the data from BLE beacons and BLE scanners to track a user's indoor location in a specific 'activity-based zone' during different ADLs (Figure 6). This approach was developed by using a k-nearest neighbor (k-NN)-based learning approach (Section 4.1). When tested on a dataset (Figure 17, Table 3) it achieved a performance accuracy of 81.36% (Figures 18 and 19).

2. Need for context-independent Indoor Localization systems: As outlined in Section 2, several recent works related to Indoor Localization systems are context-based and are only functional in the specific environments for which they were developed [26–30]. These specific environments include—factories [26], indoor parking [27], hospitals [28], industry-based settings [29], and academic environments [30]. For instance, the methodology proposed in [29] is not functional in any of the settings described in [26–28,30]. The future of interconnected Smart Cities would consist of a host of indoor environments in the living and functional spaces of humans, which would be far more diverse, different, and complicated as compared to the environments described in [26–30]. The challenge is thus to develop a means for Indoor Localization that is not environment dependent and can be seamlessly deployed in any IoT-based setting irrespective of the associated context parameters and their attributes. The work proposed in this paper addresses this challenge by proposing a novel context independent approach that can interpret the accelerometer and gyroscope data from diverse behavioral patterns to detect the 'zone-based' indoor location of a user in any IoT-based environment (Section 4.2). This proposed approach (Figure 9) can study, analyze, and interpret the distinct behavioral patterns, in terms of the associated accelerometer and gyroscope data, local to each such 'zone', in the confines of any given IoT-based space without being affected by the changes or variations in the context parameters or environment variables. It uses a Random Forest-based learning approach for the training and the same was evaluated on a dataset (Figure 26, Table 4). The performance accuracy of this method for detecting a user's location in each of these 'zones', that were present in this dataset [66], was found to be 81.13% (Figures 27 and 28). Here, the 'zone-based' mapping of a user's location refers to mapping the user in one of the multiple 'activity-based zones' that any given IoT-based environment can be classified into based on the specific activity being performed by the user. The accelerometer and gyroscope data are user behavior dependent and not context parameter dependent and neither is this approach of spatially mapping a given IoT-based space into 'activity-bases zones' dependent on any specific set of context parameters, as explained in Section 4.2. This upholds the context independent nature of this methodology. In other words, this proposed methodology can be seamlessly applied to any IoT-based environment, including all the environments described in [26–30], as well as in any other IoT-based setting that involves different forms of user interactions on context parameters or environment variables, which can be characterized by the changes in the associated behavioral data.
3. The RMSE of the existing Indoor Localization systems [33–43] are still high and greater precision and accuracy for detection of indoor location is the need of the

hour. Several performance metrics have been used by researchers for studying the characteristics of Indoor Localization systems, Indoor Positioning Systems, and Location-Based Services. However, ISO/IEC18305:2016, an international standard for evaluating localization and tracking systems [31], which is one of the recent works in this field, lists several metrics and the associated formulae for evaluating the performance characteristics of such systems. These include the formulae for determination of the RMSE in the X-direction, Y-direction, and in the X-Y plane. When the RMSE is determined in the X-Y plane, it is referred to as Horizontal Error as per the definitions of the standard [31]. We have presented and discussed the associated formulae in Equations (1)–(3). Upon reviewing the recent works [33–43] related to this field, as presented in Section 2, it can be observed that the RMSE of the works are still significantly high in view of the average dimensions of an individual's living space. As per [44,45], (1) the average dimensions of newly built one-bedroom apartments and two-bedroom apartments in United States in 2018 were 757 square feet (70.3276 square meters) and 1138 square feet (105.7236 square meters), respectively. In view of these dimensions of these apartments, it can be concluded that higher precision is needed for the future of Indoor Localization systems. Such systems should have much lower values of RMSE in X and Y directions as well as their overall Horizontal Error should be low. The work presented in this paper addresses this research challenge by proposing a methodology to detect the spatial coordinates of a user's indoor position based on the associated user interactions with the context parameters and the user-centered local spatial context, by using a reference system. In Section 4.3 we have presented the steps for development of this approach for Indoor Localization and the results of the same are discussed in Section 5.3. While RMSE is sometimes calculated by using vector analysis where a single value of RMSE is calculated instead of RMSE along X and Y directions, but as ISO/IEC18305:2016 [31] presents two separate formulae (Equations (1) and (2)) for calculation of RMSE in X-direction and RMSE in Y-direction and a third formula (Equation (3)) for Horizontal Error calculation, so, we calculated RMSE in X and Y directions separately and then calculated the Horizontal Error as per Equations (1)–(3), respectively. As can be seen from the results (Table 7), the performance characteristics of our approach are—RMSE in X-direction: 5.85 cm, RMSE in Y-direction: 5.36 cm, and Horizontal Error: 7.93 cm. As can be seen from [33–43], RMSE is usually represented in meters, so upon converting these metrics from Table 7 to meters (correct to 2 decimal places) the corresponding values are: RMSE in X-direction: 0.06 m, RMSE in Y-direction: 0.05 m, and Horizontal Error: 0.08 m. The RMSE of these existing works [33–43], in increasing to decreasing order are shown in Table 16.

Table 16. Summary of the various Indoor Localization approaches that used RMSE for evaluation of the performance metrics.

RMSE Value (in Meters)	Work(s)
0.32	Bolic et al. [34]
1.00	Chen et al. [41]
1 to 2	Angermann et al. [35]
1.20	Klingbeil et al. [38]
1.28	Chen at al. [43]
1.40	Correa et al. [33]
1.53	Evennou et al. [36]
2.90	Li et al. [42]
3.10	Liu et al. [40]
4.30	Wang et al. [37]
4.55	Pei et al. [39]

From Table 16, it can be concluded that Bolic et al.'s work [34] has the best performance accuracy out of all the works reviewed in [33–43] with the RMSE being 0.32 m. Upon comparing the performance metrics of our approach (Table 7) with Bolic et al.'s work [34], it can be easily concluded that our work outperforms the same in terms of performance accuracy as the RMSE values (RMSE in X-direction: 0.06 m, RMSE in Y-direction: 0.05 m, and Horizontal Error: 0.08 m) of our methodology are significantly lower. As our work outperforms Bolic et al.'s work, which has the best accuracy out of all the works reviewed in [33–43], so, it can also be concluded that our work outperforms all the other works as well [33,35–43], in terms of the RMSE method of performance evaluation, as recommended by ISO/IEC18305:2016 [31].

4. Need for an optimal machine learning-based approach for Indoor Localization: A range of machine learning approaches—Random Forest, Artificial Neural Network, Decision Tree, Support Vector Machine, k-NN, Gradient Boosted Trees, Deep Learning, and Linear Regression, have been used by several researchers [9–25] for development of various types of Indoor Localization systems for IoT-based environments. While each of these systems seem to perform reasonably well but none of these works attempted to develop an optimal machine learning model for Indoor Localization systems. Additionally, due to variations in the data source, differences in the types of data, varied methods of data collection, different training set to test set ratios, dissimilar data preprocessing steps, as well as because of differences in the simulated or real-world environments in which these respective systems were developed, implemented, and deployed, the performance metrics of these systems cannot be directly compared to deduce the optimal approach. These works [9–25] along with the machine learning approaches that were used in each are outlined in Table 17.

Table 17. Summary of the various machine learning approaches that have been investigated by researchers in this field.

Learning Approach Used	Work(s)
Random Forest	Varma et al. [19], Gao et al. [20]
Artificial Neural Network	Khan et al. [16], Labinghisa et al. [17], Qin et al. [18]
Decision Tree	Musa et al. [9], Yim et al. [10]
Support Vector Machine	Sjoberg et al. [11], Zhang et al. [12]
k-NN	Zhang et al. [13], Ge et al. [14], Hu et al. [15]
Gradient Boosted Trees	Wang et al. [25]
Deep Learning	Zhang et al. [23], Poulose et al. [24]
Linear Regression	Jamâa et al. [21], Barsocchi et al. [22]

There is a need to address this research challenge of identifying the optimal machine learning methodology for Indoor Localization. The work presented in this paper addresses this challenge. In Section 6—we developed, implemented, and tested the performance characteristics of different learning models to perform Indoor Localization by using the same dataset [67], the same data preprocessing steps, the same training and test ratios, and the same methodology, which we presented in Section 4.3. The learning models that we developed and studied included—Random Forest, Artificial Neural Network, Decision Tree, Support Vector Machine, k-NN, Gradient Boosted Trees, Deep Learning, and Linear Regression. These models were developed to detect the spatial coordinates of a user's indoor location as per the methodology outlined in Section 4.3. RapidMiner was used to develop these machine learning models, and the corresponding RapidMiner "processes" are shown in Figures 12 and 39–45, respectively. We evaluated the performance characteristics of these models based on three performance metrics as outlined in ISO/IEC18305:2016—an international standard for evaluating localization and tracking systems [31]. These include—RMSE in X-direction, RMSE in Y-direction, and Horizontal Error (Equations (1)–(3)). The performance characteristics of these respective machine learning models are shown in Tables 7–14. In Table 15 and Figure 46, we present the comparisons amongst these learning models to deduce the optimal machine learning approach for development of an Indoor Localization system. Based on the findings presented in Table 15 and Figure 46, the following can be observed:

i. Out of all these learning approaches, the Random Forest-based learning approach has the least Horizontal Error of 7.93 cm. In an increasing to decreasing order, the Horizontal Errors of these machine learning models can be arranged as: Horizontal Error (Random Forest) < Horizontal Error (k-NN) < Horizontal Error (Decision Tree) < Horizontal Error (Deep Learning) < Horizontal Error (Artificial Neural Network) < Horizontal Error (Support Vector Machine) < Horizontal Error (Linear Regression) < Horizontal Error (Gradient Boosted Trees).

ii. Out of all these learning approaches, the RMSE in X-direction is the least for the Random Forest-based learning approach, which is equal to 5.85 cm. In an increasing to decreasing order, the RMSE in X-direction of these machine learning models can be arranged as: RMSE in X-direction (Random Forest) < RMSE in X-direction (k-NN) < RMSE in X-direction (Decision Tree) < RMSE in X-direction (Support Vector Machine) < RMSE in X-direction (Artificial Neural Network) < RMSE in X-direction (Linear Regression) < RMSE in X-direction (Gradient Boosted Trees) < RMSE in X-direction (Deep Learning).

iii. Out of all these learning models, the RMSE in Y-direction of the k-NN-based learning approach is the lowest and the RMSE in Y-direction of the Random Forest-based learning approach is the second lowest. Their respective values being 2.96 cm and 5.36 cm, respectively. In an increasing to decreasing order, the RMSE in Y-direction of these machine learning models can be arranged as: RMSE in Y-direction (k-NN) < RMSE in Y-direction (Random Forest) < RMSE in Y-direction (Decision Tree) < RMSE in Y-direction (Deep Learning) < RMSE in Y-direction (Artificial Neural Network) < RMSE in Y-direction (Support Vector Machine) < RMSE in Y-direction (Linear Regression) RMSE in Y-direction (Gradient Boosted Trees)

iv. From (i) and (ii), it can be deduced that for the RMSE in X-direction and for the Horizontal Error (Equations (1) and (3)) methods of performance evaluation, the Random Forest-based learning approach outperforms all the other learning approaches—Artificial Neural Network, Decision Tree, Support Vector Machine, k-NN, Gradient Boosted Trees, Deep Learning, and Linear Regression. Even though the k-NN-based learning approach performs better than the Random Forest-based learning approach for determination of the RMSE in Y-direction, as can be seen from (iii), however, the difference between RMSE in Y-direction for the k-NN based learning approach and the RMSE in Y-direction for the Random Forest based learning approach is not high. To add, for the other two performance metrics—RMSE in X-direction and Horizontal Error, the k-NN based learning approach does not perform as good as the Random Forest-based learning approach and its error values are much higher. Thus, based on these findings and the discussions, which are presented in an elaborate manner in Section 6, it can be concluded that a Random Forest-based learning approach is the optimal machine learning model for development of Indoor Localization systems, Indoor Positioning Systems, and Location-Based Services.

8. Conclusions and Scope for Future Work

The future of technology-laden living and functional environments, for instance, Smart Homes, Smart Cities, Smart Workplaces, Smart Industries, and Smart Vehicles, would involve Human–Computer, Human–Robot, Human–Machine, and other forms of human interactions with technology-laden gadgets, systems, or devices. Although Global Positioning System (GPS) and Global Navigation Satellite Systems (GNSS) have significantly revolutionized navigation research by being able to track people, objects, and assets in real-time, such technologies are still ineffective in indoor settings [1]. Indoor Localization has multiple applications in the context of such forms of human interactions with technology. As per [3], the market opportunities of Indoor Localization related systems are expected to be in the order of USD 10 billion by 2024 due to the diverse societal needs that such systems can address. There can be multiple use cases and applications of Indoor Localization systems that can be investigated and studied. This paper focuses on one specific application domain—Ambient Assisted Living (AAL) of elderly people in the future of Internet of Things (IoT)-based living environments, such as Smart Homes and Smart Cities. The work presented in this paper addresses multiple research challenges and makes several scientific contributions to this field by integrating the latest advancements from Big Data, Machine Learning, Indoor Localization, Ambient Assisted Living, Internet of Things, Activity Centric Computing, Human–Computer Interaction, Pattern Recognition, Assisted Living Technologies, and their related application domains.

First, to address the research challenge that the AAL-based systems and technologies [46–51] for activity recognition, activity analysis, and fall detection, currently lack the ability to track the indoor location of the user; this paper proposes a novel Big-Data driven methodology that studies the multimodal components of user interactions and analyzes the data from BLE beacons and BLE scanners to track a user's indoor location in a specific 'activity-based zone' during Activities of Daily Living. This approach was developed by using a k-nearest neighbor (k-NN)-based learning approach. When tested on a dataset this methodology achieved a performance accuracy of 81.36%.

Second, to address the limitation in several Indoor Localization systems [26–30], that they are context-based and are only functional in the specific environments in which they were developed; this paper proposes a context independent approach that can interpret the accelerometer and gyroscope data from diverse behavioral patterns to detect the 'zone-based' indoor location of a user in any IoT-based environment. Here, the 'zone-based' mapping of a user's location refers to mapping the user in one of the multiple 'activity-based zones' that any given IoT-based environment can be classified into, based on the specific activity being performed by the user. This methodology was developed by using the Random Forest-based learning approach. When tested on a dataset this novel methodology achieved a performance accuracy of 81.13%.

Third, to address the challenge that the RMSE of the existing Indoor Localization systems are still high [33–43] and greater precision and accuracy for detection of indoor location is the need of the hour; this paper proposes a methodology to detect the spatial coordinates of a user's indoor position based on the associated user interactions with the context parameters and the user-centered local spatial context, by using a reference system. The performance characteristics of this system were evaluated as per three metrics stated in ISO/IEC18305:2016 [31], which is an international standard for testing Localization and Tracking Systems. These metrics included root mean squared error (RMSE) in X-direction, RMSE in Y-direction, and the Horizontal Error which were found to be 5.85 cm, 5.36 cm, and 7.93 cm, respectively. A comparison study of this approach with similar researches [33–43] in this field showed that our system outperformed all these works that had used a similar approach of performance evaluation.

Finally, in view of the fact that multiple machine learning-based approaches have been used by researchers [9–25] and there is a need to identify the optimal machine learning model that can be used to develop the future of Indoor Localization systems, Indoor Positioning Systems, and Location-Based Services; the paper presents a comprehensive comparative study of different machine learning approaches that include—Random Forest, Artificial Neural Network, Decision Tree, Support Vector Machine, k-NN, Gradient Boosted Trees, Deep Learning, and Linear Regression. The performance characteristics of each of these learning methods were studied by evaluating the RMSE in X-direction, the RMSE in Y-direction, and the Horizontal Error as per ISO/IEC18305:2016 [31]. The results and findings of this study show that the Random Forest approach can be considered as the optimal learning method for development of such technologies for all practical purposes.

To the best knowledge of the authors, no similar work has been done yet and no work in the field of Indoor Localization thus far has achieved such a superior performance accuracy (RMSE for detection of X coordinate: 5.85 cm, RMSE for detection of Y coordinate: 5.36 cm, and Horizontal Error: 7.93 cm) as presented in this work. Future work would involve—(1) Implementing and deploying all these proposed approaches for Indoor Localization in real-time in different IoT-based environments by using the Context-Driven Human Activity Recognition Framework [64]. For real-time implementation of all these proposed approaches, we plan on conducting experiments as per Institutional Review Board (IRB) approved protocols by setting up an experiment procedure for data collection and analysis. The specific functionalities and characteristic features of the different methodologies that we have outlined in Section 4.1, Section 4.2, Section 4.3 would then be implemented in real-time. Thereafter, the performance characteristics from the real-time data would be studied and compared with the findings presented in Section 5.1, Section 5.2, Section 5.3 (2) Extending the functionalities of the two 'zone'-based Indoor Localization approaches and evaluating their performance characteristics by using the RMSE approach as well as by using some of the other performance metrics defined in ISO/IEC18305:2016 [31]. This would be performed either by analyzing the real-time data collected from (1) or by using a different dataset that consists of user interaction data related to different ADLs and the spatial coordinates of the user's varying position recorded during the dynamic user interactions associated with these different activities.

Author Contributions: Conceptualization, N.T. and C.Y.H.; methodology, N.T.; software, N.T.; validation, N.T.; formal analysis, N.T.; investigation, N.T.; resources, N.T.; data curation, N.T.; visualization, N.T.; data analysis and results, N.T.; writing—original draft preparation, N.T.; writing—review and editing, N.T. and C.Y.H.; supervision, C.Y.H.; project administration, C.Y.H.; funding acquisition, Not Applicable. All authors have read and agreed to the published version of the manuscript.

Funding: This research received no external funding.

Institutional Review Board Statement: Not applicable.

Informed Consent Statement: Not applicable.

Data Availability Statement: Publicly available datasets were analyzed in this study. This data can be found at: https://doi.org/10.17632/sy3kcttdtx.1 and https://www.kaggle.com/liwste/indoor-positioning (accessed on 13 February 2021).

Conflicts of Interest: The authors declare no conflict of interest.

References

1. Langlois, C.; Tiku, S.; Pasricha, S. Indoor Localization with Smartphones: Harnessing the Sensor Suite in Your Pocket. *IEEE Consum. Electron. Mag.* **2017**, *6*, 70–80. [CrossRef]
2. Zafari, F.; Papapanagiotou, I.; Devetsikiotis, M.; Hacker, T.J. An iBeacon based proximity and indoor localization system. *arXiv* **2017**, arXiv:1703.07876.
3. Dardari, D.; Closas, P.; Djuric, P.M. Indoor Tracking: Theory, Methods, and Technologies. *IEEE Trans. Veh. Technol.* **2015**, *64*, 1263–1278. [CrossRef]
4. Thakur, N. Framework for a Context Aware Adaptive Intelligent Assistant for Activities of Daily Living. Master's Thesis, University of Cincinnati, Cincinnati, OH, USA, 2019. Available online: http://rave.ohiolink.edu/etdc/view?acc_num=ucin1553528536685873 (accessed on 10 December 2020).
5. Thakur, N.; Han, C.Y. An Improved Approach for Complex Activity Recognition in Smart Homes. Available online: https://link.springer.com/chapter/10.1007/978-3-030-22888-0_15 (accessed on 3 March 2021).
6. United Nations: 2020 Report on Ageing. Available online: http://www.un.org/en/sections/issuesdepth/ageing/ (accessed on 22 November 2020).
7. United States Census Bureau Report: An Aging World: 2015. Available online: https://www.census.gov/library/publications/2016/demo/P95-16-1.html (accessed on 17 December 2020).
8. Key Facts of Dementia. Available online: https://www.who.int/news-room/fact-sheets/detail/dementia (accessed on 24 November 2020).
9. Musa, A.; Nugraha, G.D.; Han, H.; Choi, D.; Seo, S.; Kim, J. A decision tree-based NLOS detection method for the UWB indoor location tracking accuracy improvement. *Int. J. Commun. Syst.* **2019**, *32*, e3997. [CrossRef]
10. Yim, J. Introducing a decision tree-based indoor positioning technique. *Expert Syst. Appl.* **2008**, *34*, 1296–1302. [CrossRef]

11. Sjoberg, M.; Koskela, M.; Viitaniemi, V.; Laaksonen, J. Indoor location recognition using fusion of SVM-based visual classifiers. In Proceedings of the 2010 IEEE International Workshop on Machine Learning for Signal Processing, Kittila, Finland, 29 August–1 September 2010; pp. 343–348. [CrossRef]
12. Zhang, S.; Guo, J.; Wang, W.; Hu, J. Indoor 2.5D Positioning of WiFi Based on SVM. In Proceedings of the 2018 Ubiquitous Positioning, Indoor Navigation and Location-Based Services (UPINLBS) Conference, Wuhan, China, 22–23 March 2018; pp. 1–7. [CrossRef]
13. Zhang, L.; Zhao, C.; Wang, Y.; Dai, L. Fingerprint-based Indoor Localization using Weighted K-Nearest Neighbor and Weighted Signal Intensity. In Proceedings of the 2nd International Conference on Artificial Intelligence and Advanced Manufacture, Association for Computing Machinery (ACM), Manchester, UK, 15–17 October 2020; pp. 185–191.
14. Ge, X.; Qu, Z. Optimization WIFI indoor positioning KNN algorithm location-based fingerprint. In Proceedings of the 2016 7th IEEE International Conference on Software Engineering and Service Science (ICSESS), Beijing, China, 26–28 August 2016; pp. 135–137.
15. Hu, J.; Liu, D.; Yan, Z.; Liu, H. Experimental Analysis on Weight KK -Nearest Neighbor Indoor Fingerprint Positioning. *IEEE Internet Things J.* **2018**, *6*, 891–897. [CrossRef]
16. Khan, I.U.; Ali, T.; Farid, Z.; Scavino, E.; Rahman, M.A.A.; Hamdi, M.; Qiao, G. An improved hybrid indoor positioning system based on surface tessellation artificial neural network. *Meas. Control.* **2020**, *53*, 1968–1977. [CrossRef]
17. Labinghisa, B.A.; Lee, D.M. Neural network-based indoor localization system with enhanced virtual access points. *J. Supercomput.* **2021**, *77*, 638–651. [CrossRef]
18. Qin, F.; Zuo, T.; Wang, X. CCpos: WiFi Fingerprint Indoor Positioning System Based on CDAE-CNN. *Sensors* **2021**, *21*, 1114. [CrossRef]
19. Varma, P.S.; Anand, V. Random Forest Learning Based Indoor Localization as an IoT Service for Smart Buildings. *Wirel. Pers. Commun.* **2020**, 1–19. [CrossRef]
20. Gao, J.; Li, X.; Ding, Y.; Su, Q.; Liu, Z. WiFi-Based Indoor Positioning by Random Forest and Adjusted Cosine Similarity. In Proceedings of the 2020 Chinese Control and Decision Conference (CCDC), Hefei, China, 22–24 August 2020; pp. 1426–1431.
21. Ben Jamâa, M.; Koubâa, A.; Baccour, N.; Kayani, Y.; Al-Shalfan, K.; Jmaiel, M. EasyLoc: Plug-and-Play RSS-Based Localization in Wireless Sensor Networks. In *Complex Networks & Their Applications IX*; Springer: Berlin, Germany, 2013; Volume 507, pp. 77–98.
22. Barsocchi, P.; Lenzi, S.; Chessa, S.; Furfari, F. Automatic virtual calibration of range-based indoor localization systems. *Wirel. Commun. Mob. Comput.* **2011**, *12*, 1546–1557. [CrossRef]
23. Zhang, Q.; Wang, Y. A 3D mobile positioning method based on deep learning for hospital applications. *EURASIP J. Wirel. Commun. Netw.* **2020**, *2020*, 1–15. [CrossRef]
24. Poulose, A.; Han, D.S. Hybrid Deep Learning Model Based Indoor Positioning Using Wi-Fi RSSI Heat Maps for Autonomous Applications. *Electronics* **2020**, *10*, 2. [CrossRef]
25. Wang, Y.; Lei, Y.; Zhang, Y.; Yao, L. A robust indoor localization method with calibration strategy based on joint distribution adaptation. *Wirel. Netw.* **2021**, 1–15. [CrossRef]
26. Lin, Y.-T.; Yang, Y.-H.; Fang, S.-H. A case study of indoor positioning in an unmodified factory environment. In Proceedings of the 2014 International Conference on Indoor Positioning and Indoor Navigation (IPIN), Busan, Korea, 27–30 October 2014; pp. 721–722.
27. Liu, J.; Chen, R.; Chen, Y.; Pei, L.; Chen, L. iParking: An Intelligent Indoor Location-Based Smartphone Parking Service. *Sensors* **2012**, *12*, 14612–14629. [CrossRef]
28. Jiang, L.; Hoe, L.N.; Loon, L.L. Integrated UWB and GPS location sensing system in hospital environment. In Proceedings of the 2010 5th IEEE Conference on Industrial Electronics and Applications, Taichung, Taiwan, 15–17 June 2010; pp. 286–289.
29. Barral, V.; Suárez-Casal, P.; Escudero, C.J.; García-Naya, J.A. Multi-Sensor Accurate Forklift Location and Tracking Simulation in Industrial Indoor Environments. *Electronics* **2019**, *8*, 1152. [CrossRef]
30. Zadeh, A.M.H.; Koo, A.C.; Abadi, H.G.N. Design of Students' Attendance system based on mobile indoor location. In Proceedings of the International Conference on Mobile Learning, Application & Services, Malacca, Malaysia, 18–20 September 2012; pp. 1–4.
31. ISO/IEC 18305:2016 Information Technology—Real Time Locating Systems—Test and Evaluation of Localization and Tracking Systems. Available online: https://www.iso.org/standard/62090.html (accessed on 13 February 2021).
32. EVARILOS—Evaluation of RF-based Indoor Localization Solutions for the Future Internet. Available online: https://www2.tkn.tu-berlin.de/tkn-projects/evarilos/index.php (accessed on 13 February 2021).
33. Correa, A.; Llado, M.B.; Morell, A.; Vicario, J.L.; Barcelo, M. Indoor Pedestrian Tracking by On-Body Multiple Receivers. *IEEE Sens. J.* **2016**, *16*, 2545–2553. [CrossRef]
34. Bolic, M.; Rostamian, M.; Djuric, P.M. Proximity Detection with RFID: A Step toward the Internet of Things. *IEEE Pervasive Comput.* **2015**, *14*, 70–76. [CrossRef]
35. Angermann, M.; Robertson, P. FootSLAM: Pedestrian Simultaneous Localization and Mapping without Exteroceptive Sensor-sHitchhiking on Human Perception and Cognition. *Proc. IEEE* **2012**, *100*, 1840–1848. [CrossRef]
36. Evennou, F.; Marx, F. Advanced Integration of WiFi and Inertial Navigation Systems for Indoor Mobile Positioning. *EURASIP J. Adv. Signal Process.* **2006**, *2006*, 86706. [CrossRef]

37. Wang, H.; Lenz, H.; Szabo, A.; Bamberger, J.; Hanebeck, U.D. WLAN-Based Pedestrian Tracking Using Particle Filters and Low-Cost MEMS Sensors. In Proceedings of the 2007 4th Workshop on Positioning, Navigation and Communication, Hannover, Germany, 22 March 2007; pp. 1–7.
38. Klingbeil, L.; Wark, T. A Wireless Sensor Network for Real-Time Indoor Localisation and Motion Monitoring. In Proceedings of the 2008 International Conference on Information Processing in Sensor Networks (ipsn 2008), St. Louis, MO, USA, 22–24 April 2008; pp. 39–50.
39. Pei, L.; Liu, J.; Guinness, R.; Chen, Y.; Kuusniemi, H.; Chen, R. Using LS-SVM Based Motion Recognition for Smartphone Indoor Wireless Positioning. *Sensors* **2012**, *12*, 6155–6175. [CrossRef]
40. Liu, J.; Chen, R.; Pei, L.; Guinness, R.; Kuusniemi, H. A Hybrid Smartphone Indoor Positioning Solution for Mobile LBS. *Sensors* **2012**, *12*, 17208–17233. [CrossRef] [PubMed]
41. Chen, Z.; Zou, H.; Jiang, H.; Zhu, Q.; Soh, Y.C.; Xie, L. Fusion of WiFi, Smartphone Sensors and Landmarks Using the Kalman Filter for Indoor Localization. *Sensors* **2015**, *15*, 715–732. [CrossRef] [PubMed]
42. Li, Y.; Zhang, P.; Lan, H.; Zhuang, Y.; Niu, X.; El-Sheimy, N. A modularized real-time indoor navigation algorithm on smartphones. In Proceedings of the 2015 International Conference on Indoor Positioning and Indoor Navigation (IPIN), Banff, AB, Canada, 13–16 October 2015; pp. 1–7.
43. Chen, Z.; Zhu, Q.; Soh, Y.C. Smartphone Inertial Sensor-Based Indoor Localization and Tracking With iBeacon Corrections. *IEEE Trans. Ind. Inform.* **2016**, *12*, 1540–1549. [CrossRef]
44. Statistics: Average Size of Newly Built One-Bedroom Apartments in the United States from 2008 to 2018. Available online: https://www.statista.com/statistics/943956/size-newly-built-one-bed-apartments-usa/ (accessed on 7 February 2021).
45. Statistics: Average Size of Newly Built Two-Bedroom Apartments in the United States from 2008 to 2018. Available online: https://www.statista.com/statistics/943958/size-newly-built-two-bed-apartments-usa/ (accessed on 7 February 2021).
46. Ranieri, C.; MacLeod, S.; Dragone, M.; Vargas, P.; Romero, R.A. Activity Recognition for Ambient Assisted Living with Videos, Inertial Units and Ambient Sensors. *Sensors* **2021**, *21*, 768. [CrossRef] [PubMed]
47. Fahada, L.G.; Tahir, S.F. Activity recognition and anomaly detection in smart homes. *J. Neurocomput.* **2021**, *423*, 362–372. [CrossRef]
48. Suriani, N.S.; Rashid, F.A.N. Smartphone Sensor Accelerometer Data for Human Activity Recognition Using Spiking Neural Network. *Int. J. Machine Learn. Comput.* **2021**, *11*, 298–303.
49. Mousavi, S.A.; Heidari, F.; Tahami, E.; Azarnoosh, M. Fall detection system via smart phone and send people location. In Proceedings of the 2020 28th European Signal Processing Conference (EUSIPCO), Amsterdam, The Netherlands, 18–22 January 2021; pp. 1605–1607.
50. Alarifi, A.; Alwadain, A. Killer heuristic optimized convolution neural network-based fall detection with wearable IoT sensor devices. *J. Meas.* **2021**, *167*, 108258. [CrossRef]
51. Al-Okby, M.F.R.; Al-Barrak, S.S. New Approach for Fall Detection System Using Embedded Technology. In Proceedings of the 24th IEEE International Conference on Intelligent Engineering Systems (INES), Reykjavík, Iceland, 8–10 July 2020; pp. 209–214.
52. Nikoloudakis, Y.; Panagiotakis, S.; Markakis, E.; Pallis, E.; Mastorakis, G.; Mavromoustakis, C.X.; Dobre, C. A Fog-Based Emergency System for Smart Enhanced Living Environments. *IEEE Cloud Comput.* **2016**, *3*, 54–62. [CrossRef]
53. Navarro, J.; Vidaña-Vila, E.; Alsina-Pagès, R.M.; Hervás, M. Real-Time Distributed Architecture for Remote Acoustic Elderly Monitoring in Residential-Scale Ambient Assisted Living Scenarios. *Sensors* **2018**, *18*, 2492. [CrossRef]
54. Nikoloudakis, Y.; Markakis, E.; Mastorakis, G.; Pallis, E.; Skianis, C. An NF V-powered emergency system for smart enhanced living environments. In Proceedings of the 2017 IEEE Conference on Network Function Virtualization and Software Defined Networks (NFV-SDN), Berlin, Germany, 6–8 November 2017; pp. 258–263.
55. Facchinetti, D.; Psaila, G.; Scandurra, P. Mobile cloud computing for indoor emergency response: The IPSOS assistant case study. *J. Reliab. Intell. Environ.* **2019**, *5*, 173–191. [CrossRef]
56. Fundació Ave Maria, A Non-Profit Organization in Spain. Available online: https://www.avemariafundacio.org/ (accessed on 3 March 2021).
57. Anderson, G.O. Technology Use and Attitude among Mid-Life and Older Americans. *AARP Res.* **2018**, 1–29. [CrossRef]
58. Vaportzis, E.; Clausen, M.G.; Gow, A.J. Older Adults Perceptions of Technology and Barriers to Interacting with Tablet Computers: A Focus Group Study. *Front. Psychol.* **2017**, *8*. [CrossRef]
59. Elguera Paez, L.; Zapata Del Río, C. Elderly Users and Their Main Challenges Usability with Mobile Applications: A Systematic Review. In *Design, User Experience, and Usability. Design Philosophy and Theory*; HCII 2019, Lecture Notes in Computer Science; Marcus, A., Wang, W., Eds.; Springer: Berlin, Germany, 2019; Volume 11583, pp. 423–438.
60. Mierswa, I.; Wurst, M.; Klinkenberg, R.; Scholz, M.; Euler, T. YALE: Rapid prototyping for complex data mining tasks. In Proceedings of the 12th ACM SIGKDD International Conference on Knowledge Discovery and Data Mining (KDD '06), Philadelphia, PA, USA, 20–23 August 2006; pp. 935–940.
61. Waikato Environment for Knowledge Analysis (WEKA). Available online: https://en.wikipedia.org/wiki/Weka_(machine_learning) (accessed on 16 February 2021).
62. MLC++. Available online: http://robotics.stanford.edu/~{}ronnyk/mlc.html (accessed on 16 February 2021).
63. Saguna, S.; Zaslavsky, A.; Chakraborty, D. Complex activity recognition using context-driven activity theory and activity signatures. *ACM Trans. Comput. Interact.* **2013**, *20*, 1–34. [CrossRef]

64. Chakraborty, S.; Han, C.Y.; Zhou, X.; Wee, W.G. A Context Driven Human Activity Recognition Framework. In Proceedings of the 2016 International Conference on Health Informatics and Medical Systems, Monte Carlo Resort, Las Vegas, NV, USA, 25–28 July 2016; pp. 96–102.
65. Tao, M. A Framework for Modeling and Capturing Social Interactions. Ph.D. Thesis, University of Cincinnati, Cincinnati, OH, USA, 2014. Available online: https://etd.ohiolink.edu/apexprod/rws_olink/r/1501/10?clear=10&p10_accession_num=ucin1423581254 (accessed on 12 February 2021).
66. Tabbakha, N.E.; Ooi, C.P.; Tan, W.H. A Dataset for Elderly Action Recognition Using Indoor Location and Activity Tracking Data. *Mendeley Data*. 2020. [CrossRef]
67. Indoor Positioning Dataset of Bluetooth Beacons Readings Indoor. Available online: https://www.kaggle.com/liwste/indoor-positioning (accessed on 29 December 2020).
68. Root Mean Square. Available online: https://en.wikipedia.org/wiki/Root_mean_square (accessed on 13 February 2021).
69. Confusion Matrix. Available online: https://en.wikipedia.org/wiki/Confusion_matrix (accessed on 13 February 2021).
70. Taipei 101. Available online: https://en.wikipedia.org/wiki/Taipei_101 (accessed on 9 February 2021).
71. Burj Khalifa. Available online: https://en.wikipedia.org/wiki/Burj_Khalifa (accessed on 9 February 2021).
72. Masud, T.; Morris, R.O. Epidemiology of falls. *Age Ageing* **2001**, *30*, 3–7. [CrossRef]

information

MDPI

Article

An Ambient Intelligence-Based Human Behavior Monitoring Framework for Ubiquitous Environments

Nirmalya Thakur * and Chia Y. Han

Department of Electrical Engineering and Computer Science, University of Cincinnati, Cincinnati, OH 45221-0030, USA; han@ucmail.uc.edu
* Correspondence: thakurna@mail.uc.edu

Abstract: This framework for human behavior monitoring aims to take a holistic approach to study, track, monitor, and analyze human behavior during activities of daily living (ADLs). The framework consists of two novel functionalities. First, it can perform the semantic analysis of user interactions on the diverse contextual parameters during ADLs to identify a list of distinct behavioral patterns associated with different complex activities. Second, it consists of an intelligent decision-making algorithm that can analyze these behavioral patterns and their relationships with the dynamic contextual and spatial features of the environment to detect any anomalies in user behavior that could constitute an emergency. These functionalities of this interdisciplinary framework were developed by integrating the latest advancements and technologies in human–computer interaction, machine learning, Internet of Things, pattern recognition, and ubiquitous computing. The framework was evaluated on a dataset of ADLs, and the performance accuracies of these two functionalities were found to be 76.71% and 83.87%, respectively. The presented and discussed results uphold the relevance and immense potential of this framework to contribute towards improving the quality of life and assisted living of the aging population in the future of Internet of Things (IoT)-based ubiquitous living environments, e.g., smart homes.

Keywords: ambient intelligence; human behavior monitoring; smart homes; activities of daily living; elderly population; machine learning; internet of things; ubiquitous computing

Citation: Thakur, N.; Han, C.Y. An Ambient Intelligence-Based Human Behavior Monitoring Framework for Ubiquitous Environments. *Information* **2021**, *12*, 81. https://doi.org/10.3390/info12020081

Academic Editor: Spyros Panagiotakis

Received: 21 January 2021
Accepted: 10 February 2021
Published: 14 February 2021

1. Introduction

The elderly population across the globe is increasing at a very fast rate. It has been estimated [1] that by the year 2050, around 20% of the world's population will be aged 60 years or more. Aging is associated with several issues and limitations that affect a person's quality of life. According to [2], in the United States, approximately 8 out of every 10 elderly people have some form of chronic diseases, and approximately 5.4 million older adults have Alzheimer's. People living longer are causing a significant increase in the old-age dependency ratio, which is the ratio of the count of elderly people to that of the working population. On a global scale, this ratio is expected to increase from 11.7% to 25.4% over the next few years [2]. In addition to this, the population of elderly people with ages 80 and above is expected to triple within the next few years [3]. This increase in the population of older adults would bring several sociological and economic needs to the already existing challenges associated with aging. This constantly increasing elderly population is expected to impact society in multiple ways, as outlined below [3]:

i. A rise in cost of healthcare: at present, the treatment of older adults' accounts for 40% of the total healthcare costs in the United States even though older adults account for around 13% of the total population.
ii. Diseases affecting greater percentage of the population: with the increasing elderly population, there will be an increased number of people with diseases like Parkinson's and Alzheimer's, for which there is yet to be a proper and definitive cure.

iii. Decreased caregiver population: the rate of increase of caregivers is not as high as the increasing rate of the elderly population.
iv. Quality of caregiving: caregivers would be required to look after multiple older adults, and quite often they might not have the time, patience, or energy to meet the expectations of caregiving or to address the emotional needs of the elderly.
v. Dependency needs: with multiple physical, emotional, and cognitive issues associated with aging, a significant percentage of the elderly population would be unable to live independently.
vi. Societal impact: the need for the development of assisted living and nursing facilities to address healthcare-related needs.

With the decreasing count of caregivers, it is necessary that the future of technology-based living environments, e.g., smart homes and smart cities use technology-based services to address these needs and create assisted living experiences for the elderly. Over the last few years, researchers [4] have focused on developing assistive systems and devices according to a new paradigm, "ambient intelligence." Ambient intelligence may broadly be defined as a computing paradigm that uses information technology and its applications to enhance user abilities and performance through interconnected systems that can sense, anticipate, adapt, predict, and respond to human behavior and needs.

Human behavior is associated with performing activities in various environments and settings. An activity may broadly be defined as an interaction between a subject and an object for the subject to achieve a desired end goal or objective. This is typically represented as "S <–> O," where S stands for the 'subject' and O stands for the 'object.' Here, the subject is the user or the individual performing the activity, and the objects can be one or more context parameters present in the confines of the user's spatial orientation that are a part of the activity. To complete any given activity, the subject performs a set of related and sequential tasks or actions on one or more objects that depends on the kind of activity to be performed. These tasks or actions, along with their associated characteristic features, represent the user interactions related to the specific activity [5].

There can be various kinds of activities that a user performs in different environments with different spatial configurations. Activities that are crucial to one's sustenance and are performed within the confines of one's living space, e.g., personal hygiene, dressing, eating, maintaining continence, and mobility, are collectively termed as activities of daily living (ADLs) [6]. Based on the interaction patterns of the subject and object during activities, there are five broad characteristics of ADLs—(1) sequential, (2) concurrent, (3) interleaved, (4) false start, and (5) social interactions [5]. When multiple ADLs occur either at the same time or in a sequence or in a combination, they may exhibit more than one of these characteristics. Figure 1 shows four typical scenarios of different ADLs—A1, A2, A3, and A4—that can occur, where a number of these characteristics were exhibited by the activity sequences and combinations.

Elderly people need assistance to carry out ADLs due to the various bodily limitations and disabilities that they face with aging. An important aspect towards creating assisted living experiences in smart homes for the aging population is to monitor their interactions with their surroundings during ADLs [7]. The semantic analysis of user interactions during any ADL involves the monitoring of the associated behavioral patterns with respect to contextual, spatial, and temporal information. This analysis helps in interpretation of user performance during ADLs, as well as allowing for the detection of any anomalies that could constitute an emergency. For example, a person lying on a bed in a bedroom for several hours at night would mean that the person is taking rest, but if the same activity of lying is tracked to be taking place at the bathroom at the same time, it could mean an emergency situation resulting from a fall or unconsciousness, which needs the attention of caregivers or medical practitioners. In addition to aiding during ADLs, human behavior monitoring allows for the early detection of various forms of cognitive impairment, dementia, Alzheimer's, and a range of other limitations associated with old age [8]. Since it is not practically possible to manually access an older adult's behavior, it is the need of the hour to develop technology-based solutions with ambient intelligence to address this challenge. This served as the main motivation for the development of this framework that leverages the potential at the intersection of multiple disciplines including human–computer interaction, the Internet of Things (IoT), ubiquitous computing, machine learning, and pattern recognition.

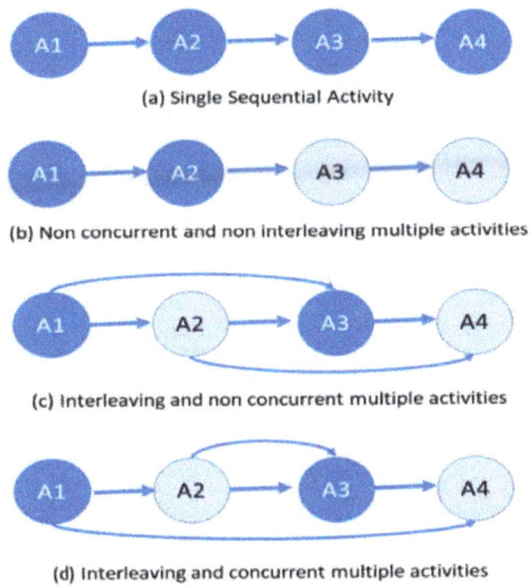

Figure 1. Representation of four typical scenarios of different activities of daily living (ADLs)—(**a**) A1, (**b**) A2, (**c**) A3, and (**d**) A4—that can occur, where different characteristics of ADLs are exhibited by the activity sequences and combinations.

To summarize, the scientific contributions of this paper are as follows:

1. It provides a novel approach to perform the semantic analysis of user interactions on the diverse contextual parameters during ADLs in order to identify a list of distinct behavioral patterns associated with different complex activities performed in an IoT-based environment. These behavioral patterns include walking, sleeping, sitting, and lying. This functionality was developed and implemented by using a k-nearest neighbor algorithm (k-NN) classifier. The performance accuracy of this approach was found to be 76.71% when it was evaluated on a dataset of ADLs.
2. It provides a novel intelligent decision-making algorithm that can analyze such distinct behavioral patterns associated with different complex activities and their relationships with the dynamic contextual and spatial features of the environment in order to detect any anomalies in user behavior that could constitute an emergency, such as a fall or unconsciousness. This algorithm was developed and implemented by using a k-NN classifier, and it achieved an overall performance accuracy of 83.87% when tested on a dataset of ADLs.

This paper is organized as follows. We present an overview of the related works in Section 2. The proposed framework is introduced and explained in Section 3. Section 4 discusses the results and performance characteristics of this framework. In Section 5, we discuss the limitations and drawbacks in the existing works and outline how our framework addresses these challenges and outperforms these existing systems in terms of their technical characteristics, functionalities, and operational features. It is followed by Section 6, where the conclusion and scope for future work are outlined.

2. Literature Review

This section outlines the recent works in the fields of human behavior research, i.e., assistive technology, Internet of Things, human–computer interaction, and their related disciplines for creating assisted living experiences in the future of technology-laden living environments, e.g., smart homes and smart cities.

A system comprised of wireless sensors to track and interpret human motion data for performing activity recognition was proposed by Azkune et al. [9]. The system consisted of an approach of activity clusters that were developed by using knowledge engineering principles. Based on these clusters, the associated patterns of human motion related to different activities could be tracked and interpreted by this system. Boualia et al. [10] proposed a Red Green Blue (RGB) frame analysis-based activity recognition framework with a specific focus on the study of human poses during different activities. The framework used a Convolutional Neural Network (ConvNet) architecture that was adapted for a regression problem and a support vector machine (SVM) classifier to detect activities. The authors evaluated the performance characteristics of their framework by testing it on two activity recognition datasets. Kasteren et al. [11] proposed a hidden Markov model-based architecture that analyzed the multimodal characteristics of sensor data for activity recognition. The authors used a recorded dataset and developed its annotation by using an off-the-shelf sensor, the Jabra BT250v. The Jabra BT250v was used to develop annotations for all the activities performed during each day, and these annotations were then used to train the hidden Markov model-based architecture for activity recognition. Cheng et al. [12] developed a framework that used concepts from computer vision, image processing, and video-data analysis to track and detect activities for both one and multiple users in the confines of a given IoT-based space. The approach combined characteristic features of motion data and user appearance information, as well as the spatiotemporal features of user behavior to train multiple learning models. The authors evaluated their approach by testing it on a dataset of activities. Skocir et al. [13] developed an artificial neural network-driven architecture that tracked human motion during different activities, with a specific focus on detecting enter and exit events in the confines of a given spatial environment, e.g., entering and exiting a room. The architecture used two IoT-based sensors with distinct functionalities to develop its foundation. One of these sensors was used to detect the presence or absence of the user, and the other sensor was used to detect whether the door was opened or closed. A dataset of different activities was used by the authors to test and discuss the performance characteristics of their approach.

The work done by Doryab et al. [14] involved the development of a task recommendation system to augment performances of medical practitioners in hospitals. This recommendation system was sensor technology-driven and focused on recommending tasks specifically related to different kinds of surgeries. The sensor data were used to detect the current action being performed by the user, and based on the same action, tasks associated with co-located activities were recommended by the system. A sensor network-driven activity assistance framework with the aim to assist users to perform different kinds of activities was proposed by Abascal et al. [15]. This work was specifically focused on helping elderly people with different kinds of impairments such as sensory, motor, or cognitive. In addition to performance characteristics, the authors also evaluated the accessibility, usability, and validity of their system. A system for the behavior monitoring of older adults in smart homes that used concepts of activity recognition and analysis was proposed by Chan et al. [16]. This system collected human motion data related to specific ADLs—walking, sleeping, and using the bathroom. The authors conducted real-time experiments in an Alzheimer's unit with a specific focus on studying and analyzing the human behavior and activities of people with Alzheimer's. Rashid et al. [17] developed a wearable neckband for human eating activity recognition and analysis. The system had a functionality to automatically update its database to adjust depending on the changing eating styles and eating habits of users. It used an artificial neural network-based approach that could detect four eating states—chewing, swallowing, talking, and idle. Siraj et al. [18]

developed a framework to recognize small activities, such as cooking, that are performed with other complex activities during a day. The authors used multiple machine learning models including those of deep learning, convolutional neural network, and gated recurrent unit to train their framework for the recognition of tasks and actions associated with the activity of cooking. They evaluated their framework on a dataset which consisted of various actions and tasks related to cooking. Mishra et al. [19] proposed a system that used video data for activity recognition and analysis. The system consisted of a spatiotemporal approach defined by considering the fuzzy lattices of the video frames. These lattices were described by kinetic energy, which was calculated by the Schrödinger wave equation. The system could detect any changes in human behavior or motion based on the change in kinetic energy associated with these lattices. In [20], Fu et al. proposed a wireless wearable sensor-driven device that could perform activity recognition. The device consisted of an air pressure sensor and an inertial measurement unit to study and analyze human behavior related to different activities. The wearable used a transfer learning approach to perform personalized activity recognition. The work done by Yared et al. [21] involved the development of an intelligent activity analysis framework to reduce accidents in the kitchen area. The authors analyzed multiple activities performed in the kitchen to identify characteristic features such as gas concentration, smoke, the temperature of utensils, and the temperature of burner that needed to be monitored to detect any accidents. The findings of this work listed a set of factors that were responsible for most kitchen accidents. Angelini et al. [22] developed a smart bracelet that could collect multiple features of a user's movement data to interpret the health status of the user. It also had the functionality to remind the user of their routine medications. The bracelet was developed to work for different kinds of indoor and outdoor activities. The authors conducted usability studies to discuss the effectiveness of this bracelet.

In the work done by Dai et al. [23], the dynamics of the motion data coming from the user's android phone were analyzed to detect falls. The authors developed a proof-of-concept model that was based on an Android phone that collected real-time behavior-related data of the user. The architecture of the system was developed in a specific way to ensure that it did not contribute to high central processing unit (CPU) usage and did not occupy a significant percentage of the computer's random-access memory (RAM). The results discussed by the authors showed that the average CPU usage was 7.41% by the system, and it occupied about 600 KB on the RAM. Kong et al. [24] proposed a depth recognition and distance-based algorithm for detecting falls. The algorithm tracked the distance between the neck of the user and the ground, and if the distance was found to decrease with a situation lasting greater than a minute, then the algorithm interpreted the situation as a fall. Shao et al. [25] proposed an approach that analyzed the characteristics of floor vibrations to detect falls. The authors performed experiments with objects and humans falling on the ground to study the characteristics of floor vibrations. The system consisted of a k-means classification approach to detect falls. Chou et al. [26] proposed an Electrocardiography (ECG)-based system for fall detection. The system consisted of a smart cane with an ECG detection circuit along with other sensors to study the behavioral patterns of the user. The authors developed and implemented a microcontroller-based circuit that could detect falls based on the data collected from the ECG circuit and the associated sensors. In [27], Keaton et al. proposed an WiFi channel state-based approach for the detection of falls in IoT-based environments. The authors developed a neural network-based learning model that could study, track, and analyze the changes in WiFi channel state data based on normal behaviors and falls. Anceschi et al. [28] proposed a machine learning-based wearable system for fall detection in a workplace environment. To develop and train the machine learning model, the authors merged four different datasets that consisted of diverse activities performed in a workplace. This device used a couple of IoT-based off-the-shelf products that worked in coordination with a microcontroller circuit to detect falls from human motion data. Mousavi et al. [29] used acceleration data available from smartphones to develop a fall detection system. This system consisted of an SVM classifier that interacted with the triaxial accelerometer data coming from a smartphone that had an IOS operating system. The system also had a feature to alert caregivers via either an SMS or email when a fall was detected.

Despite these recent advances in this field, there are still several limitations and challenges. For instance, (1) a number of these works have superficially focused on activity recognition without an analysis of the fine grain characteristics of activities and the associated dynamics of human behavior; (2) several activity analysis approaches are confined to specific tasks and cannot always be applied seamlessly to other activities; (3) a number of these methodologies have been developed and implemented for specific settings with a fixed set of context parameters and environment variables, and their real world deployment is difficult because the real world environments are different compared to such settings; (4) the video-based systems may have several challenges related to the categorization and transcription of data, the selection of relevant fragments, the selection of camera angle, and the determination of the number of frames; (5) some of the fall detection technologies are built for specific operating systems, devices, or gadgets and cannot be implemented on other platforms; (6) some of these systems have a dependency on external parameters, such as floor vibrations, that can affect the readings and performance characteristics; and (7) some of the systems are affected by user diversity such as the user's height and weight. To add to the above, some of these works have focused on activity recognition and analysis, while others have focused on fall detection. None of these works have focused on both of these challenges at the same time. Thus, it can be concluded that it is the need of the hour to leverage the immense potential at the intersection of ambient intelligence and the IoT to develop a framework that can not only track, study, analyze, and anticipate human behavior but also detect any anomalies, such as a fall or unconsciousness, that could constitute an emergency. It is also necessary that such systems are developed in way so that they are not environment-specific and can be seamlessly implemented and deployed in any IoT-based real world setting. This framework aimed to address these challenges by developing an approach for the analysis of human behavior at a fine-grain level with respect to the associated dynamic contextual, spatial, and temporal features to detect any anomalies that could constitute an emergency. The work involved the integration of advancements and technologies from multiple disciplines. This framework is introduced in Section 3, and a further discussion of how the salient features of this framework address these challenges and the drawbacks in the existing systems is presented in Section 5.

3. Proposed Work

In this section, we first present the steps towards the development of the functionality in our framework for the semantic analysis of user interactions on the context parameters during ADLs in order to identify a list of common behavioral patterns associated with different complex activities performed in any given IoT-based environment. In a real-world scenario, human activities are highly complex and involve multiple forms of user interactions that include a myriad of tasks and their dynamic characteristics, performed on the context parameters, based on the associated need related to the activity. Such complex real-world activities are referred to as complex activities. A complex activity can be broken down into atomic activities, context attributes, core atomic activities, core context attributes, other atomic activities, other context attributes, start atomic activities, end atomic activities, start context attributes, and end context attributes [30]. Here, atomic activities refer to the macro and micro level tasks and sub-tasks associated with the complex activity, and the environment parameters on which these atomic activities are performed are collectively known as context attributes. Those specific atomic activities that are crucial to a complex activity and without which the complex activity can never be completed are referred to as core atomic activities, and their associated context attributes are known as core context attributes. The atomic activities that are necessary to start a given complex activity are known as start atomic activities, and the atomic activities that are necessary to successfully end a given complex activity are known as end atomic activities. The context parameters on which these two types of atomic activities take place are known as start context attributes and end context attributes, respectively. All the atomic activities other than the core atomic activities are known as other atomic activities, and their associated context attributes are known as other context attributes. The semantic analysis of user interactions during complex activities involves analyzing all these characteristic features of activities with respect to contextual, spatial, and temporal information. The following are the steps for the development of this functionality in the proposed framework:

i. Deploy both wireless and wearable sensors to develop an IoT-based interconnected environment.
ii. Set up a data collection framework to collect the big data from these sensors during different ADLs performed in the confines of a given IoT-based space.
iii. Use context-based user interaction data obtained from the wireless sensors to spatially map a given environment into distinct 'zones,' in terms of context attributes associated with distinct complex activities. Here, we define a 'zone' as a region in the user's spatial orientation where distinct complex activities take place. For instance, in the cooking zone, the complex activity of cooking could take place, but other complex activities like sleeping or taking a shower could not.
iv. Analyze the atomic activities performed on different context attributes for a given complex activity, along with their characteristic features.
v. Track user behavior in terms of joint point movements and joint point characteristics [31] for each atomic activity associated with any given complex activity.
vi. Analyze the user behavior, atomic activities, and context attributes to form a general definition of a complex activity in each context-based spatial 'zone.'
vii. Repeat (vi) for all the complex activities with respect to the context attributes as obtained from (iii) for a given IoT-based environment.
viii. Analyze the activity definitions to find atomic activities and their characteristic features for all the complex activities associated with the different 'zones.'
ix. Study the activity definitions to record the human behavior for all the atomic activities obtained from (viii).
x. Analyze the behavior definitions in terms of joint point movements and characteristics to develop a knowledge base of common behaviors associated with all the complex activities in the different 'zones.'
xi. Develop a dataset that consists of all these behavioral patterns and the big data from user interactions for each of these 'zones' in a given IoT-based environment.
xii. Preprocess the data to detect and eliminate outliers and any noise prior to developing a machine learning model.
xiii. Split the data into training and test sets and then test the machine learning model on the test set to evaluate its performance characteristics.

Upon the development of the above-discussed functionality in our framework, we implemented the following steps to develop the proposed intelligent decision-making algorithm that can detect emergencies or anomalies in user behavior based on studying the multimodal components of user interactions during complex activities in each 'zone.' Each 'zone' is associated with distinct complex activities that are further associated with a set of atomic activities, context attributes, core atomic activities, core context attributes, other atomic activities, other context attributes, start atomic activities, end atomic activities, start context attributes, and end context attributes. An analysis of the user behavior in terms of joint point characteristics [31] allows for the detection and analysis of these behavioral patterns and their relationships with the dynamic spatial features of the environment to detect any anomalies in user behavior that could constitute an emergency. For instance, the atomic activity of lying at night in the sleeping or bedroom zones could be interpreted as the person taking rest. However, the detection of the same atomic activity in the bathroom at the same time could indicate an emergency that could have resulted from a fall or unconsciousness. Such a situation would need the attention of caregivers or medical practitioners. The proposed intelligent decision-making algorithm was built on this concept for the detection of emergencies during complex activities, and the following are the steps for the development of this functionality:

i. Classify the complex activities from this dataset as per their relationships with atomic activities, context attributes, other atomic activities, other context attributes, core atomic activities, core context attributes, start atomic activities, end atomic activities, start context attributes, and end context attributes to develop semantic characteristics of complex activities.
ii. Track user movements to detect start atomic activities and start context attributes.
iii. If these detected start atomic activities and start context attributes match with the semantic characteristics of complex activities in the database, run the following algorithm: emergency detection from semantic characteristics of complex activities (EDSCCA).
iv. If these detected start atomic activities and start context attributes do not match with the semantic characteristics of complex activities in the knowledge base, then track the atomic activities, context attributes, other atomic activities, other context attributes, core atomic activities, core context attributes, start atomic activities, end atomic activities, start context attributes, and end context attributes to develop a semantic definition for a complex activity (SDCA).
v. If an SDCA is already present in the knowledge base, go to (vi), else update the database with the SDCA.
vi. Develop a dataset that consists of all these semantic definitions for complex activities and the big data from user interactions associated with them.
vii. Preprocess the data to detect and eliminate outliers and any noise prior to developing a machine learning model.
viii. Split the data into training and test sets and then test the machine learning model on the test set to evaluate its performance characteristics.

Next, we outline the steps for developing the proposed EDSCCA algorithm:

i. Track if the start atomic activity was performed on the start context attribute.
ii. Track if the end atomic activity was performed on the end context attribute.
iii. If (i) is true and (ii) is false:
   a. Track all the atomic activities, context attributes, other atomic activities, other context attributes, core atomic activities, and core context attributes.
   b. For any atomic activity or other atomic activity that does not match its associated context attribute, track the features of the user behavior.
   c. If the user behavior features indicate lying and no other atomic activities are performed, the inference is an emergency.
iv. If (i) is true and (ii) is true:
   a. The user successfully completed the activity without any emergency detected, so the inference is no emergency.
v. If (i) is false and (ii) is true:
   a. Track all the atomic activities, context attributes, other atomic activities, other context attributes, core atomic activities, and core context attributes.
   b. For any atomic activity or other atomic activity that does not match its associated context attribute, track the features of the user behavior.
   c. If the user behavior features indicate lying and no other atomic activities performed, the inference is an emergency.
vi. If (i) is false and (ii) is false:
   a. No features of human behavior were associated with the observed activities or, in other words, the user did not perform any activity, so the inference is no emergency.

We used one of our previous works [31] that presented a framework for human behavior representation in the context of ADLs based on joint point characteristics. These joint point characteristics primarily include joint point distances and joint point speeds. By studying these joint point characteristics associated with diverse behavioral patterns, this framework tracks the dynamic changes in the skeleton that point to interpretations of human pose and posture. The dynamics of human pose and posture are then used by the framework to analyze human behavior and its associated features during multimodal interactions in the context of ADLs. This concept is outlined in Figure 2. According to this methodology, each point on the skeletal tracking, as obtained from a Microsoft Kinect sensor, is assigned a joint number and a definition based on the kind of underlining movements associated with that joint point. The associated joint point characteristics, in terms of the individual joint point speeds and the distance between two or more joint points, undergo changes based on the behavioral patterns of the user. We applied this concept to analyze the ADLs in terms of the atomic activities, context attributes, other atomic activities, other context attributes, core atomic activities, and core context attributes in order to identify the list of behavioral patterns associated with each of these ADLs. This analysis also involved modelling all possible instances of each complex activity while assigning weights to the individual atomic activities, context attributes, other atomic activities, other context attributes, core atomic activities, and core context attributes based on probabilistic reasoning. This was done by using Equations (1)–(3), which were proposed in [32].

$$\alpha = a_tC0 + a_tC1 + a_tC2 + \ldots \ldots a_tCa_t = 2^{at} \quad (1)$$

$$\beta = (a_t - c_t)C0 + (a_t - ct)C1 + (a_t - c_t)C2 + \ldots + (a_t - c_t)C(a_t - c_t) = 2^{(at-ct)} \quad (2)$$

$$\gamma = 2at - 2^{(at-ct)} = 2^{(at-ct)} * (2^{ct} - 1) \quad (3)$$

where α represents all possible ways by which any complex activity can be performed including false starts; β represents all the ways of performing any complex activity where the user always reaches the end goal; γ represents all the ways of performing any complex activity where the user never reaches the end goal; Ati represents all the atomic activities related to the complex activity, where i is a positive integer; Cti represents all the context attributes related to the complex activity, where i is a positive integer; AtS represents a list of all the Ati that are start atomic activities; CtS represents a list of all the Cti that are start context attributes; AtE represents a list of all the Ati that are end atomic activities; CtE represents a list of all the Cti that are end context attributes; γAt represents a list of all the Ati that are core atomic activities; ρCt represents a list of all the Cti that are core context attributes; a_t represents the number of Ati related to the complex activity; b_t represents the number of Cti related to the complex activity; c_t represents the number of γAt related to the complex activity; and d_t represents the number of ρCt related to the complex activity.

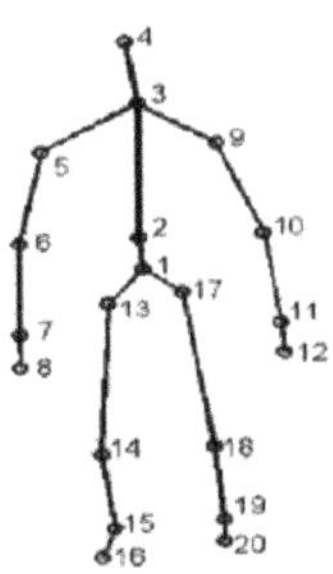

Joint	Definition	Joint	Definition
1	Center of Hip	11	Right Wrist
2	Spine	12	Right Hand
3	Center of Shoulder	13	Left Hip
4	Head	14	Left Knee
5	Left Shoulder	15	Left Ankle
6	Left Elbow	16	Left Foot
7	Left Wrist	17	Right Hip
8	Left Hand	18	Right Knee
9	Right Shoulder	19	Right Ankle
10	Right Elbow	20	Right Foot

Figure 2. The methodology to represent skeletal tracking in terms of joint points and their associated definitions [31].

The work in [32] presented a mathematical foundation for modelling all possible user interactions related to atomic activities, context attributes, other atomic activities, other context attributes, core atomic activities, and core context attributes associated with any given complex activity. The objective of the work in [32] was to develop a knowledge base that would consist of all possible tasks and actions performed on context parameters, related to any given complex activity, arising from universal diversity and the variations in the context parameters based on the associated spatial and temporal characteristics of user interactions. In this work, these equations were developed by integrating complex activity analysis [30], the principles of the binomial theorem [33], and permutation and combination principles. These equations represent the diverse ways by which a complex activity may be performed. Equation (1) represents all possible ways by which a complex activity can be modelled, including distractions, false starts, one or more missed Ati, one or more missed Cti, one or more missed AtS, one or more missed CtS, one or more missed AtE, one or more missed CtE, one or more missed γAt, and one or more missed ρCt. Equation (2) represents all those scenarios where the user reached the end goal or, in other words, the user performed all the γAt on the ρCt related to a given complex activity. Equation (3) represents all those scenarios where the user did not perform one or more γAt on the ρCt related to a given complex activity, as well as one or more missed AtS, one or more missed CtS, one or more missed AtE, and one or more missed CtE. Weights were assigned to the individual Ati and Cti by probabilistic reasoning principles, as outlined in [30]. The weights indicate the relevance or importance of the task or action towards helping the user reach the end goal or desired outcome. A higher value of the weight indicates a greater relevance, and a lower value of the weight indicates a lesser relevance of the associated Ati and Cti. The γAt and ρCt are assigned the highest weights as compared to the other Ati and Cti. The weights associated with all the Ati and Cti can be analyzed to determine the threshold weight of the complex activity, which determines whether a given complex activity was properly performed. Here, properly performed refers to whether the user was able to successfully reach the end goal or outcome associated with a given complex activity. The threshold weight varies based on the nature and number of AtS, CtS, AtE, CtE, γAt, and ρCt related to a complex activity. Each instance of a complex activity, denoted by Equation (1), is also assigned a different weight based on the number of AtS, CtS, AtE, CtE, γAt, and ρCt, as well as the nature and sequence in which these actions were performed. When this weight exceeds the threshold weight, it indicates that the user reached the end goal, and such activity instances are represented by Equation (2). Table 1 outlines the analysis for a typical ADL, eating lunch, as described by this methodology. In Table 2, we represent the analysis of this complex activity as per Equations (1)–(3) to study the characteristics of the associated Ati, Cti, AtS, CtS, AtE, CtE, γAt, ρCt, a_t, b_t, c_t, and d_t.

Table 1. Analysis of the complex activity of eating lunch in terms of joint point characteristics [31].

Atomic Activities	Context Attributes	Joint Points Pairs That Experience Change
At1: Standing (0.08)	**Ct1**: Lights on (0.08)	No change
At2: Walking towards dining table (0.20)	**Ct2**: Dining area (0.20)	(13,17), (14,18), (15,19), and (16,20)
At3: Serving food on a plate (0.25)	**Ct3**: Food present (0.25)	(7, 11) and (8,12)
At4: Washing hand/using hand sanitizer (0.20)	**Ct4**: Plate present (0.20)	(7, 11) and (8,12)
At5: Sitting down (0.08)	**Ct5**: Sitting options available (0.08)	No change
At6: Starting to eat (0.19)	**Ct6**: Food quality and taste (0.19)	(6,3), (7,3), (8,3), (6,4), (7,4), (8,4) or (10,3), (11,3), (12,3), (10,4), (11,4), and (12,4)

As can be seen from Table 2, where $\alpha = 64$, there are 64 different ways by which this complex activity can be performed. However, the value of $\gamma = 60$ means that 60 out of these 64 ways would not lead to the end goal or the desired outcome. The remaining activity instances indicated by $\beta = 4$ refers to those instances when the user would always reach the end goal of this complex activity. One such instance is shown in Table 1.

Table 2. Analyzing multiple characteristics of a typical complex activity—eating lunch.

Complex Activity Characteristics	Value(s)
Ati, all the atomic activities related to the complex activity	At1, At2, At3, At4, At5, and At6
Cti, all the context attributes related to the complex activity	Ct1, Ct2, Ct3, Ct4, Ct5, and Ct6
AtS, list of all the Ati that are start atomic activities	At1 and At2
CtS, list of all the Cti that are start context attributes	Ct1 and Ct2
AtE, list of all the Ati that are end atomic activities	At5 and At6
CtE, list of all the Cti that are end context attributes	Ct5 and Ct6
γAt, list of all the Ati that are core atomic activities	At2, At3, At4, and At6
ρCt, list of all the Cti that are core context attributes	Ct2, Ct3, Ct4, and Ct6
a_t, number of Ati related to the complex activity	6
b_t, number of Cti related to the complex activity	6
c_t, number of γAt related to the complex activity	4
d_t, number of ρCt related to the complex activity	4
α, all possible ways by which any complex activity can be performed including false starts	64
β, all the ways of performing any complex activity where the user always reaches the end goal	4
γ, all the ways of performing any complex activity where the user never reaches the end goal	60

To develop this framework, we used an open-source dataset [34] that contains the big data of user interactions recorded during multiple ADLs in an IoT-based environment. The complex activities and their associated characteristics in this dataset can be distinctly mapped to four spatial 'zones'—kitchen, bedroom, office, and toilet—in the simulated and interconnected IoT-based environment. The big data in this dataset consisted of data attributes that provided the location, or the 'zone'-related data associated with all these ADLs. These data were used to analyze the indoor location of the user with respect to the context attributes of interest for a given complex activity in the IoT-based environment. The context attributes associated with different instances of each of these ADLs were studied by the approach discussed in Tables 1 and 2. The dataset also consisted of accelerometer and gyroscope data that were collected from wearables and that represented diverse behavioral patterns during different instances of each of the ADLs performed in each of these spatial 'zones.' These data were used to study, analyze, and interpret the multimodal characteristics of human behavior distinct to different complex activities. Here, as per the data and their characteristics present in the dataset, we defined lying and being unable to get up in any other location other than a bedroom as an emergency. This definition of an emergency can also be modified, e.g., to detect a long lie, as per the complex activities and their semantic characteristics for a given IoT-based environment.

We used RapidMiner, previously known as Yet Another Learning Environment (YALE) [35], for the development of this framework. RapidMiner is a software tool that consists of several built-in functions known as 'operators' that can be used to implement a range of computational functions including machine learning, artificial intelligence, and natural language processing algorithms. The tool also allows for the seamless customization of these 'operators' as per the needs of the model being developed. Multiple 'operators' can be put together in the tool to develop an application, which is known as a 'process.' There are two versions of RapidMiner available—the free version and the paid version. The free version has a processing limit of 10,000 rows. The dataset that we used for this study did not exceed 10,000 rows, so this limitation of the free version did not affect our results and findings. The version of RapidMiner that we used was 9.8.000, and the same was run on a Microsoft Windows 10 computer with an Intel (R) Core(TM) i7-7600U CPU @ 2.80 GHz, 2 core(s) and 4 logical processor(s) for the development and implementation of the proposed framework.

4. Results

In this section, we present the results obtained from the proposed framework by using the dataset [34]. The big data present in the dataset represented various kinds of ADLs—sleeping, changing clothes, relaxing, cooking, eating, working, and defecating, as well as emergency situations in the kitchen, bedroom, office, and toilet. The emergency corresponded to the user lying on the ground in any location other than the bedroom, which could have resulted from a fall or unconsciousness. As per the methodology discussed in Figure 2 and Tables 1 and 2, we developed definitions of all the complex activities that occurred in a given IoT-space. Then, we developed a process in RapidMiner to identify and interpret the list of common behavioral patterns associated with each of these ADLs in this dataset, performed in the spatial locations or 'zones'—bedroom, kitchen, office, and toilet. We used the 'Dataset' operator to import this dataset into RapidMiner. The 'Data Preprocessing' operator was used to preprocess the data and to study the various characteristics of human behavior as outlined in Section 3. The data processing involved the studying, analysis, and interpretation of the dynamic characteristics of human behavior data associated with the diverse complex activities performed in each of the spatial 'zones' represented in the dataset. The dataset that we used for these pre-processing steps consisted of 295 rows. First, we studied the different ADLs performed in each of these 'zones'—bedroom, kitchen, office, and toilet. This is shown in Figure 3, where the location or 'zone' is plotted on the x-axis, and the different ADLs are represented on the y-axis. As there were nine different ADLs, so we represented each ADL with a different color; this color coding is mentioned in the figure. Each occurrence of an ADL in a specific 'zone' is represented with a bubble corresponding to that zone. For instance, in the toilet zone, the activities of defecating and emergency were observed, so these two activities were tracked using distinct colors for this 'zone.'

Figure 3. Analysis of different ADLs performed in the different spatial locations or 'zones' in a given Internet of Things (IoT)-based environment.

After detecting and studying the different ADLs local to each 'zone,' we studied the associated atomic activities, context attributes, other atomic activities, other context attributes, core atomic activities, and core context attributes associated with each of these ADLs to study the common behavioral patterns local to each ADL in each zone that we observed from Figure 3. This analysis is shown in Figure 4, where the x-axis represents the location, and the common behavioral patterns of lying, standing, sitting, and walking are represented on the y-axis. As there were multiple ADLs to which these common behavioral patterns belonged, so we represented each ADL by using a different color. Each occurrence of an ADL in a specific 'zone' is represented with a bubble corresponding to that zone. For instance, from Figure 3, we can observe that in the toilet zone, the activities of defecating and emergency occur multiple times. The behavioral patterns associated with these activities are sitting and lying, so these behaviors were represented using different colors, as shown in Figure 4.

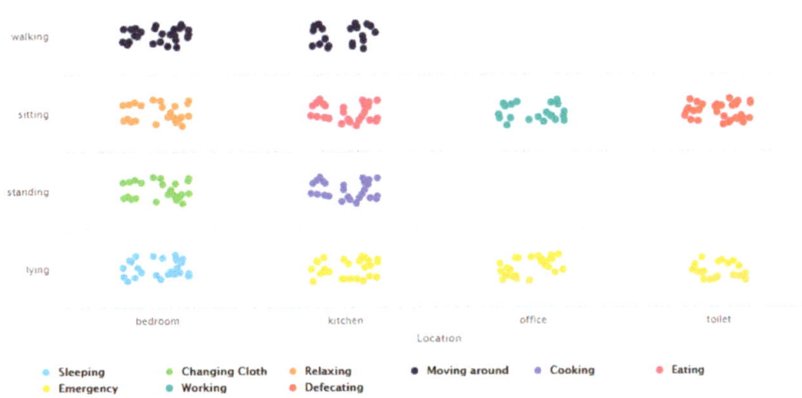

Figure 4. Representation of common and distinct behavioral patterns associated with the different ADLs performed in the different spatial locations or 'zones' in a given IoT-based environment.

After studying these activity patterns distinct to different ADLs local to each zone, we studied the characteristics of the human behaviors at a fine-grain level associated with each of these ADLs. This was done by analyzing the accelerometer and gyroscope data corresponding to occurrences of each of the common behavioral patterns—lying, standing, sitting, and walking—for different ADLs in each of these spatial 'zones.' The study and analysis of the accelerometer and gyroscope data for these common behavioral patterns for all these ADLs performed in the kitchen, bedroom, office area, and toilet are shown in Figures 5–8, respectively. In each of these figures, the common behavioral patterns are plotted on the x-axis. The y-axis represents the accelerometer data and gyroscope in the x, y, and z directions, each of which is plotted with a distinct color.

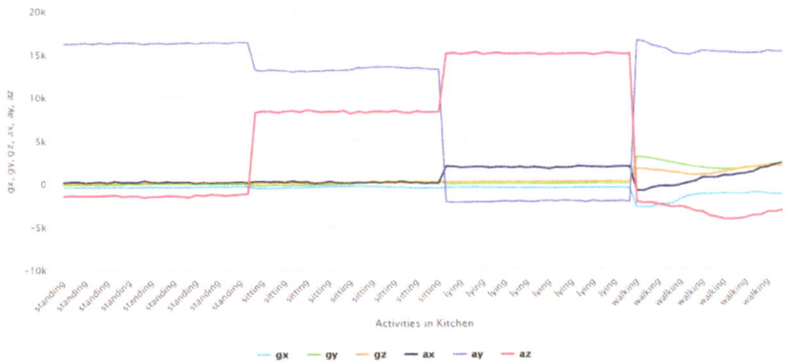

Figure 5. The study and analysis of the accelerometer and gyroscope data for the common behavioral patterns—lying, standing, sitting, and walking—for all ADLs performed in the kitchen. Due to paucity of space, analyses of a some of the ADLs are shown here.

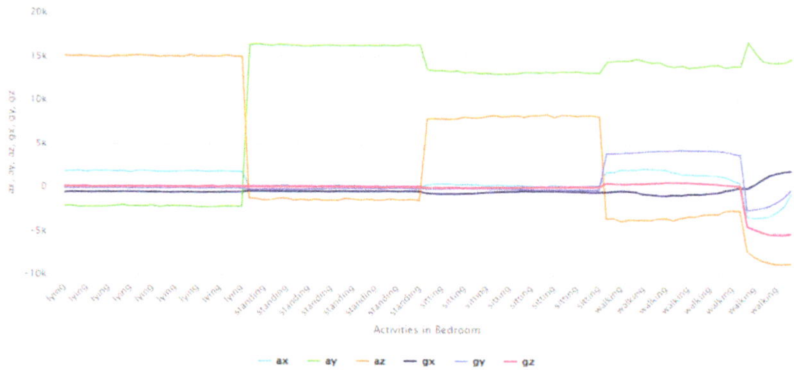

Figure 6. The study and analysis of the accelerometer and gyroscope data for the common behavioral patterns—lying, standing, sitting, and walking—for all ADLs performed in the bedroom. Due to paucity of space, analyses of a some of the ADLs are shown here.

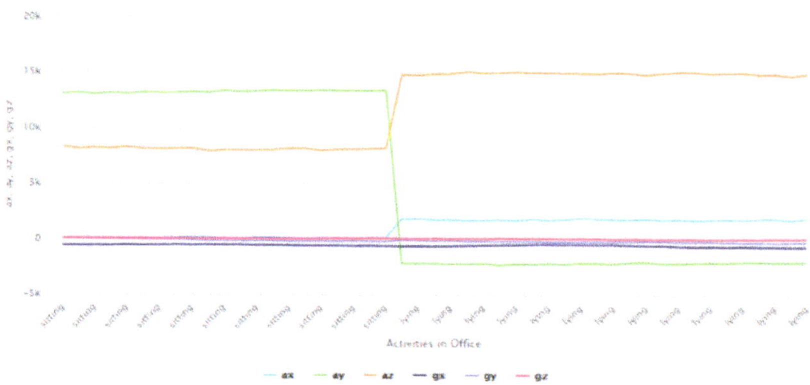

Figure 7. The study and analysis of the accelerometer and gyroscope data for the common behavioral patterns—lying, standing, sitting, and walking—for all ADLs performed in the office area. Due to paucity of space, analyses of a some of the ADLs are shown here.

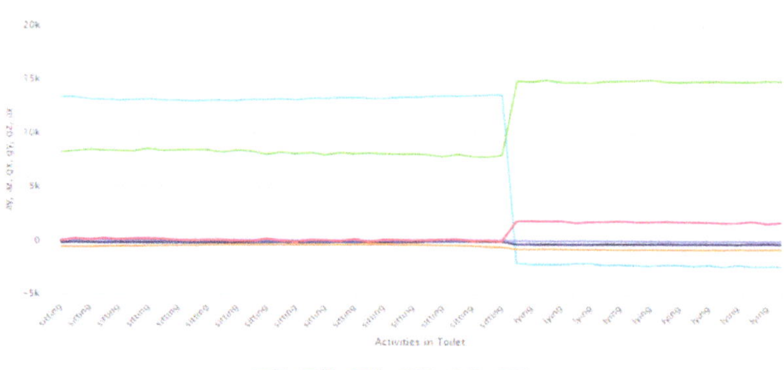

Figure 8. The study and analysis of the accelerometer and gyroscope data for the common behavioral patterns—lying, standing, sitting, and walking—for all ADLs performed in the toilet. Due to paucity of space, analyses of a some of the ADLs are shown here.

After performing this analysis, we used the 'Split Data' operator to split the data into training and test sets; 75% of the data were used for training, and the remaining 25% were used for testing. We used a k-NN classifier to develop the machine-learning functionality of our framework. k-NN [36] is a non-parametric machine learning classifier. k-NN works by comparing an unknown data sample to 'k' closest training examples in the dataset to classify the unknown data into one of these samples. Here, closeness refers to the distance in a space represented by 'p,' where 'p' is the number of attributes in the training set. There are various approaches for the calculation of this distance. For the development of the proposed approach, we used the Euclidean distance approach in RapidMiner [35]. The Euclidean distance [37] between two points 'm' and 'n' is computed as shown in Equation (4):

$$\mathrm{d}(\mathrm{m},\mathrm{n}) = \sqrt{\sum_{i=1}^{p}(m_i - n_i)^2} \quad (4)$$

where m and n are two points in the Euclidean space, d (m,n) represents the distance between the two points m and n in the Euclidean space, m_i represents the vector in the Euclidean space that connects the point m to the origin, n_i represents the vector in the Euclidean space that connects the point n to the origin, and p represents the p-space.

The k-NN model that we developed consisted of 11 nearest neighbors. The model was developed using 222 examples consisting of three dimensions of each of the activity classes representing lying, standing, walking, and sitting. We tested the classifier by using the 'Apply Model' operator and evaluated its performance characteristics by using the 'Performance' operator. This RapidMiner process is shown in Figure 9, and the order in which the 'operators' associated with this RapidMiner process were executed when the process was compiled and run is shown in Figure 10. Thereafter, we studied the effectiveness and performance characteristics of our framework to detect these behavioral patterns—walking, sleeping, sitting, and lying—in different spatial locations. The RapidMiner process studied each row of the test dataset, which constituted a user interaction with the context parameters and detected the associated behavioral patterns.

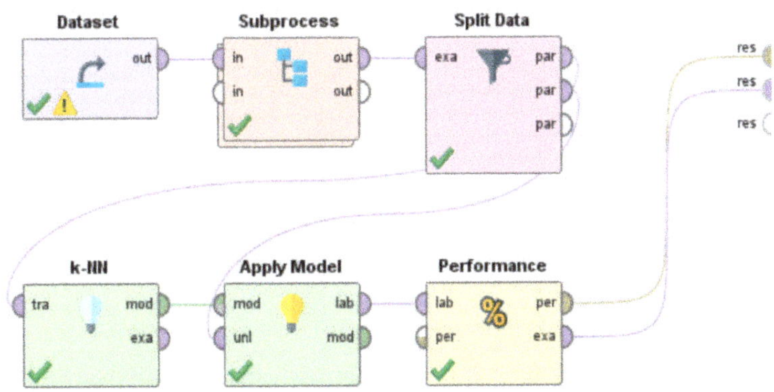

Figure 9. RapidMiner process for the development of the functionality of the framework to perform the semantic analysis of user interactions on the diverse context parameters during ADLs to identify a list of distinct behavioral patterns associated with different complex activities performed in a given IoT-based environment.

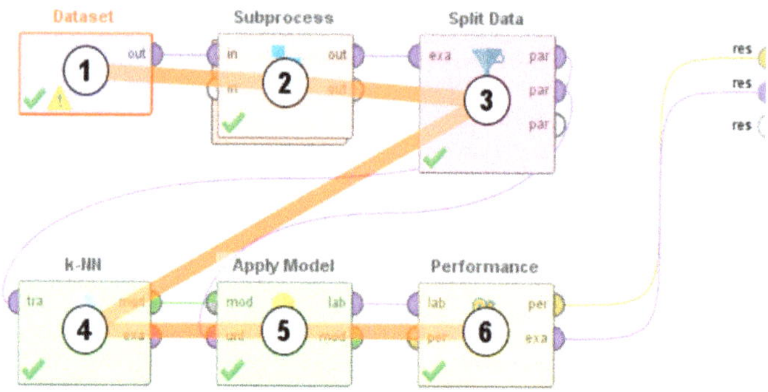

Figure 10. The order of execution of all the operators upon the compilation and execution of the RapidMiner process shown in Figure 9.

The output of this RapidMiner process was in the form of a table where each row consisted of the attributes as outlined in Table 3. Here, the degree of certainty expresses the certainty of prediction of the associated behavioral pattern of the user by the developed k-NN-based machine learning model. To predict the same, the k-NN model in RapidMiner assigned a confidence value to each of these behavioral patterns, and the behavior with the highest confidence was the final prediction of the model for that specific user interaction. For instance, in row number 2, the confidence values associated with lying, standing, sitting, and walking are 0.818, 0.182, 0, and 0, respectively, so the prediction of the model was lying. This output table had 73 rows, but only the first 13 rows are shown in Figure 11.

Table 3. Description of the attributes of the output of the RapidMiner process shown in Figure 11.

Attribute Name	Description
Row No	The row number in the output table
Activity	The actual behavioral pattern associated with a given ADL
Prediction (Activity)	The predicted behavioral pattern associated with a given ADL
Confidence (lying)	The degree of certainty that the user was lying during this ADL
Confidence (standing)	The degree of certainty that the user was standing during this ADL
Confidence (sitting)	The degree of certainty that the user was sitting during this ADL
Confidence (walking)	The degree of certainty that the user was sitting during this ADL

Row No.	Activity	prediction(Activity)	confidence(lying)	confidence(standing)	confidence(sitting)	confidence(walking)
1	lying	lying	0.723	0.277	0	0
2	lying	lying	0.818	0.182	0	0
3	lying	lying	0.645	0.355	0	0
4	standing	sitting	0.089	0.272	0.639	0
5	standing	standing	0.447	0.553	0	0
6	standing	lying	0.637	0.363	0	0
7	standing	lying	0.630	0.370	0	0
8	sitting	sitting	0	0	1	0
9	sitting	sitting	0	0	1.000	0
10	sitting	sitting	0.088	0	0.912	0
11	sitting	sitting	0.172	0	0.828	0
12	sitting	sitting	0	0	1	0
13	sitting	sitting	0.087	0	0.913	0

Figure 11. The output table of the RapidMiner process shown in Figure 3 for the detection of distinct behavioral patterns associated with the different ADLs. This output table had 73 rows, but only the first 13 rows are shown here.

The performance accuracy of this model was evaluated by using a confusion matrix, where both the overall performance and the individual class precision values were computed. Figures 12 and 13 show the tabular representation and plot view of the confusion matrix, respectively. A confusion matrix [38] is a method of evaluating and studying the performance characteristics of a machine learning-based algorithm. The number of instances of a data label in the predicted class is represented by each row of the matrix, and the number of instances of a data label in the actual class is represented by each column of the matrix. The matrix can also be inverted to have the rows represent the columns and vice versa. Such a matrix allows for the calculation of multiple performance characteristics associated with the machine learning model. These include overall accuracy, individual class precision values, recall, specificity, positive predictive values, negative predictive values, false positive rates, false negative rates, and F-1 scores. To evaluate the performance characteristics of our proposed approach, we focused on two of these performance metrics—the overall accuracy and the individual class precision values, which are calculated by the formula as shown in Equations (5) and (6), respectively:

$$\text{Acc} = \frac{\text{True(P)} + \text{True(N)}}{\text{True(P)} + \text{True(N)} + \text{False(P)} + \text{False(N)}} \quad (5)$$

$$\text{Pr} = \frac{\text{True(P)}}{\text{True(P)+False(P)}} \quad (6)$$

where Acc is the overall accuracy of the machine-learning model, Pr is the class precision value, True(P) means true positive, True(N) means true negative, False(P) means false positive, and False(N) means false negative.

accuracy: 76.71%

	true lying	true standing	true sitting	true walking	class precision
pred. lying	19	8	3	0	63.33%
pred. standing	1	3	0	0	75.00%
pred. sitting	4	1	22	0	81.48%
pred. walking	0	0	0	12	100.00%
class recall	79.17%	25.00%	88.00%	100.00%	

Figure 12. The performance accuracy (studied via a confusion matrix—tabular view) of the RapidMiner process shown in Figure 9 for the detection of distinct behavioral patterns associated with the different ADLs performed in the different spatial locations in a given IoT-based environment.

As can be seen from Figures 12 and 13, this machine learning model achieved an overall performance accuracy of 76.71%, with respective class precision values for lying, standing, sitting and walking of 63.33%, 75.00%, 81.48%, and 100.00%. Our understanding is that out of lying, standing, sitting and walking, only walking constitutes a movement of the user from one location to the other, which is distinct compared to the other behaviors on the dataset—lying, sitting, and standing. This makes the associated user interactions and behavior-related data very different from the other behaviors. Thus, the detection of walking by the machine learning model could achieve 100.00% accuracy.

Thereafter, we developed the other functionality of our framework—the intelligent decision-making algorithm that can analyze these behavioral patterns and their relationships with the dynamic spatial features of the environment to detect any anomalies in user behavior that could constitute an emergency, as outlined in Section 2. This functionality of our framework was developed as a RapidMiner 'process' that is shown in Figure 14, and the order in which the various operators of this RapidMiner process were executed upon the compilation of the same is shown in Figure 15. For the purpose of evaluating the efficacy of this framework, we were interested in developing a binary classifier that could classify a situation as an 'emergency' or 'non-emergency.' Thus, all instances of activities other than an emergency were labelled as 'non-emergency' in this dataset for the development of this RapidMiner 'process.' The 'Dataset' operator allowed for the importation of the data into the RapidMiner platform for developing this 'process.' The 'Set Role' operator was used to inform RapidMiner of the data attribute and its characteristics that should be predicted. In this case, it was either 'emergency' or 'non-emergency.' The 'Data Processing' operator was used to implement the knowledge base and make the model aware of the rest of the relationships and dependencies amongst the data attributes as per the characteristics of our framework and the proposed EDSCCA. The 'Data Preprocessing' operator also consisted of the of the 'Split Data' operator, which was used to split the data into training and test sets. We used 75% of the data for training and 25% of the data for testing after the removal of the outliers, as per the data preprocessing steps outlined in Section 3. Next, we used a k-NN classifier to develop this binary classification model. This k-NN classifier was also developed based on the Euclidean distance approach represented in Equation (4). This classification model consisted of five nearest neighbors and 186 examples with eight dimensions of the two classes—emergency and non-emergency. The 'Apply Model' operator was used to apply this learning model to the test data. The 'Performance' operator was used to evaluate the performance characteristics of the learning model. For the performance metrics, we used the confusion matrix to study the overall accuracy of the model, as well as the individual class precision values.

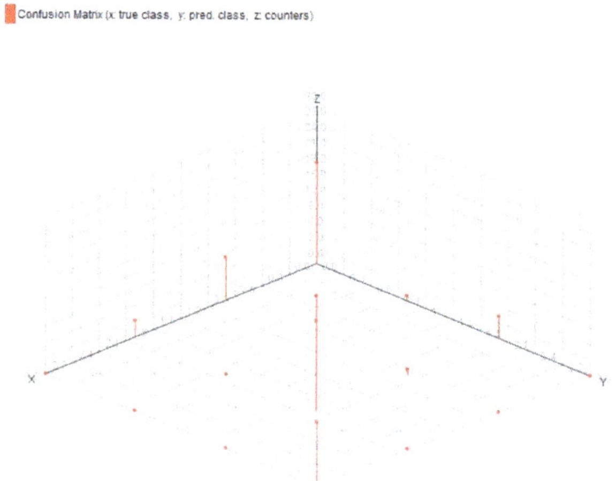

Figure 13. The performance accuracy (studied via a confusion matrix—plot view) of the RapidMiner process shown in Figure 9 for the detection of distinct behavioral patterns associated with different ADLs performed in the different spatial locations in a given IoT-based environment.

This RapidMiner process studied each row of the dataset, which consisted of different behavioral patterns associated with an ADL, to classify the associated behavior as emergency or non-emergency.

The output of this RapidMiner process was in the form of a table where each row consisted of the attributes outlined in Table 4. Here, the degree of certainty expresses the certainty of prediction of emergency or non-emergency by the developed k-NN-based machine learning model. To predict the same, the k-NN classification model in RapidMiner assigned a confidence value to each of these behavioral patterns, and the behavior with the highest confidence value was the final prediction of the model for that specific user interaction. For instance, in row number 2, the confidence values associated with non-emergency and emergency are 0.811 and 0.189, respectively, so the prediction of the model was non-emergency for the specific user interaction represented by this row. This output table had 62 rows, but only the first 13 rows are shown in Figure 16.

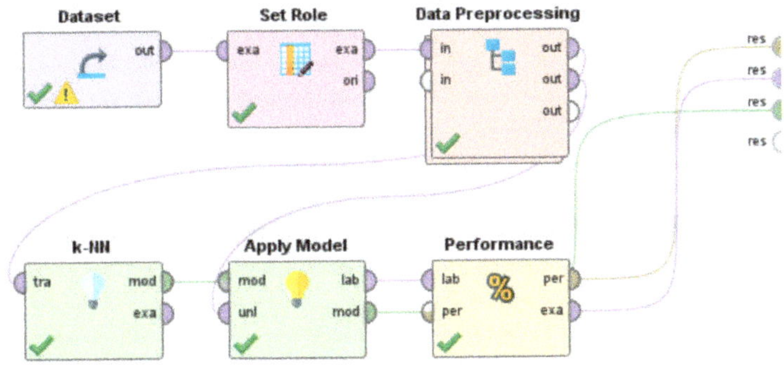

Figure 14. RapidMiner process for the development of the intelligent decision-making algorithm of the framework that can analyze distinct behavioral patterns and their relationships with the dynamic contextual and spatial features of the environment to detect any anomalies in user behavior that could constitute an emergency.

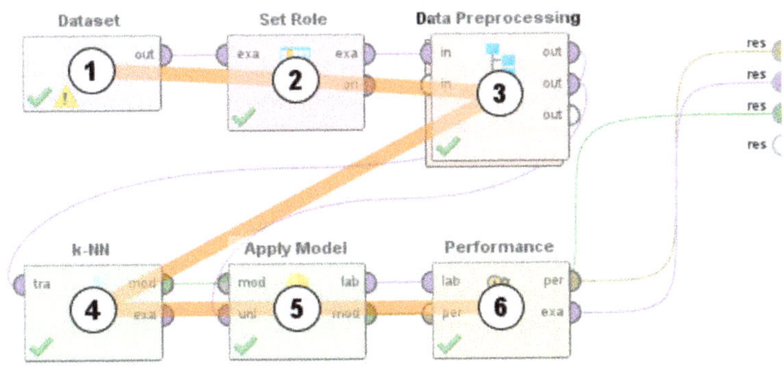

Figure 15. The order of execution of all the operators upon the compilation and execution of the RapidMiner process shown in Figure 14.

Table 4. Description of the attributes of the output of the RapidMiner process shown in Figure 16.

Attribute Name	Description
Row No	The row number in the output table
Complex Activity	The actual user behavior (either emergency or non-emergency) associated with a given complex activity (ADL)
Prediction (Complex Activity)	The predicted user behavior (either emergency or non-emergency) associated with a given complex activity (ADL)
Confidence (Non-Emergency)	The degree of certainty that the user behavior associated with a given complex activity did not constitute an emergency
Confidence (Emergency)	The degree of certainty that the user behavior associated with a given complex activity constituted an emergency

Row No.	Complex Activity	prediction(Complex Activity)	confidence(Non-Emergency)	confidence(Emergency)
1	Non-Emergency	Non-Emergency	0.801	0.199
2	Non-Emergency	Non-Emergency	0.811	0.189
3	Non-Emergency	Non-Emergency	0.616	0.384
4	Non-Emergency	Emergency	0.220	0.780
5	Non-Emergency	Emergency	0.403	0.597
6	Non-Emergency	Non-Emergency	0.815	0.185
7	Non-Emergency	Non-Emergency	1	0
8	Non-Emergency	Emergency	0.393	0.607
9	Non-Emergency	Non-Emergency	1	0
10	Non-Emergency	Non-Emergency	1	0
11	Non-Emergency	Non-Emergency	1	0
12	Non-Emergency	Non-Emergency	1	0
13	Non-Emergency	Non-Emergency	1	0

Figure 16. The output table of the intelligent decision-making algorithm of the framework, developed as a RapidMiner process, that can analyze distinct behavioral patterns and their relationships with the dynamic contextual and spatial features of the environment to detect any anomalies in user behavior that could constitute an emergency. This output table had 62 rows, but only the first 13 rows are shown here.

The performance characteristics of this framework were evaluated in the form a confusion matrix, as shown in Figures 17 and 18, with Figure 17 representing the tabular view and Figure 18 representing the plot view of the confusion matrix. By using the confusion matrix, both the overall performance and individual class precision performance values were computed.

As can be observed from Figures 17 and 18, the framework achieved an overall performance accuracy of 83.87%, with the sub-class precision for the detection of 'non-emergency' being 85.42% and the sub-class precision for the detection of 'emergency' being 78.57%.

accuracy: 83.87%

	true Non-Emergency	true Emergency	class precision
pred. Non-Emergency	41	7	85.42%
pred. Emergency	3	11	78.57%
class recall	93.18%	61.11%	

Figure 17. The performance accuracy (studied via a confusion matrix—tabular view) of the RapidMiner process shown in Figure 14 that involves the development of the intelligent decision-making algorithm of the framework that can analyze distinct behavioral patterns and their relationships with the dynamic contextual and spatial features of the environment to detect any anomalies in user behavior that could constitute an emergency.

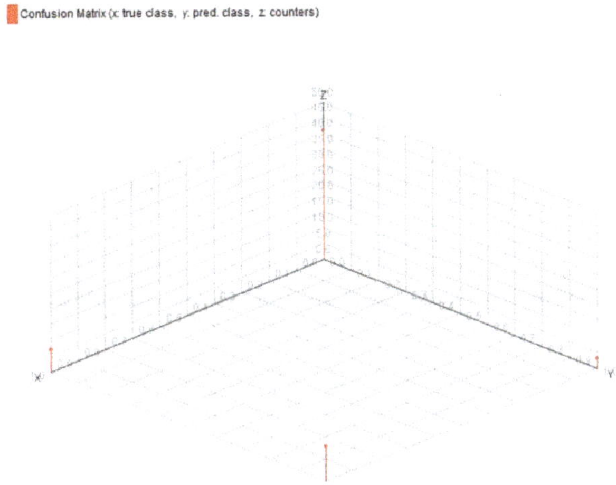

Figure 18. The performance accuracy (studied via a confusion matrix—plot view) of the RapidMiner process shown in Figure 14 that involves the development of the intelligent decision-making algorithm of the framework that can analyze distinct behavioral patterns and their relationships with the dynamic contextual and spatial features of the environment to detect any anomalies in user behavior that could constitute an emergency.

5. Comparative Discussion

Despite several advances and emerging technologies in the fields of human activity recognition, human behavior analysis, and their related application domains, the existing systems [9–29] have several limitations and drawbacks, as outlined in Section 2. This framework, which integrates the latest advancements and technologies in human–computer interaction, machine learning, Internet of Things, pattern recognition, and ubiquitous computing, aims to take a rather comprehensive approach to addressing these challenges in this field. In this section, we discuss these specific challenges and outline how our framework addresses the same and outperforms these existing systems in terms of their technical characteristics, functionalities, and operational features. This is presented as follows:

1. Several researchers in this field have only focused on activity recognition and that too at a superficial level. Various methodologies such as sensor technology-driven [9], RGB frame-based [10], hidden Markov model-based [11], and computer vision-based [12] methodologies have been proposed by researchers, but the main limitation of such systems is their inability to analyze complex activities at a fine-grain level to interpret the associated dynamics of user interactions and their characteristic features. Our framework addresses this challenge by being able to perform the semantic analysis of user interactions with diverse contextual parameters during ADLs. By semantic analysis, we refer to the functionalities of our framework to (1) analyze complex activities in terms of the associated postures and gestures, which are interpreted in terms of the skeletal joint point characteristics (Figure 1); (2) interpret the interrelated and interdependent relationships between atomic activities, context attributes, core atomic activities, core context attributes, other atomic activities, other context attributes, start atomic activities, end atomic activities, start context attributes, and end context attributes associated with any complex activity (Table 1); (3) detect all possible dynamics of user interactions and user behavior that could be associated with any complex activity (Table 2); (4) identify a list of distinct fine-grain level behavioral patterns—walking, sleeping, sitting, and lying—associated with different complex activities (Figures 9 and 11), which achieved a performance accuracy of 76.71% when tested on a dataset of ADLs (Figures 12 and 13); and (5) use an intelli-

gent decision-making algorithm that can analyze these distinct behavioral patterns and their relationships with the dynamic contextual and spatial features of the environment to detect any anomalies in user behavior that could constitute an emergency (Figures 14 and 16), which achieved an overall performance accuracy of 83.87% when tested on a dataset of ADLs (Figures 17 and 18).

2. Some of the recent works that have focused on activity analysis were limited to certain tasks and could not be generalized for different activities. For instance, in [17], the work focused on eating activity recognition and analysis; in [13], the activity analysis was done to detect enter and exit motions only in a given IoT-based space. In [18], the methodology focused on the detection of simple and less complicated activities, such an cooking, and [22] presented a system that could remind its users to take their routine medications. The analysis of such small tasks and actions are important, but the challenge in this context is the fact that these systems are specific to such tasks and cannot be deployed or implemented in the context of other activities. With its functionalities to perform complex activity recognition and analysis of skeletal joint point characteristics, our framework can analyze and interpret any complex activity and its associated tasks and actions, thereby addressing this challenge. When tested on a dataset, our framework was able to recognize and analyze all nine complex activities—sleeping, changing cloth, relaxing, moving around, cooking, eating, emergency, working, and defecating—that were associated with this dataset. It is worth mentioning here that our framework cannot only recognize these specific nine complex activities, because its characteristics allow it to recognize and analyze any set of complex activities represented by the big data associated with user interactions in a given IoT-based context, which could be a from a dataset or from a real-time sensor-based implementation of the IoT framework.
3. A number of these methodologies have focused on activities in specific settings and cannot be seamlessly deployed in other settings consisting of different context parameters and environment variables. For instance, in [14,16], the presented systems are specific to hospital environments, the methodology presented in [21] is only applicable to a kitchen environment, and the approach in [28] is only applicable to a workplace environment. While such systems are important for safe and assisted living experiences in these local spatial contexts, their main drawback is the fact that these tools are dependent on the specific environmental settings for which they have been designed. Our framework develops an SDCA by analyzing the multimodal components of user interactions on the context parameters, from an object centered perspective, as outlined in Section 3. This functionality allows our framework to detect and interpret human activities, their associated behavioral patterns, and the user interaction features in any given setting consisting of any kind of context attributes and environment variables.
4. Video-based systems for activity recognition and analysis, such as [12,19] may have several drawbacks associated with their development, functionalities, and performance metrics. According to [39], video 'presents challenges at almost every stage of the research process.' Some of these are the categorization and transcription of data, the selection of relevant fragments, the selection of camera angle, and the determination of the number of frames. By not using viewer-centered image analysis but by using object centered data directly from the sensors, our proposed framework bypasses all these challenges.
5. Some of the frameworks that have focused on fall detection are dependent on a specific operating system or platform or device. These include the smartphone-based fall detection approach proposed in [23] that uses an Android operating system, the work presented in [29] that uses an IOS operating system, the methodology proposed in [26] that requires a smart cane, and the approach in [15] that requires a handheld device. To address universal diversity and ensure the wide-scale user acceptance of such technologies, it is important that such fall detection systems are

platform-independent and can run seamlessly on any device that uses any kind of operating system. Our framework does not have this drawback because it does not need an Android or IOS operating system or any specific device for running. Even though it uses RapidMiner as a software tool to develop its characteristic features, RapidMiner is written in Java—which is platform-independent. RapidMiner allows for the exportation of any process in the form of the associated Java code. Java applications are known as write once run anywhere (WORA). This essentially means that when a Java application is developed and compiled on any system, the Java compiler generates a bytecode or class file that is platform-independent and can be run seamlessly on any other system without re-compilation by using a Java virtual machine (JVM). Additionally, RapidMiner also consists of multiple extensions that can be added to a RapidMiner process and used to seamlessly integrate a RapidMiner process with other applications or software based on the requirements.

6. Several fall detection systems are dependent on external parameters that cannot be controlled and could affect the performance characteristics. For instance, Shao et al. [25] proposed a fall detection methodology based on measuring the vibrations of the floor. Several factors such as the weight of the user, the material of the floor, the condition of the floor, and other objects placed on the floor can impact the intensity of vibrations that could affect the performance of the system. Kong et al.'s [24] system used the distance between the neck of the user and the ground to detect falls. The performance of such a system could be affected by the height of the user, the posture of the user, and any elevations on the ground such as high objects or stairs. The work proposed in [27] by Keaton et al., which used WiFi channel state data to detect falls, could be affected by external factors that tend to influence the WiFi channel state data. Similarly, the methodology developed in [20] worked by using an air pressure sensor, the readings of which could be affected by environmental factors and external phenomena. Such influences or effects of external conditions could have a negative effect on the operational and performance characteristics of the system, and it could even lead to false alarms, thus causing alert fatigue [40] in caregivers and medical personnel. Such false alarms and alert fatigue can decrease the quality of care, increase response time, and make caregivers and medical personnel insensitive to the warnings of such fall detection systems. The challenge is therefore to ensure that such fall detection systems can seamlessly function without being dependent on external factors that could affect its operation or performance metrics. Our framework uses concepts of complex activity recognition [30] and two related works [31,32], as well as taking the context-driven approach outlined in Section 3, for the analysis of diverse components of user interactions performed on context parameters to interpret the dynamics of human behavior and their relationships with the contextual and spatial features of an environment to detect any anomalies that could constitute an emergency. The performance, operation, and functionality of such an approach is independent of the effect of any external factors or conditions, such as floor vibrations, WiFi channel state data, and the distance between the user and the ground.

6. Conclusions and Scope for Future Work

Ambient intelligence in the future of smart homes and smart cities has the potential to address the multiple elderly needs during ADLs due to the behavioral, physical, mental, psychological, and other forms of impairments or limitations that they face with increasing age. A key to developing ambient intelligence in order to address and access these needs lies in monitoring human behavior while analyzing the multimodal components of user interactions with the dynamic contextual, spatial, and temporal features of a given IoT-based ubiquitous environment in which these activities are performed. Therefore, this work provides an interdisciplinary framework that takes a comprehensive approach to study, track, monitor, and analyze human behavior during ADLs. Based on the understanding of the behaviors associated with ADLs, abnormal behaviors leading to situations that might have resulted in health-threatening situations, such as from a fall or unconsciousness, that would need the immediate attention of caregivers or medical practitioners can be detected, and necessary actions can be taken accordingly.

The framework has two novel functionalities that were implemented and tested with an existing dataset. First, it is able to analyze multimodal components of user interactions to identify a list of distinct behavioral patterns associated with each ADL. Using the given dataset, the results showed that it achieved an overall performance accuracy of 76.71%. Second, it uses an intelligent decision-making algorithm that can analyze these behavioral patterns and their relationships with the dynamic contextual and spatial features of the environment to detect any anomalies in user behavior that could constitute an emergency, such as from a fall or unconsciousness. This algorithm achieved an overall performance accuracy of 83.87% when tested on a dataset consisting of multiple ADLs.

To the best of the authors' knowledge, no similar work has been done yet. The presented and discussed results uphold the immense potential and relevance of the framework for the development of ambient intelligence in the future of ubiquitous living environments, e.g., smart homes, to address multiple needs associated with the aging population. Our framework addresses several limitations and challenges in this field, but at this point, its functionality is limited to one user in the confines of a given IoT-based space. Future work along these lines would involve extending the functionality of the framework to incorporate multiple users. We also plan to implement this framework in real-time by setting up an IoT-based environment and incorporating relevant practices and measures to address the healthcare- and safety-related needs of the elderly.

Author Contributions: Conceptualization, N.T. and C.Y.H.; methodology, N.T.; software, N.T.; validation, N.T.; formal analysis, N.T.; investigation, N.T.; resources, N.T.; data curation, N.T.; visualization, N.T.; data analysis and results, N.T.; writing—original draft preparation, N.T.; writing—review and editing, C.Y.H. and N.T.; supervision, C.Y.H.; project administration, C.Y.H.; funding acquisition, Not Applicable. All authors have read and agreed to the published version of the manuscript.

Funding: This research received no external funding.

Institutional Review Board Statement: Not applicable.

Informed Consent Statement: Not applicable.

Data Availability Statement: Publicly available datasets were analyzed in this study. This data can be found here: https://doi.org/10.17632/sy3kcttdtx.1, accessed on 13 February 2021.

Conflicts of Interest: The authors declare no conflict of interest.

References

1. World Population Ageing: 1950–2050. ISBN-10: 9210510925. Available online: http://globalag.igc.org/ruralaging/world/ageingo.htm (accessed on 17 October 2020).
2. Esnaola, U.; Smithers, T. Whistling to Machines. In *Ambient Intelligence in Everyday Life, Lecture Notes in Computer Science Series*; Springer: Berlin/Heidelberg, Germany, 2006; Volume 3864, pp. 198–226.
3. Rashidi, P.; Mihailidis, A. A Survey on Ambient-Assisted Living Tools for Older Adults. *IEEE J. Biomed. Health Inform.* **2013**, *17*, 579–590. [CrossRef] [PubMed]
4. Sadri, F. Ambient intelligence: A survey. *Acm Comput. Surv.* **2011**, *36*, 1–66. [CrossRef]
5. Thakur, N. Framework for a Context Aware Adaptive Intelligent Assistant for Activities of Daily Living. Master's Thesis, University of Cincinnati, Cincinnati, OH, USA, 2019. Available online: http://rave.ohiolink.edu/etdc/view?acc_num=ucin1553528536685873 (accessed on 10 December 2020).
6. Thakur, N.; Han, C.Y. *A Review of Assistive Technologies for Activities of Daily Living of Elderly, Book Chapter in: Assisted Living: Current Issues and Challenges*; Nova Science Publishers: Hauppauge, NY, USA, 2020; pp. 61–84. ISBN 978-1-53618-446-4.
7. Katz, S.; Downs, T.D.; Cash, H.R.; Grotz, R.C. Progress in development of the index of ADL. *Gerontologist* **1970**, *10*, 20–30. [CrossRef] [PubMed]
8. Debes, C.; Merentitis, A.; Sukhanov, S.; Niessen, M.; Frangiadakis, N.; Bauer, A. Monitoring Activities of Daily Living in Smart Homes: Understanding human behavior. *IEEE Signal. Process. Mag.* **2016**, *33*, 81–94. [CrossRef]
9. Azkune, G.; Almeida, A.; López-de-Ipiña, D.; Liming, C. Extending knowledge driven activity models through data-driven learning techniques. *Expert Syst. Appl.* **2016**, *42*, 3115–3128. [CrossRef]

10. Neili Boualia, S.; Essoukri Ben Amara, N. Deep Full-Body HPE for Activity Recognition from RGB Frames Only. *J. Inform.* **2021**, *8*, 2.
11. Van Kasteren, T.; Noulas, A.; Englebienne, G.; Kröse, B. Accurate activity recognition in a home setting. In Proceedings of the 10th International Conference on Ubiquitous Computing, Seoul, Korea, 21–24 September 2008; pp. 1–9.
12. Cheng, Z.; Qin, L.; Huang, Q.; Jiang, S.; Yan, S.; Tian, Q. Human Group Activity Analysis with Fusion of Motion and Appearance Information. In Proceedings of the 19th ACM International Conference on Multimedia, Scottsdale, AZ, USA, 28 November–1 December 2011; pp. 1401–1404.
13. Skocir, P.; Krivic, P.; Tomeljak, M.; Kusek, M.; Jezic, G. Activity detection in smart home environment. *Procedia Comput. Sci.* **2016**, *96*, 672–681. [CrossRef]
14. Doryab, A.; Bardram, J.E. Designing Activity-aware Recommender Systems for Operating Rooms. In Proceedings of the 2011 Workshop on Context-awareness in Retrieval and Recommendation, Association for Computing Machinery, New York, NY, USA, 13 February 2011; pp. 43–46.
15. Abascal, J.; Bonail, B.; Marco, A.; Sevillano, J.L. AmbienNet: An Intelligent Environment to Support People with Disabilities and Elderly People. In Proceedings of the 10th International ACM SIGACCESS Conference on Computers and Accessibility, Halifax, NS, Canada, 13–15 October 2008; pp. 293–294.
16. Chan, M.; Campo, E.; Bourennane, W.; Bettahar, F.; Charlon, Y. Mobility Behavior Assessment Using a Smart-Monitoring System to Care for the Elderly in a Hospital Environment. In Proceedings of the 7th International Conference on Pervasive Technologies Related to Assistive Environments, Island of Rhodes, Greece, 27–30 May 2014; Article 51. pp. 1–5.
17. Rashid, N.; Dautta, M.; Tseng, P.; Al Faruque, M.A. HEAR: Fog-Enabled Energy-Aware Online Human Eating Activity Recognition. *IEEE Internet Things J.* **2021**, *8*, 860–868. [CrossRef]
18. Siraj, M.S.; Shahid, O.; Ahad, M.A.R. Cooking Activity Recognition with Varying Sampling Rates Using Deep Convolutional GRU Framework. In *Human Activity Recognition Challenge*; Springer: Berlin/Heidelberg, Germany, 2021; pp. 115–126. ISBN 978-981-15-8269-1.
19. Mishra, O.; Kavimandan, P.S.; Tripathi, M.M.; Kapoor, R.; Yadav, K. Human Action Recognition Using a New Hybrid Descriptor. In *Advances in VLSI, Communication, and Signal Processing*; Springer: Berlin/Heidelberg, Germany, 2021; pp. 527–536. ISBN 978-981-15-6840-4.
20. Fu, Z.; He, X.; Wang, E.; Huo, J.; Huang, J.; Wu, D. Personalized Human Activity Recognition Based on Integrated Wearable Sensor and Transfer Learning. *J. Sens.* **2021**, *21*, 885. [CrossRef] [PubMed]
21. Yared, R.; Abdulrazak, B.; Tessier, T.; Mabilleau, P. Cooking risk analysis to enhance safety of elderly people in smart kitchen. In Proceedings of the 8th ACM International Conference on Pervasive Technologies Related to Assistive Environments, Corfu, Greece, 1–3 July 2015; Article 12. pp. 1–4.
22. Angelini, L.; Nyffeler, N.; Caon, M.; Jean-Mairet, M.; Carrino, S.; Mugellini, E.; Bergeron, L. Designing a Desirable Smart Bracelet for Older Adults. In Proceedings of the 2013 ACM Conference on Pervasive and Ubiquitous Computing, Zurich, Switzerland, 8–12 September 2013; pp. 425–433.
23. Dai, J.; Bai, X.; Yang, Z.; Shen, Z.; Xuan, D. PerFallD: A Pervasive Fall Detection System Using Mobile Phones. In Proceedings of the 8th IEEE International Conference on Pervasive Computing and Communications Workshops, Mannheim, Germany, 29 March–2 April 2010; pp. 292–297.
24. Kong, X.; Meng, Z.; Meng, L.; Tomiyama, H. A Neck-Floor Distance Analysis-Based Fall Detection System Using Deep Camera. In *Advances in Artificial Intelligence and Data Engineering*; Springer: Berlin/Heidelberg, Germany, 2021; pp. 1113–1120. ISBN 978-981-15-3514-7.
25. Shao, Y.; Wang, X.; Song, W.; Ilyas, S.; Guo, H.; Chang, W.-S. Feasibility of Using Floor Vibration to Detect Human Falls. *Int. J. Environ. Res. Public Health* **2021**, *18*, 200. [CrossRef] [PubMed]
26. Chou, H.; Han, K. Developing a smart walking cane with remote electrocardiogram and fall detection. *J. Intell. Fuzzy Syst.* **2021**, 1–14, (Pre-press).
27. Keaton, M.; Nordstrom, A.; Sherwood, M.; Meck, B.; Henry, G.; Alwazzan, A.; Reddy, R. WiFi-based In-home Fall-detection Utility: Application of WiFi Channel State Information as a Fall Detection Service. In Proceedings of the 2020 IEEE International Conference on Engineering, Technology and Innovation (ICE/ITMC), Cardiff, UK, 15–17 June 2020; pp. 1–6.
28. Anceschi, E.; Bonifazi, G.; De Donato, M.C.; Corradini, E.; Ursino, D.; Virgili, L. SaveMeNow.AI: A Machine Learning Based Wearable Device for Fall Detection in a Workplace. In *Enabling AI Applications in Data Science*; Springer: Berlin/Heidelberg, Germany, 2021; pp. 493–514. ISBN 978-3-030-52067-0.
29. Mousavi, S.A.; Heidari, F.; Tahami, E.; Azarnoosh, M. Fall detection system via smart phone and send people location. In Proceedings of the 2020 28th European Signal. Processing Conference (EUSIPCO), Amsterdam, The Netherlands, 18–22 January 2021; pp. 1605–1607.
30. Saguna, S.; Zaslavsky, A.; Chakraborty, D. Complex Activity Recognition Using Context-Driven Activity Theory and Activity Signatures. *Acm Trans. Comput. Hum. Interact.* **2013**, *20*, 32. [CrossRef]
31. Thakur, N.; Han, C.Y. Towards A Language for Defining Human Behavior for Complex Activities. In Proceedings of the 3rd International Conference on Human Interaction and Emerging Technologies, Paris, France, 27–29 August 2020; pp. 309–315.
32. Thakur, N.; Han, C.Y. Towards a Knowledge Base for Activity Recognition of Diverse Users. In Proceedings of the 3rd International Conference on Human Interaction and Emerging Technologies, Paris, France, 27–29 August 2020; pp. 303–308.

33. Biggs, N.L. The roots of combinatorics. *Hist. Math.* **1979**, *6*, 109–136. [CrossRef]
34. Tabbakha, N.E.; Ooi, C.P.; Tan, W.H. A Dataset for Elderly Action Recognition Using Indoor Location and Activity Tracking Data. Available online: https://doi.org/10.17632/sy3kcttdtx.1 (accessed on 10 August 2020).
35. Mierswa, I.; Wurst, M.; Klinkenberg, R.; Scholz, M.; Euler, T. YALE: Rapid prototyping for complex data mining tasks. In Proceedings of the 12th ACM SIGKDD international conference on Knowledge discovery and Data Mining (KDD' 06), Pennsylvania, PI, USA, 20–23 August 2006; pp. 935–940.
36. K-Nearest Neighbors Algorithm. Available online: https://en.wikipedia.org/wiki/K-nearest_neighbors_algorithm (accessed on 1 February 2021).
37. Euclidean Distance. Available online: https://en.wikipedia.org/wiki/Euclidean_distance (accessed on 1 February 2021).
38. Confusion Matrix. Available online: https://en.wikipedia.org/wiki/Confusion_matrix (accessed on 1 February 2021).
39. Luff, P.; Heath, C. Some 'technical challenges' of video analysis: Social actions, objects, material realities and the problems of perspective. *Qual. Res.* **2012**, *12*, 255–279. [CrossRef]
40. Alarm Fatigue. Available online: https://en.wikipedia.org/wiki/Alarm_fatigue (accessed on 2 February 2021).

information

MDPI

Article

A Method of Human Activity Recognition in Transitional Period

Lei Chen [1], Shurui Fan [1,*], Vikram Kumar [2,*] and Yating Jia [1]

1 Tianjin Key Laboratory of Electronic Materials Devices, School of Electronics and Information Engineering, Hebei University of Technology, Tianjin 300401, China; 201931903018@stu.hebut.edu.cn (L.C.); ya1552715935@126.com (Y.J.)
2 Indian Institute of Information Technology, Una 177220, India
* Correspondence: fansr@hebut.edu.cn (S.F.); vikram@iiitu.ac.in (V.K.); Tel.: +86-139-0201-3210 (S.F.); +91-98163-17024 (V.K.)

Received: 10 July 2020; Accepted: 26 August 2020; Published: 28 August 2020

Abstract: Human activity recognition (HAR) has been increasingly used in medical care, behavior analysis, and entertainment industry to improve the experience of users. Most of the existing works use fixed models to identify various activities. However, they do not adapt well to the dynamic nature of human activities. We investigated the activity recognition with postural transition awareness. The inertial sensor data was processed by filters and we used both time domain and frequency domain of the signals to extract the feature set. For the corresponding posture classification, three feature selection algorithms were considered to select 585 features to obtain the optimal feature subset for the posture classification. And We adopted three classifiers (support vector machine, decision tree, and random forest) for comparative analysis. After experiments, the support vector machine gave better classification results than other two methods. By using the support vector machine, we could achieve up to 98% accuracy in the Multi-class classification. Finally, the results were verified by probability estimation.

Keywords: activity recognition; posture transitions; inertial sensor; feature selection; support vector machine

1. Introduction

The human activity and posture transformation recognition is useful to provid users with valuable situational awareness, thus become one of the hotspots in many fields such as medical care, human-computer interaction, film and television production, and motion analysis [1]. The two dominant approaches for human activity classification used in literature are Vision-based systems and Wearable Sensor-based systems. Vision-based systems are widely used to detection of human parts and identification of daily activities [2]. These systems process the collected visual data for activity classification.

Wearable Sensor based systems consist of multiple inertial sensors connected to a human sensor network. After receiving and executing system commands, the raw human body data would be given feedback [3,4]. Inertial measurement (accelerometers and gyroscopes) units are used to measure the triaxle angular velocity and the triaxle acceleration signals generated during human body movement [5]. Sensors available in smartphones, such as temperature sensors and pressure sensors, are useful to know the surroundings [6]. The data collected from the sensors attached to the user and sensors installed in the surroundings are proceed to provide situational awareness to the user [7]. One of the problems of using accelerometer to detect the motion of an object is that it often affected by the gravitational field in the measurement, and its value ($g = 9.81\ m/s^2$) is relatively high. However, many studies have found that gravity factors can be separated from body motion by filtering. When using three-axis accelerometer, the induced gravity vector can also help determine the direction of the object relative to the gravity axis [8]. The gyroscope measures the direction indirectly; that is, it first estimates the angular velocity, and integrates the angular velocity to obtain the direction. However, a reference initial angular position is needed to obtain the direction from the gyroscope [9]. Gyroscopes are also prone to noise, resulting in different offsets which can be eliminated by filtering.

At present, many scholars have studied the problem of human behavior recognition based on video data [10]. In [11], the authors proposed depth video-based HAR system to utilize skeleton joints features indoors. They used processed depth maps to track human silhouettes and produce body joints information in the of skeleton, then the hidden Markov model was trained by features calculated from the joint information. The trained model was adopted to recognize various human activities with a mean rate of 84.33% for nine daily routine activities of the elderly. Basbiker M, etc. [12] developed an intelligent human recognition system. In multiple stages of the system, a series of digital image processing technologies were used to extract the human activity feature data from the frame sequence, and a robust neural networks was established to classify the activity models by using a multi-layer feedforward perceptor network. However, the vision-based HAR is limited by spatial location, and video data is relatively complex. It is easier to cause privacy leakage. In contrast, data based on inertial measurement unit can avoid these problems very well, thus it is becoming a new trand of HAR.

The human activity recognition system has three types of feature extraction methods: temporal features, frequency features, and a combination of the two [13]. The authors of [14] put forward an algorithm named S-ELM-KRSL, which is more suitable for processing large-scale data with noises or outliers to identify the motion sequence of body. After experiment, the scheme could detect symptoms of mild cognitive impairment and dementia with satisfactory accuracy. In [15], Zhu, etc. proposed a semi-supervised deep learning approach using temporal Ensembling of deep long short-term memory to extract high-level features for human activity recognition. They investigated temporal Ensembling with some randomness to enhance the generalization of the neural networks. Besides the use of ensemble approach based on both labeled and unlabeled data, they also combined the supervised and unsupervised losses and demonstrated the effectiveness of the semi-supervised learning scheme in experimental results. The authors of [16] brought up a novel ensemble extreme learning machine (ELM) algorithm, in which Gaussian random projection is employed to initialize the input weights of base ELMs and more diversities had been generated to boost the performance of ensemble learning. The algorithm demonstrated recognition accuracies of 97.35% and 98.88% on two datasets. However, the training time of the algorithm is slightly longer. In [17], a feature selection algorithm based on fast correlation filtering was developed to achieve data preprocessing and demonstrated that the classification accuracy can reach up to 100%. However, the classification model only used the AIRS2 algorithm which may not be suitable for other classifier. Feature selection is based on well-defined evaluation criteria to select the original feature set, which eliminates small correlations and unnecessary features. The selected features don't change the original representation of the feature set, and feature selection helps online classification to be more flexible [18].

Most human behavior recognition systems developed in the past ignored posture transitions because the incidence of posture transitions is lower and the duration is shorter than other basic physical activities [19]. However, the above assumptions depend on different applications and are not applicable when multiple activities must be performed in a short period of time. On the other hand, in many practical scenarios, such as fitness or disability monitoring systems, determining posture transitions is critical because in these cases the user performs multiple tasks in a short period of time [20]. In fact, in the case of human behavior recognition system and transient posture perception, the classification will change slightly, and the absence of specified posture transformation may lead to poor system performance [21].

A posture transition is a finite duration event determined by its start and end times. In general, the time required for posture transitions between different individuals is different. The posture transition is limited by the other two activities and represents the transition period between the two activities [22]. Basic activities like standing and walking can be extended for a longer period of time than posture transitions. The data collection of the two types of activities is also different. The posture transformation needs to be repeated to obtain a separate sample. Since the basic activities are continuous, multiple window samples can be obtained from a single test according to the limitation of its time range [23].

The other works related to this paper are referred in [24,25]. We have researched a large number of features on HAR assisted by an inertial measurement unit in the past. The various activity features are classified hierarchical, and six basic activities can be identified with an average accuracy of 96.4%. However, the transition period of activities was out of account.

This paper focuses on Human Activity Recognition with postural transition awareness. In this paper, the motion of the human body was sensed by an accelerometer and a gyroscope of the inertial measurement unit. The magnitude and direction of the acceleration can be measured by vertically arranging the sensors in three-dimensional space. It can also be built on a single chip, and it is now common to use three-axis accelerometers in some commercial electronic devices [26]. First, we analyzed the six-axis signal data acquired by the inertial measurement unit, and thenpreprocessed to obtain a variety of signals that can represent the action. The various signals obtained from the preprocessing were extracted in the time domain and the frequency domain using various standard and original measurement methods to characterize each active sample. Thereafter, we perform feature selection according to the specific classification condition by using various feature selection algorithms. A variety of machine learning methods are used to classify and selected the one with the highest classification accuracy. Finally, we use support vector machine to classify the posture. Different kernel functions and specific parameters are used to optimize the model.

Figure 1 shows the framework followed in this paper for Activity Recognition. The framework consists of four modules: Data preprocessing, Feature Extraction and Selection, Classifier Selection, and Classifier Evaluation. The details of each module are given in next sections. In Section 2, we described the Data preprocessing, Feature extraction, and Data selection. Section 3 is focused on the Classifier Selection. In Section 4, we discussed Classifier Selection and Results. We concluded the paper in Section 5.

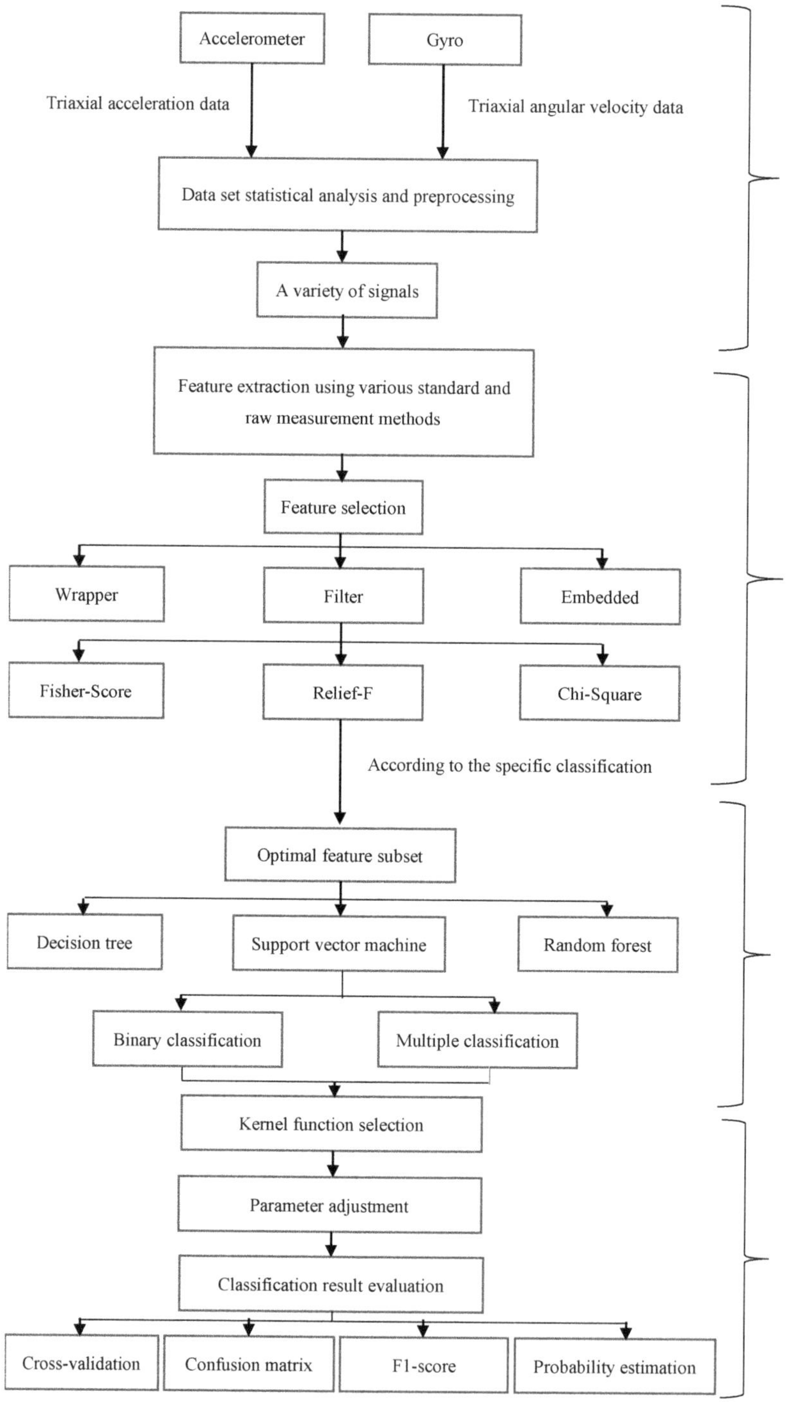

Figure 1. System Framework.

2. Data Preprocessing and Feature Selection

2.1. Data Preprocessing

The role of this module is to process the activity data received from the sensors and extract the variety of signals useful for activity recognition.

In this paper, we used the second generation human behavior recognition database available in the University of California Irvine (UCI) public platform [27]. The data set includes 6 basic activities: 3 static poses (standing, sitting, lying) and 3 dynamic poses (walking, downstairs, upstairs) for 30 different volunteers (everyone, aged between 19 and 48, who was instructed to follow the activity protocol when wearing an SGSII Smartphone at the waist as shown in Table 1), each volunteer was asked to do it twice. In addition, all possible pose transitions that occur between the existing three static poses are also available, including: standing-sitting (St-Si), sitting-standing (Si-St), sitting-lying (Si-Li), lying-sitting (Li-Si), standing-lying (St-Li), and lying-standing (Li-St). The frequency of the IMU was 100 Hz.

Table 1. Human activity recognition experiment protocol.

Serial Number	Static Poses	Time (s)	Serial Number	Dynamic Poses	Time (s)
0	Start (standing)	0	8	Walk (1)	15
1	Stand (1)	15	9	Walk (2)	15
2	Sit (1)	15	10	Downstairs (1)	12
3	Stand (2)	15	11	Upstairs (1)	12
4	Lay down (1)	15	12	Downstairs (2)	12
5	Sit (2)	15	13	Upstairs (2)	12
6	Lay down (2)	15	14	Downsairs (3)	12
7	Stand (3)	15	15	Upstairs (3)	12
			16	Stop	0

Table 1 shows all the activity tasks in order, and the corresponding time. In the process of experiment, every posture transformation performed twice by each volunteer. 60 labels were generated for each posture transformation which is accounting for 9% of all recorded experimental data. The duration of each posture tranformation is different, and even reverse transitions (for example, Stand-Sit and Sit-Stand). The average duration of posture transition is 3.7 s, while the basic activity is about 20.1 s. The signals collected from one volunteer were extracted and the data of 12 movements (6 basic movements and 6 posture transformation) were statistically analyzed as shown in Figure 2.

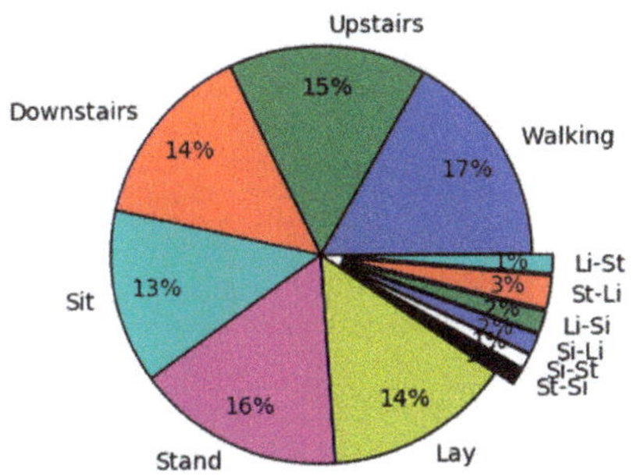

Figure 2. The statistics of posture data.

We process the original sensor signals obtained from the accelerometer (ar (t)) and the gyroscope (wr (t)) in three steps. First, we used a third-order median filter and a third-order low filter with a cutoff frequency of 20 Hz. Second, Battworth filter is applied for (transfer function is H1 (ω)) noise reduction, high-pass filter with a cutoff frequency of 0.3 Hz (transfer function is H2 (ω)) to eliminate the influence of DC bias in the gyroscope. Third, the acceleration signal is divided into gravity g (t) and object motion acceleration a (t).

The sensor data is plotted as Figures 3 and 4. The red line is the acceleration signal in the *X*-axis, the green line is the acceleration signal in *Y*-axis, and the blue line is the acceleration signal in *Z*-axis. It is evident from Figures 3 and 4 that the sensor data in the attitude transition phase changes significantly. The units used for the accelerations are g's, while the gyroscope units are rad'seg. The horizontal axis describes the sampling points which is corresponding to the time. All the preprocessed signals are summarized in Table 2.

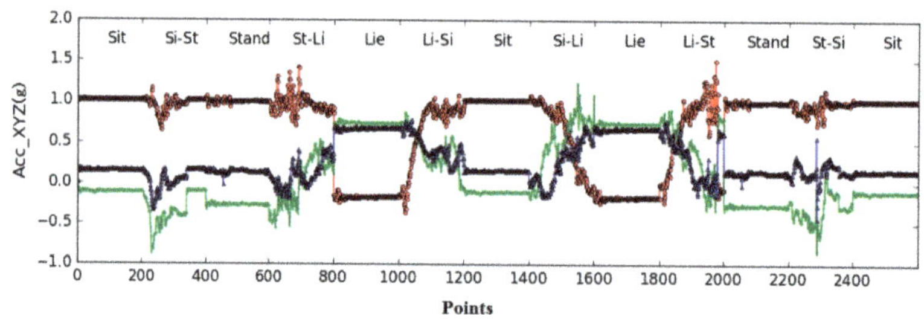

Figure 3. Acceleration *X*, *Y*, *Z* axis data.

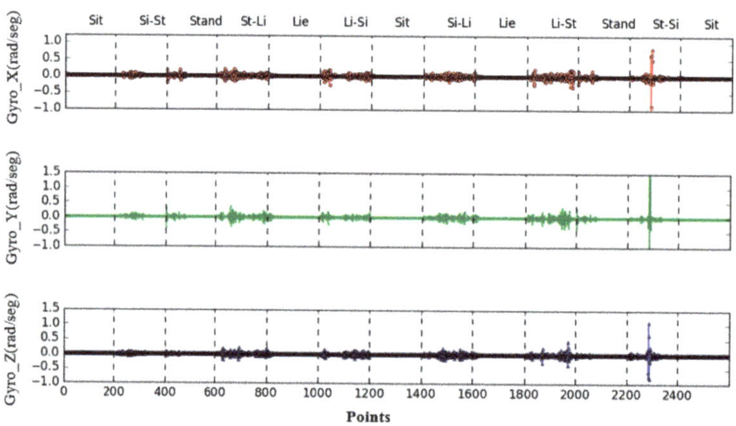

Figure 4. Angular velocity *X*, *Y*, *Z* axis data.

Table 2. Sensor inertial signal preprocessing.

Name	Quantity	Formula
Acceleration signal	tAcc (X,Y,Z)	$a_\tau(t) = H_1(a_r(t))$
Body acceleration signal	tAccBody (X,Y,Z)	$a(t) = H_2(a_\tau(t))$
Gravity signal	tGravity (X,Y,Z)	$g(t) = a_\tau(t) - a(t)$
Angular velocity signal	tGyro (X,Y,Z)	$\omega(t) = H_2(H_1\omega_r(t))$
Acceleration differential signal	tAccJerk (X,Y,Z)	$diff(a_\tau(t))$
Angular velocity diff- erential signal	tGyroJerk (X,Y,Z)	$diff(\omega(t))$
Acceleration amplitude signal	tAccMag	$\|a_\tau(t)\|$
Angular velocity amp- litude signal	tGyroMag	$\|\omega(t)\|$
Gravity amplitude signal	tGravityMag	$\|g(t)\|$
Acceleration and gravity angle signal	tAccAng	$\angle(a_\tau(t), g(t))$
Angular velocity and gravity angle signal	tGyroAng	$\angle(\omega(t), g(t))$
Acceleration frequency domain signal	fAcc (X,Y,Z)	$fft(a_\tau(t))$
Angular velocity fre- quency domain signal	fGyro (X,Y,Z)	$fft(\omega(t))$
Acceleration differential frequency domain signal	fAccJerk (X,Y,Z)	$fft(diff(a_\tau(t)))$

2.2. Feature Extraction

We used both the time and the frequency domain to extract the features. Table 3 shows the various measures and formulas used for generating feature sets on a fixed width window of length N, and there is 50% overlap between the two windows. The length of the window used in experiment is 2.56 s, since a person typically takes 1.5 steps per second on average, each window requires at least one full walking cycle.

Table 3. Feature Vector.

Function	Function Description	Formula
Mean (v)	Sample mean	$\overline{v} = \frac{1}{N}\sum_{i=1}^{N} v_i$
Var (v)	Sample variance	$\frac{1}{N-1}\sum_{i=1}^{N} (v_i - \overline{v})^2$
RMS (v)	Root mean square	$RMS = \left(\frac{1}{N}\sum_{i=1}^{N} v_i^2\right)^{1/2}$
Energy (v)	Average of the sum of squares	$P(v) = \frac{1}{N}\sum_{i=1}^{N}(v_i)^2$
Entropy (v)	Information entropy	$E = -\sum_{\mathrm{i=1}}^{\mathrm{N}} v_i \log v_i$
Distance (v)	Euclidean distance	$L2 = \frac{1}{N-1}\sqrt{\sum_{i=2}^{N}(v_{i-1} - v_i)^2}$
MaxfreqInd (v)	Maximum frequency component	$\mathrm{argmax}(v_i)$
MeanFreq (v)	Frequency signal weighted average	$\frac{\sum_{i=1}^{N}(iv_i)}{\sum_{j=1}^{N} v_j}$
EnergyBand (v,a,b)	Spectral energy in the [a,b] band	$\frac{1}{\mathrm{a-b+1}}\sum_{i=a}^{b} {v_i}^2$

In our past work, we extracted a total of 585 features to describe each active window [25]. From the various features tabulated in Table 3, some new features are taken into account. These features are extracted from each axis of the acceleration signal and the angular velocity signal. The statistical features in Table 3 are also applicable to the x-axis, y-axis, z-axis, Mag, differential, and tilt angle of acceleration and angular velocity. Table 3 shows the feature representation form calculated by generating the metrics of the data set and the window signal of length 128. Taking the Mean (v) as an example to perform feature calculation on different processed signals and corresponding feature descriptions. Table 4 shows the characterization of the average value.

Table 4. Signal processing methods for feature average.

Characterization	Explanation
tAcc-X-Mean	The x-axis body acceleration signal after noise removal is averaged according to the window length
tAcc-Y-Mean	The y-axis body acceleration signal after noise removal is averaged according to the window length
tAcc-Z-Mean	The z-axis body acceleration signal after noise removal is averaged according to the window length
tGyro-X-Mean	The x-axis angular velocity signal after noise removal is averaged according to the window length
tGyro-Y-Mean	The y-axis angular velocity signal after noise removal is averaged according to the window length
tGyro-Z-Mean	The z-axis angular velocity signal after noise removal is averaged according to the window length
tGravityAcc-X-Mean	The gravity component of the x-axis acceleration signal is averaged according to the length of the window
tGravityAcc-Y-Mean	The gravity component of the y-axis acceleration signal is averaged according to the length of the window
tGravityAcc-Z-Mean	The gravity component of the z-axis acceleration signal is averaged according to the length of the window

Table 4. *Cont.*

Characterization	Explanation
tAccJerk-X-Mean	The derivative of the *x*-axis body acceleration signal is averaged according to the length of the window
tAccJerk-Y-Mean	The derivative of the *y*-axis body acceleration signal is averaged according to the length of the window
tAccJerk-Z-Mean	The derivative of the *z*-axis body acceleration signal is averaged according to the length of the window
tGyroJerk-X-Mean	The derivative of the gravity component of the *x*-axis acceleration signal is averaged according to the length of the window
tGyroJerk-Y-Mean	The derivative of the gravity component of the *y*-axis acceleration signal is averaged according to the length of the window
tGyroJerk-Z-Mean	The derivative of the gravity component of the *z*-axis acceleration signal is averaged according to the length of the window
tAccMag-Mean	The amplitude of the triaxial body acceleration signal is averaged according to the length of the window
tGyroMag-Mean	The amplitude of the three-axis angular velocity signal is averaged according to the window length
tGravityAccMag-Mean	The amplitude of the gravity component of the three-axis acceleration signal is averaged according to the length of the window
tAccAng-Mean	The angle between the acceleration signal and the direction of gravity is averaged according to the length of the window
tGyroAng-Mean	The angle between the angular velocity signal and the direction of gravity is averaged according to the length of the window

2.3. Feature Selection

The objective of this step is to select the significant features from the feature set obtained in the feature extraction module to the training model [28,29]. The feature selection methods adopted by most researchers include Filter, Embedded, Wrapper. In this step, we used the filtering methods in the feature selection algorithm. The basic principle of feature selection algorithm is shown in Figure 5.

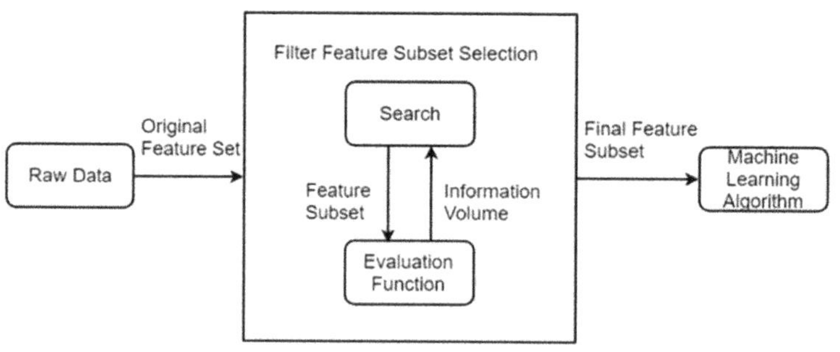

Figure 5. Filtered feature selection algorithm.

The algorithm uses divergence or correlation indicators to score each feature, and selects features with scores greater than a threshold or selects the top K features with the largest scores. Specifically, calculate the divergence of each feature, remove the features whose divergence is less than the threshold/select the top k features with the largest score; calculate the correlation between each feature and the label, and remove the features/selection with a correlation less than the threshold the top k features with the largest scores.

The advantages of the filtered feature selection algorithm are mainly versatility, low complexity, and fast running speed [30]. In this paper, three filtering feature selection algorithms, Relief-F, Fisher-Score, and Chi-Square, were applied to select the features.

The purpose of selected feature set is to classify the posture transformation between six basic movements (walking, going upstairs, downstairs, sitting, standing, and lying) and to achieve this, we selected 585 features. First, feature selection is made for the two categories: one is six basic actions, and another is six posture transformations. The results are shown in Figure 6. Secondly, the multiple classifications are characterized. The six basic movements are six categories, and another is all posture transformations. The results are shown in Figure 7.

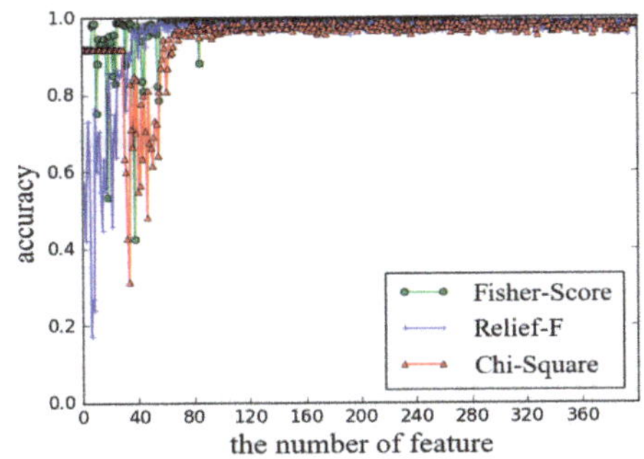

Figure 6. Three feature selection algorithm results in two categories.

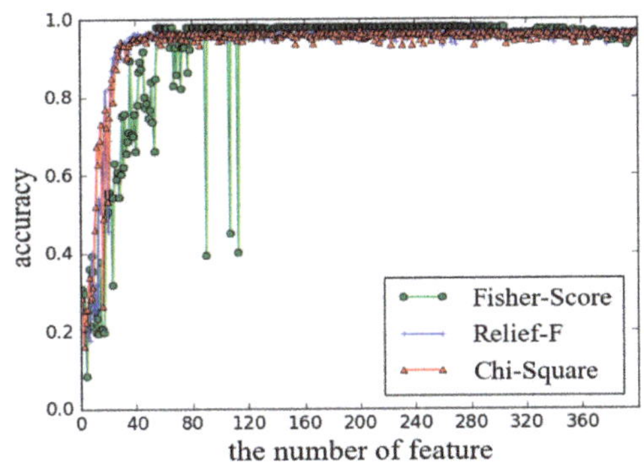

Figure 7. Three feature selection algorithm results in multiple classification.

In Figures 6 and 7, the abscissa refers to the number of features selected by the three feature selection algorithms, and the ordinate refers to the classification accuracy. It can be seen from Figures 6 and 7 that the classification accuracy increases gradually with increase in the number of selected features and approaches to 1. The ordering of the abscissa features in the three feature selection algorithms is sorted according to the scores of the features in the three algorithm principles.

In order to further select features of smaller dimensions to classify human poses with higher accuracy, we first input the first feature selected by each algorithm, that is, the three features into the classifier for training, obtain a classification model, and test it. If the test accuracy does not reach the ideal value, the first two features selected by each feature selection algorithm are selected for classification training, and so on, the feature combination with the highest classification accuracy is selected.

Finally, the features with highest score got from three feature selection methods were selected in the two categories: the maximum value in the fAcc (X) sequence, the frequency signal kurtosis in the fAcc (Y) sequence, and the sample range of the fAcc (X) sequence. In order to ensure classification accuracy in multiple classifications, 30 features (The top ten features selected by each feature selection method) were selected as shown in Table 5.

Table 5. Selected features.

SF. no.	Feature Description	Symbol Used
1, 2	sequence mean	tAcc (Y), tGravity (Y)
3, 4	sequence median	tAcc (Y), tGravity (Y)
5, 6	maximum value in the sequence	fAcc (Y), fGyro (X)
7	standard deviation	tAccJerk (X)
8	frequency signal skew	fAcc (Y)
9	range	fGyro (X)
10, 11, 12, 13	quartile of the sequence	tAccJerk (X), tGyroJerk (Z), tAccMag, tGravityMag
14, 15, 16, 17	10th percentile	tGyroJerk (Z), tAccMag, tGravityMag, tGyroAng
18, 19, 20	25th percentile	tAccMag, tGravityMag, fGyro (X)
21, 22	50th percentile	tAcc (Y), Gravity (Y)
23, 24, 25, 26	75th percentile	tGravity (Y), tGyroJerk (Z), tAccMag, tGravityMag
27, 28, 29, 30	90th percentile	tGyroJerk (Z), tAccMag, tGravityMag, tGyroAng

3. Classifier Selection

We used Support Vector Machine (SVM), which is a supervised machine learning algorithm developed in the last century and often used in statistical classification problems [31]. It was more often applied to the two-classification problem. The basic model is a linear classifier, which is transformed into a convex quadratic programming problem by maximizing the interval [32]. SVM is effective in high-dimensional space and suitable for situations where the dimensions are larger than the samples. Different kernel functions can be formulated for different scenarios. Linear separable samples can be classified by linear function. In diverse dimensions, the classifier shows different forms, such as a straight line for two-dimensions as shown in Figure 8, a plane for three-dimension and hyperplane for high-dimensional space.

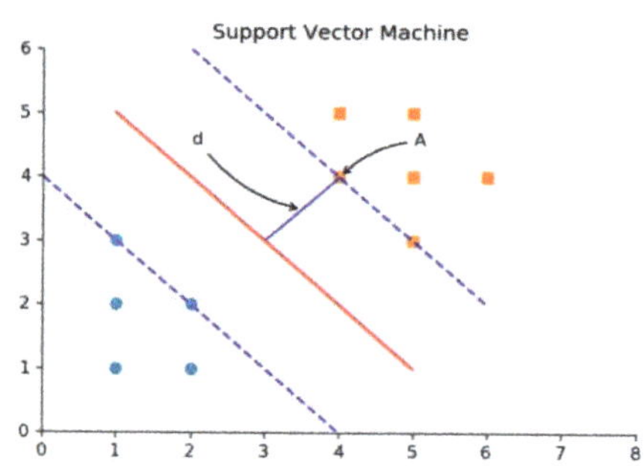

Figure 8. Support vector machine in two-dimensional space.

The decision tree is a tree that is constructed according to different strategies. By training the input data, the decision tree can be constructed, which can classify the unknown data efficiently, that is, predict the future based on the known [33]. It is a tree structure algorithm composed of root node, internal node, and leaf node. The core idea of the decision tree algorithm is to select attributes based on information gain and select the attribute with the largest information gain as the root [34]. The root is the top classification condition, each node of the tree acts as a test point on the property. The leaf node represents each category number, and the branch is on behalf of the output of each criteria. A binary tree has two branches on each node, while a node in a multi-tree has more than two branches.

The random forest algorithm is mainly based on the model aggregation idea, and has high precision in the classification and regression of high dimensional uncertainties [35]. The key idea under the random forest classifier is to grow a large number of unbiased decision trees from the guided samples, where each tree is voted for an activity class, and the random forest finally selects the most voted classification in the forest [36]. The random forest starts by selecting guide samples from the original training data. Then learning each guide sample through the decision tree. Only a small number of variables are available for binary partitioning on each node.

In the previous section, three filtering feature selection algorithms were used to select three features for the two-category case, and 30 features were selected for the multi-classification case. Next, for the different classification cases, three features and 30 features were respectively applied to the three classifiers, and the test set classification accuracy is shown in Tables 6 and 7. According to the analysis of the classification results, there is no significant difference between the classification accuracy of the three sets of testers. We found that the results of the SVM are better than the other two. Precision, recall and F1-score is the evaluation index of the classification results. Avg/total calculates the mean value of entirety, which represents the overall situation of evaluation index. We used the features selected by Fisher-Score, Relief-F and Chi-Square to train the SVM, and the training set accuracy is shown in Table 8.

Table 6. Three classifier experimental results of two categories.

	Test Set Classification Accuracy		Precision	Recall	F1-Score
SVM	1.0	Class 1	1.00	1.00	1.00
		Class 2	1.00	1.00	1.00
		Avg/total	1.00	1.00	1.00
Decision tree	0.9767	Class 1	0.97	1.00	0.99
		Class 2	1.00	0.75	0.86
		Avg/total	0.98	0.98	0.98
Random forest	0.9827	Class 1	0.97	1.00	0.99
		Class 2	1.00	0.75	0.86
		Avg/total	0.98	0.98	0.98

Table 7. Three classifier experimental results of multiple classification

	Test set Classification Accuracy		Precision	Recall	F1-Score
SVM	0.9827	Avg/total	0.99	0.98	0.98
Decision tree	0.9792	Avg/total	0.98	0.98	0.98
Random forest	0.9801	Avg/total	0.98	0.98	0.98

Table 8. Training set results of SVM.

	Fisher-Score	Relief-F	Chi-Square
Accuracy	0.954	0.988	0.976

4. Classification Results Analysis and Improvement

4.1. Classifier Parameter Selection

In this Module, we used the support vector machine as a common classifier to classify the pose. The role of the kernel function is to map the input space to a high-dimensional space with certain rules, and construct an optimal separation hyperplane in it, and finally achieve the effect of separating nonlinear data [37]. We mainly used linear and Radial Basis Function.

If we learn and test the classifier model on the same subset of data, it will lead over-fitting phenomenon which can be avoided by cross-validation.

The data of 30 volunteers in the original data set were divided: the data of the first 15 people were used as the feature selection set, the data from th 16th to 26th person were used as the training set of the classifier, and the others were used as the test set of the classifier.

4.1.1. Classifier Linear Kernel Parameter Selection

A commonly used parameter in a linear kernel is the penalty factor C. When the value of C is large, the misclassification is less, the fitting to the sample is better, but it is easy to cause overfitting [38]. Although the possibility of misclassification becomes larger and the fit to the sample is degraded, the prediction effect may be more desirable due to the influence of noise between the samples [39].

First, based on the three features selected in the previous section, the linear kernel support vector machine is used to solve the two-class problem in behavior recognition. Figure 9 shows the selection process for parameter C in the two classifications. Next, based on the 30 features selected in the previous section, we used the linear kernel support vector machine to solve the seven classification problem in behavior recognition. Figure 10 shows the selection process of parameter C in the multi-class.

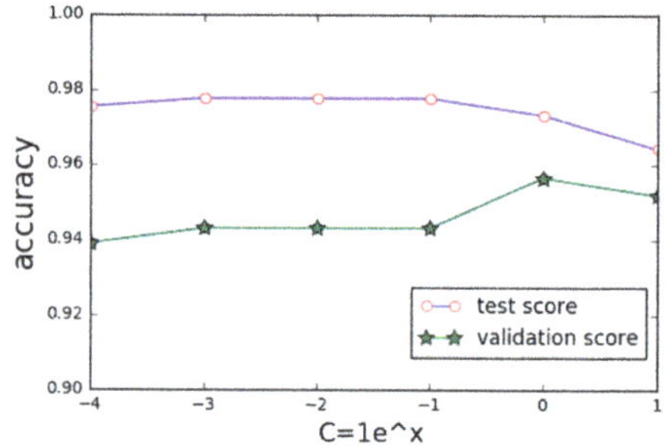

Figure 9. Selection of penalty factor C in the two categories.

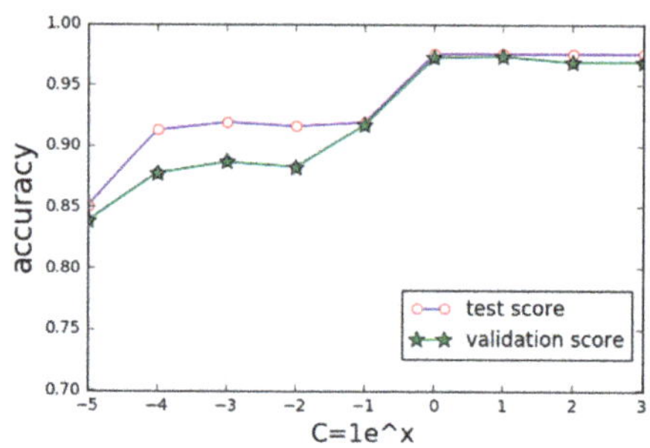

Figure 10. Selection of penalty factor C in multiple classification.

In Figures 9 and 10, the upper line represents the test set classification accuracy, and the lower line represents the cross-validation average. The abscissa shows the change of the penalty factor C, and the ordinate indicates the classification accuracy. It can be seen that with the increase of the penalty factor C, the classification accuracy and cross-validation average of the test set increase, but when the value of C is too large, the classification accuracy decreases slightly. In the process of processing the data, the larger the value of C, the more the error cannot be tolerated, and the time required for data processing will be longer. However, if the value of C is too small, we cannot guarantee that the parameter can be applied to other data sets. However, It still has a better effect. Therefore, considering the comprehensive consideration, we used the penalty factor value equals to 1. The 27th–29th people in the database were used for cross-validation to calculate the average precision value, mean value and standard deviation. We noticed that the classification accuracy of the test set is 0.973, the average cross-validation is 0.956, and the standard deviation of cross-validation is 0.042, which can achieve the desired effects. The factor C has a value of 1, and the classification accuracy of the test set is 0.975, the cross-validation average is 0.972, and the cross-validation standard deviation is 0.033, which can achieve the desired effect.

4.1.2. Classifier RBF Kernel Parameter Selection

The radial basis function (RBF) is a localized kernel function whose role is to map samples to high dimensional space. There are two main parameters in the classifier of RBF: the penalty factors C and σ [40]. The parameter σ reflects the clustering of the points after the mapping. The smaller the parameter σ, the distance between the mapped points tends to be equal, and the classification of the points will be finer, which will easily lead to overfitting. The larger the parameter σ, the coarser the classification will be, making it impossible to distinguish the data.

In the process of selecting the penalty factor C and the parameter σ, when the value of C is too large, over-fitting is easy to occur. When the value of σ is too small, the more support vectors are, the finer the classification is, and over-fitting easily occurs. And the increasing of the number of support vectors affects the speed of training and prediction [41]. The cross-validation is also used to determine whether the classification result has been over-fitted.

First, based on the three features selected in the previous section, the classifier of the radial basis kernel was used to solve the two-class problem in behavior recognition. Figure 11 shows the selection process of parameters C and σ in dichotomies. We used radial basis kernel support vector machine to solve the seven classification problem in behavior recognition based on the 30 features selected in the previous section. Figure 12 shows the selection process for parameters C and σ in the multi-category.

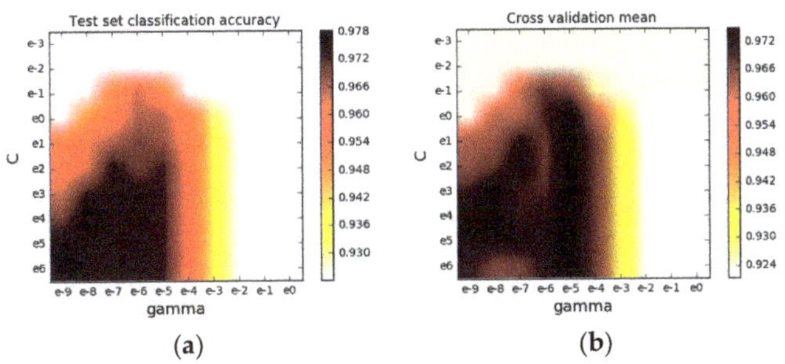

Figure 11. Selection of parameters C and σ in two categories: (**a**) Test set classification accuracy; (**b**) Cross validation mean.

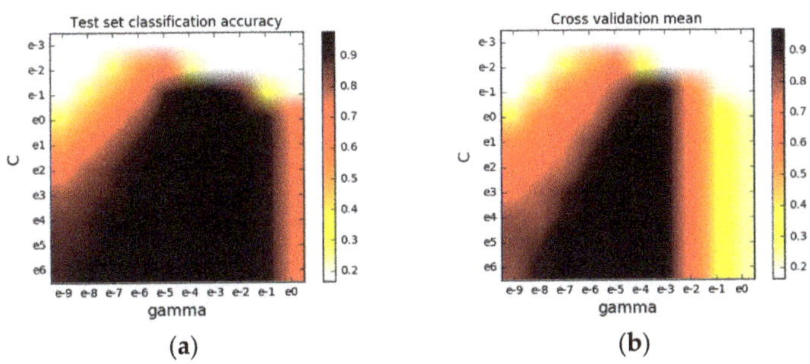

Figure 12. Selection of parameters C and σ in multiple classification: (**a**) Test set classification accuracy; (**b**) Cross validation mean.

There are two subgraphs in Figures 11 and 12. The abscissa shows the change of the parameter σ and the ordinate shows the change of the parameter C. While Figures 11 and 12a shows the classification accuracy of the test set, and Figures 11 and 12b represents the cross validation average. The darker the color, the larger the value. When the penalty factor C is too small and the parameter value σ is too large, the classification accuracy may not reach the ideal value. However, excessive pursuit of classification accuracy may cause computational complexity. Considering comprehensively, when the penalty factor

C is selected as 100 and the parameter is selected as 0.00001 in the second classification, the classification accuracy of the test set is 0.973, the cross-validation average is 0.975, and the cross-validation standard deviation is 0.011, which can achieve the desired effect, the penalty factor C in the seven classification. When the parameter is selected and the parameter is 0.001, the classification accuracy of the test set is 0.978, the average cross-validation is 0.938, and the cross-validation standard deviation is 0.057, which can achieve the desired effect.

4.2. Probability Estimation

Commonly used SVM can only generate categories without probability. The probability estimation can be used to transform the classification result of the support vector machine, that is, the probability that a sample belongs to each category [42].

The probabilistic calibration used in this study is isotonic regression, which is a nonparametric method. The core idea is to fit the deviation between the current classifier output and the real results. Isotonic regression is suitable for cases with large sample sizes, and over-fitting is prone to occur when the sample size is small. The Brier score can be used to evaluate the results of the probabilistic calibration. The Brier score is a loss, so the smaller score is better [43]. In all categories in which N predictions are aggregated, the Brier score measures the mean square error between the predicted probability and the actual probability assigned to the category. Therefore, for a set of predictionsmeans the lower the Brier score, the better the prediction calibration effects.

In this paper, we used data of five volunteers on which we used the support vector machine to learn and classify, and then uses isotonic regression to probabilistically estimate the data compiled by the volunteers. Due to individual differences, they completed each activity in different time actually. In order to maintain the integrity of a whole set of actions, result of one volunteer was presented only in Figure 13. In Figure 13, the abscissa is the test set data corresponding to different postures randomly selected from the volunteer data, and the ordinate is the predicted probability value obtained by estimating the probability of the data. The seven different colored lines represent the probability that the data is predicted into seven categories.

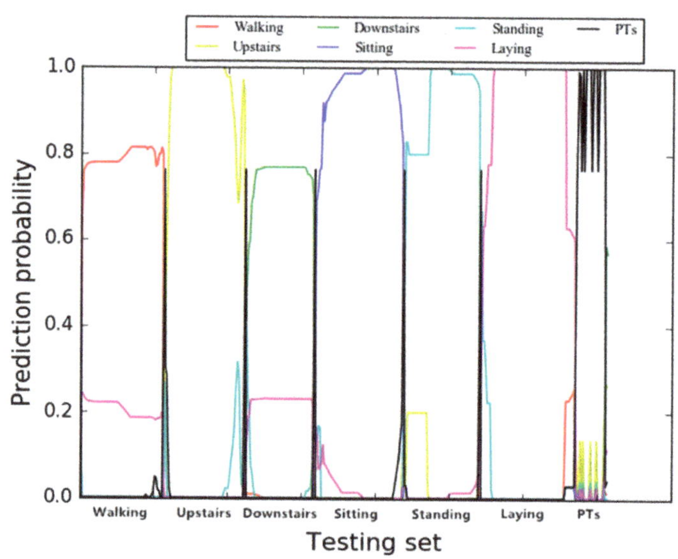

Figure 13. Seven classification probability estimation result.

The Brier score is then used to evaluate the results of theprobability estimates. The average results of five volunteers are shown in Table 9. The column labels in the table represent which actions the selected data comes from, the row labels represent the seven categories, and the values in the table are the obtained Brier scores. The Brier score on the diagonal in the table is relatively small, so the result of the probability estimation achieves the desired effects. Comparing with the experiments adopted SVM in the literature [44], SVM with kernel parameter selection adjustment has a significantly higher effectiveness and accuracy in identifying "walking", "upstairs", and "downstairs".

Table 9. Brier score evaluation result of probability estimation.

Brier-Score	Walking	Upstairs	Downstairs	Sitting	Standing	Laying	Posture Transitions
Walking	0.0105	0.2801	0.2432	0.2746	0.2332	0.2702	0.1980
Upstairs	0.2741	0.0107	0.2648	0.3023	0.2332	0.3344	0.2365
Downstairs	0.2532	0.2890	0.0204	0.2833	0.2374	0.2562	0.2087
Sitting	0.2820	0.3172	0.2741	0.0116	0.2714	0.3172	0.2293
Standing	0.2046	0.2231	0.1968	0.2341	0.0082	0.2701	0.1753
Laying	0.2820	0.3321	0.2840	0.3187	0.2622	0.0181	0.2543
Posture transitions	0.1861	0.2142	0.1602	0.2194	0.1793	0.2433	0.0356

5. Conclusions

In recent years, research on behavioral recognition methods for transitional attitude perception has become more and more widely used in many fields such as medical care. Based on the evaluated human behavior recognition data set it is found that the three-axis acceleration values of different static actions are significantly different, the three-axis angular velocity values are basically the same, and the posture conversion data between static actions changes significantly. It is undeniable that the data of the static posture is not always stable, as it cannot be guaranteed that the volunteer was completely still while sitting (or standing or lying) during the experiment.

We used Fisher-Score, Relief-F, and Chi-Square to select 585 features to obtain relatively good features set for classification. The features with higher scores were calculated using methods such as maximum value, minimum value, variance, skewness, kurtosis, and information entropy. The investigation shows that support vector machine gives better results than decision tree and random forest. In the second classification, the classification accuracy of the linear kernel ($C = 1$) is 97%, and the classification accuracy of the RBF kernel ($C = 1$, $\sigma = 0.001$) in the multi-class is 98%. Probability estimation overcomes some of the shortcomings of SVM and can directly output the probability that the data belongs to each category, thus making the results more intuitive.

Author Contributions: Funding acquisition, S.F.; resources, S.F.; project administration, S.F.; methodology, L.C.; writing—original draft preparation, L.C.; Visualization V.K.; Validation Y.J. All authors have read and agreed to the published version of the manuscript.

Funding: This work was funded by (Key Research and Development Plan Project of Hebei province, China) grant number (19210404D); (Key research and development project from Hebei Province, China) grant number (20351802D); (Key Research Project of Science and Technology from Ministry of Education of Hebei Province, China) grant number (ZD2019010); (Graduate Innovation Funding Project of Hebei Province) grant number (CXZZSS2020031).

Conflicts of Interest: The authors declare no conflict of interest.

References

1. Nweke, H.F.; Teh, Y.W.; Mujtaba, G.; Al-Garadi, M.A. Data fusion and multiple classifier systems for human activity detection and health monitoring: Review and open research directions. *Inf. Fusion* **2019**, *46*, 147–170. [CrossRef]
2. Hõrak, H. Computer Vision-Based Unobtrusive Physical Activity Monitoring in School by Room-Level Physical Activity Estimation: A Method Proposition. *Information* **2019**, *10*, 269. [CrossRef]

3. Xu, K.; Lu, Y.; Takei, K. Multifunctional Skin-Inspired Flexible Sensor Systems for Wearable Electronics. *Adv. Mater. Technol.* **2019**, *4*, 4. [CrossRef]
4. Gao, W.; Ota, H.; Kiriya, D.; Takei, K.; Javey, A. Flexible Electronics toward Wearable Sensing. *Acc. Chem. Res.* **2019**, *52*, 523–533. [CrossRef] [PubMed]
5. Xu, H.; Pan, Y.; Li, J.; Nie, L.; Xu, X. Activity Recognition Method for Home-Based Elderly Care Service Based on Random Forest and Activity Similarity. *IEEE Access* **2019**, *7*, 16217–16225. [CrossRef]
6. Sony, S.; LaVenture, S.; Sadhu, A. A literature review of next-generation smart sensing technology in structural health monitoring. *Struct. Control Health Monit.* **2019**, *26*, e2321. [CrossRef]
7. Li, J.H.; Tian, L.; Wang, H.; An, Y.; Wang, K.; Yu, L. Segmentation and Recognition of Basic and Transitional Activities for Continuous Physical Human Activity. *IEEE Access* **2019**, *7*, 42565–42576. [CrossRef]
8. Liu, Y.; Wang, X.; Zhai, Z.; Chen, R.; Zhang, B.; Jiang, Y. Timely daily activity recognition from headmost sensor events. *ISA Trans.* **2019**, *94*, 379–390. [CrossRef]
9. Ma, C.; Li, W.; Cao, J.; Du, J.; Li, Q.; Gravina, R. Adaptive sliding window based activity recognition for assisted livings. *Inf. Fusion* **2020**, *53*, 55–65. [CrossRef]
10. Zhang, S.; Wei, Z.; Nie, J.; Huang, L.; Wang, S.; Li, Z. A Review on Human Activity Recognition Using Vision-Based Method. *J. Healthc. Eng.* **2017**, *2017*, 343. [CrossRef]
11. Kim, K.; Jalal, A.; Mahmood, M. Vision-Based Human Activity Recognition System Using Depth Silhouettes: A Smart Home System for Monitoring the Residents. *J. Electr. Eng. Technol.* **2019**, *14*, 2567–2573. [CrossRef]
12. Babiker, M.; Khalifa, O.O.; Htike, K.K.; Hassan, A.; Zaharadeen, M. Automated daily human activity recognition for video surveillance using neural network. In Proceedings of the 2017 IEEE 4th International Conference on Smart Instrumentation, Measurement and Application (ICSIMA), Putrajaya, Malaysia, 28–30 November 2017; pp. 1–5.
13. De Leonardis, G.; Rosati, S.; Balestra, G.; Agostini, V.; Panero, E.; Gastaldi, L.; Knaflitz, M. Human Activity Recognition by Wearable Sensors: Comparison of different classifiers for real-time applications. In Proceedings of the 2018 IEEE International Symposium on Medical Measurements and Applications (MeMeA), Rome, Italy, 11–13 June 2018.
14. Chen, M.J.; Li, Y.; Luo, X.; Wang, W.P.; Wang, L.; Zhao, W.B. A Novel Human Activity Recognition Scheme for Smart Health Using Multilayer Extreme Learning Machine. *Cyber Enabled Intell.* **2019**, *6*, 239–258. [CrossRef]
15. Zhu, Q.; Chen, Z.; Soh, Y.C.; Yeng, C.S. A Novel Semisupervised Deep Learning Method for Human Activity Recognition. *IEEE Trans. Ind. Inform.* **2019**, *15*, 3821–3830. [CrossRef]
16. Chen, Z.; Jiang, C.; Xie, L. A Novel Ensemble ELM for Human Activity Recognition Using Smartphone Sensors. *IEEE Trans. Ind. Inform.* **2019**, *15*, 2691–2699. [CrossRef]
17. Ridok, A.; Mahmudy, W.F.; Rifai, M. An improved artificial immune recognition system with fast correlation based filter (FCBF) for feature selection. In Proceedings of the 2017 Fourth International Conference on Image Information Processing (ICIIP), Shimla, India, 21–23 December 2017; pp. 1–6.
18. Truong, P.H.; You, S.; Ji, S.-H.; Jeong, G.-M. Wearable System for Daily Activity Recognition Using Inertial and Pressure Sensors of a Smart Band and Smart Shoes. *Int. J. Comput. Commun. Control* **2019**, *14*, 726–742. [CrossRef]
19. Jiang, S.; Lv, B.; Guo, W.C.; Zhang, C.; Wang, H.T.; Sheng, X.J.; Shull, P.B. Feasibility of Wrist-Worn, Real-Time Hand, and Surface Gesture Recognition via sEMG and IMU Sensing. *IEEE Trans. Ind. Inform.* **2018**, *14*, 3376–3385. [CrossRef]
20. Lu, J.; Tong, K. Robust Single Accelerometer-Based Activity Recognition Using Modified Recurrence Plot. *IEEE Sens. J.* **2019**, *19*, 6317–6324. [CrossRef]
21. Hsu, Y.; Yang, S.; Chang, H.; Lai, H. Human Daily and Sport Activity Recognition Using a Wearable Inertial Sensor Network. *IEEE Access* **2018**, *6*, 31715–31728. [CrossRef]
22. Aminikhanghahi, S.; Cook, D.J. Enhancing activity recognition using CPD-based activity segmentation. *Pervasive Mob. Comput.* **2019**, *53*, 75–89. [CrossRef]
23. Gani, M.O.; Fayezeen, T.; Povinelli, R.J.; Smith, R.O.; Arif, M.; Kattan, A.J.; Ahamed, S.I. A light weight smartphone based human activity recognition system with high accuracy. *J. Netw. Comput. Appl.* **2019**, *141*, 59–72. [CrossRef]
24. Fan, S.R.; Jia, Y.T.; Liu, J.H. Feature selection based on three-axis acceleration sensor for human body attitude recognition. *J. Appl. Sci.* **2019**, *37*, 427–436.

25. Fan, S.R.; Jia, Y.T.; Jia, C.Y. A Feature Selection and Classification Method for Activity Recognition Based on an Inertial Sensing Unit. *Information* **2019**, *10*, 290. [CrossRef]
26. Liu, B.; Cai, H.; Ju, Z.; Liu, H. Multi-stage adaptive regression for online activity recognition. *Pattern Recognit.* **2020**, *98*, 107053. [CrossRef]
27. Jorge-Lr, O.; Luca, O.; Albert, S.; Xavier, P.; Davide, A. Transition-aware human activity recognition using smartphones. *Neurocomputing* **2016**, *171*, 754–767.
28. Haryanto, A.W.; Mawardi, E.K.; Muljono. Influence of Word Normalization and Chi-Squared Feature Selection on Support Vector Machine (SVM) Text Classification. In Proceedings of the 2018 International Seminar on Application for Technology of Information and Communication, Semarang, Indonesia, 21–22 September 2018; pp. 229–233. [CrossRef]
29. Wang, A.; Chen, G.; Wu, X.; Liu, L.; An, N.; Chang, C.-Y. Towards Human Activity Recognition: A Hierarchical Feature Selection Framework. *Sensors* **2018**, *18*, 3629. [CrossRef] [PubMed]
30. Dai, H. Research on SVM improved algorithm for large data classification. In Proceedings of the 2018 IEEE 3rd International Conference on Big Data Analysis (ICBDA), Shanghai, China, 9–12 March 2018; pp. 181–185.
31. Huo, Z.; Zhang, Y.; Shu, L.; Gallimore, M. A New Bearing Fault Diagnosis Method Based on Fine-to-Coarse Multiscale Permutation Entropy, Laplacian Score and SVM. *IEEE Access* **2019**, *7*, 17050–17066. [CrossRef]
32. Zhou, B.; Wang, H.; Hu, F.; Feng, N.S.; Xi, H.L.; Zhang, Z.H.; Tang, H. Accurate recognition of lower limb ambulation mode based on surface electromyography and motion data using machine learning. *Comput. Methods Programs Biomed.* **2020**, *193*, 105486. [CrossRef]
33. Sagi, O.; Rokach, L. Explainable decision forest: Transforming a decision forest into an interpretable tree. *Inf. Fusion* **2020**, *61*, 124–138. [CrossRef]
34. Clutterbuck, G.L.; Auld, M.L.; Johnston, L.M. High-level motor skills assessment for ambulant children with cerebral palsy: A systematic review and decision tree. *Dev. Med. Child Neurol.* **2020**, *62*, 693–699. [CrossRef]
35. Ishwaran, H.; Lu, M. Standard errors and confidence intervals for variable importance in random forest regression, classification, and survival. *Stat. Med.* **2019**, *38*, 558–582. [CrossRef]
36. Probst, P.; Wright, M.N.; Boulesteix, A.-L. Hyperparameters and tuning strategies for random forest. *Wiley Interdiscip. Rev. Data Min. Knowl. Discov.* **2019**, *9*, e1301. [CrossRef]
37. Nanda, M.A.; Seminar, K.B.; Nandika, D.; Maddu, A. A Comparison Study of Kernel Functions in the Support Vector Machine and Its Application for Termite Detection. *Information* **2018**, *9*, 5. [CrossRef]
38. Chun-Yi, T.; Wen-Hsiu, C. Multiclass object classification using covariance descriptors with kernel SVM. *J. Comput.* **2018**, *29*, 244–249.
39. Han, Y.; Li, J.; Xing, H.W.; Yang, A.-M.; Pan, Y.-H. Demonstration of SVM Classification Based on Improved Gauss Kernel Function. *Adv. Intell. Syst. Comput.* **2018**, *613*, 189–195. [CrossRef]
40. Shi, L.J.; Sun, B.B.; Ibrahim, D.S. An active learning reliability method with multiple kernel functions based on radial basis function. *Struct. Multidiscip. Optim.* **2019**, *60*, 211–229. [CrossRef]
41. Zhao, T.; Pei, J.H.; Chen, H. Multi-layer radial basis function neural network based on multi-scale kernel learning. *Appl. Soft Comput.* **2019**, *82*, 105541. [CrossRef]
42. Konopko, K.; Janczak, D. Classification method based on multidimensional probability density function estimation dedicated to embedded systems. *IFAC Pap.* **2018**, *51*, 318–323. [CrossRef]
43. Goldenholz, D.M.; Goldenholz, S.; Romero, J.; Moss, R.; Sun, H.Q.; Westover, B. Development and Validation of Forecasting Next Reported Seizure Using e-Diaries. *Ann. Neurol.* **2020**, *88*, 588–595. [CrossRef]
44. Shi, J.; Zuo, D.; Zhang, Z. Transition Activity Recognition System Based on Standard Deviation Trend Analysis. *Sensors* **2020**, *20*, 3117. [CrossRef]

information

Article

Federated Blockchain Learning at the Edge

James Calo [1,*] and Benny Lo [2]

1 Department of Computing, The Hamyln Centre, Imperial College London, London SW7 2AZ, UK
2 Department of Surgery and Cancer, The Hamyln Centre, Imperial College London, London SW7 2AZ, UK; benny.lo@imperial.ac.uk
* Correspondence: jam414@ic.ac.uk

Abstract: Machine learning, particularly using neural networks, is now widely adopted in practice even with the IoT paradigm; however, training neural networks at the edge, on IoT devices, remains elusive, mainly due to computational requirements. Furthermore, effective training requires large quantities of data and privacy concerns restrict accessible data. Therefore, in this paper, we propose a method leveraging a blockchain and federated learning to train neural networks at the edge effectively bypassing these issues and providing additional benefits such as distributing training across multiple devices. Federated learning trains networks without storing any data and aggregates multiple networks, trained on unique data, forming a global network via a centralized server. By leveraging the decentralized nature of a blockchain, this centralized server is replaced by a P2P network, removing the need for a trusted centralized server and enabling the learning process to be distributed across participating devices. Our results show that networks trained in such a manner have negligible differences in accuracy compared to traditionally trained networks on IoT devices and are less prone to overfitting. We conclude that not only is this a viable alternative to traditional paradigms but is an improvement that contains a wealth of benefits in an ecosystem such as a hospital.

Keywords: IoT; machine learning; neural networks; federated learning; blockchain; learning on the edge

Citation: Calo, J.; Lo, B. Federated Blockchain Learning at the Edge. *Information* **2023**, *14*, 318. https://doi.org/10.3390/info14060318

Academic Editors: Spyros Panagiotakis and Evangelos K. Markakis

Received: 7 April 2023
Revised: 18 May 2023
Accepted: 23 May 2023
Published: 30 May 2023

1. Introduction

The number of and use cases for IoT devices are increasing exponentially [1]; this is due in part to the increase in their ability to handle complex tasks that were once only possible on full-sized computers. One of the core use cases is within the medical field, referred to as the IoMT, with the market share of these devices on the rise. In 2017, the IoMT market was worth USD 28 billion and is projected to be USD 135 billion by 2025 [2]. Therefore, a solution that leverages this infrastructure while utilizing technologies proven to be effective is essential.

IoT architecture contains two main boundaries, cloud computing and edge computing, which differ primarily by the location where the computations take place and additionally differ in computing power. Additionally, there are three primary layers: cloud, fog and mist [3], as depicted in Figure 1. These layers are analogous to their namesakes, clouds are large and furthest away from the ground, fog is lighter and hovers between the ground and the clouds and mist is a thin layer of water molecules suspended just above the ground. The distance from the ground can be thought of as the distance from the users and the size of the water molecules are analogous to the computing resources at each layer.

Cloud computing is still the most prevalent [4,5] , allowing complex computations to be run on dedicated machines with the results streamed to IoT devices on demand, significantly reducing the need for IoT devices to perform complex calculations and instead act as a go-between for the client and server. This is effective but requires a connection to work; any disruption, due to congestion or network cutout, will cause a delay or even complete loss of results. This is unacceptable in many contexts such as during surgery or within a self-driving vehicle.

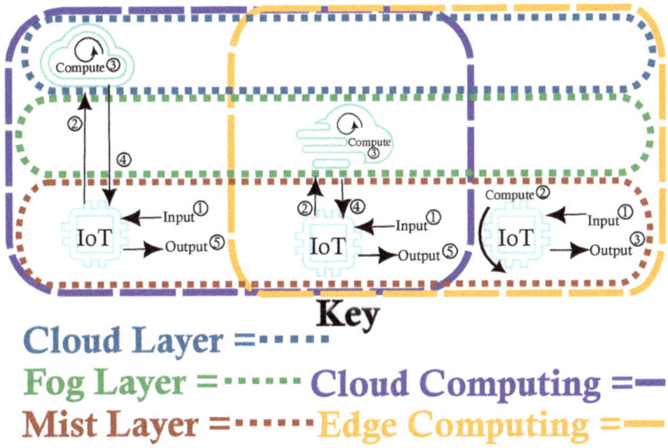

Figure 1. IoT Architecture Archetypes: Cloud and Fog both require an external server for handling the computations requested by the IoT device. The primary difference is the server's location, which is local for fog and external for cloud, and communication protocol. Mist, on the other hand, requires no external communication and all computations are performed on the device itself.

Fog computing [6,7] attempts to overcome these issues by exchanging the cloud server for a local server in the same location as the IoT devices; for example, a fog server may be housed in a hospital so that every internal IoMT device can connect directly to it as opposed to an external connection such as via the internet as is commonly the case in cloud computing. However, fog computing may still experience connection issues and requires space for dedicated hardware in close proximity. This is not ideal for situations where users either cannot be in close proximity or servers cannot be installed in the required location, such is the case with IoT sensors used in remote locations such as the Antarctic. Fog computing is a hybrid approach and is thus in the intersection of cloud computing and edge computing.

Therefore the most useful archetype is mist computing [8], where all computations are run on the IoT device itself or another (local) participant. Whilst this too requires a connection and is therefore prone to the same issues as fog computing, all participating devices are autonomous and can act individually as required. Unfortunately, this is also the most challenging archetype as the available resources are heavily limited and often require more specific solutions dependent upon the abilities of the individual device. However, whilst mist computing itself is completely within the edge computing boundary, it is frequently used in concert with the other layers. Therefore, when we refer to edge computing we refer to using only the mist layer, with each IoT device communicating only to each other without the requirement of dedicated (and more powerful) external servers.

Finally, there are additional paradigms that use software approaches to allow for additional benefits; dew computing, for example, caches and mirrors data on the cloud or fog in order to respond as if the data were local but requires syncing with the true servers. The most prevalent examples of dew computing are in cloud file storage where a local computer may contain cached copies of the files which are then synced with the cloud whenever changes are made or a refresh is requested.

Edge and fog computing, as a whole, have made great strides. Mobile Edge Computing (MEC), for example, allows for cloud computing capabilities with vastly reduced latency, high bandwidth and real-time access to network information [9]; therefore big data techniques can be applied with real-time feedback. However, this system does not actually leverage edge computing, instead using a fog computing architecture and as such must overcome issues such as reliance on users being in proximity to the fog servers, ensuring the connection does not interfere with the user's ability to access the internet (mobile data).

In order to overcome these issues, there are a small number of IoT systems that utilize true edge computing; however, these solutions either do not leverage machine learning at all or cannot be trained at the edge. For example, research by A.A. Abdellatif et al. developed an edge-based classifier for seizure detection and compressed sending of results [10,11]. Whilst this is a great step towards edge computing there is no machine learning being leveraged and so these types of systems are not able to use some of the recent advances in the field of machine learning for health. On the other hand, IoT devices using machine learning do exist and bringing machine learning, particularly neural networks and their collective family (convolutional, deep, etc.), onto the edge is a wide area of research [12]. TensorFlow has the TensorFlow Lite [13,14] framework which is designed to convert a trained neural network to run on edge devices; it is a very popular toolkit and is used in the book, TinyML [15] for running neural networks on embedded devices. However, whilst it can be used for on-device training, to our best knowledge, there are no systems that employ this.

Given the accelerated demand for faster training of more complex neural networks, there has been a focus on dedicated hardware such as GPUs and application-specific integrated circuits such as Google's Tensor Processing Unit (TPU) and field-programmable gate arrays (FPGAs), which are inherently cohesive with the matrix multiplication operations that underpin neural networks.

Whilst improvements to these specialized hardware components have accelerated our machine learning abilities, due to their high cost, large footprint and high energy draw, they are often missing from IoT devices, where smaller and less obtrusive devices are preferred. Even IoT devices that do contain specialized hardware, such as modern mobile phones which may have GPUs, require smaller, embedded, versions that are significantly less powerful than their full-sized counterparts. Furthermore, the less hardware required the lower the baseline battery draw is, a vital factor in IoT.

As a result, learning on a CPU on the edge is paramount! Removing dependency on specific hardware and providing all benefits via software allows existing devices to use this framework and keep the footprint of new devices small and focused on efficiency. Intel developed the Intel Xeon Scalable processor, which is geared towards learning and inference, focusing on three main features: the computation and memory capacity of the CPU, software optimizations for the CPU within machine learning tasks, specifically DNNs, and the use of distributed training algorithms for supervised deep learning workloads [16]. All three of these features translate to edge computing perfectly; for example, the memory capacity allows for better batching of input data and by not using a GPU there is no overhead between moving data from the CPU to GPU and back again. Furthermore, edge computing and IoT are inherently distributed and as such learning can be offloaded and computed at the source or destination, potentially increasing the efficiency of the whole system allowing each device to act autonomously and only interact with other devices when needed [7]. Furthermore, since training occurs upon observation of new data, which may not be consistent, inference, via the previous globally updated network, may be used simultaneously resulting in an online, continuous learning paradigm where networks are incrementally improved while running; this could be further improved by leveraging unsupervised techniques [17] or by using the inferring network as the supervisor.

Therefore, a privacy-ensuring, low-powered and generalizable method is required for training neural networks running at the edge. Federated learning allows multiple networks (of the same architecture) to train simultaneously on non-iid data without ever storing them and aggregate their knowledge into a global network, thereby learning at the edge. However, vanilla federated learning requires a centralized server for the aggregation; when privacy is involved, trust can be an issue. Therefore, we propose a hybrid approach utilizing the decentralization of a blockchain to create a framework that runs efficiently on multiple IoT devices across a P2P network, removing the need for a trusted centralized server, thereby improving privacy and adding robustness to data integrity. Furthermore, this enables the learning process to be distributed across participating devices (which federated learning does not implicitly do) leveraging the ubiquity of IoT devices to offset their lack of power by increasing the number of devices that are able to run in parallel.

1.1. Contributions

In this paper, we utilized our previous federated blockchain learning framework [18] to propose a pure mist solution to train a consortium of neural networks on a single Android phone to perform image classification on the CIFAR-10 dataset [19] Therefore, the main contributions of this work are as follows:

1. We developed a novel IoT federated learning framework, using Tensorflow Lite and our previously developed blockchain framework, to perform training at the edge (LotE) that is fully decentralized, leveraging our blockchain framework, ensuring that the data are private and secured against malicious attacks and requiring no trust between participants. This system requires no intermediary servers, which results in a mist-only architecture.
2. Using this framework, we build a configurable system for the training of neural networks on IoT devices and tested it on the CIFAR-10 dataset using a physical Pixel 4 Android smartphone running Android 13 with a Qualcomm Snapdragon 855 Octa-core CPU (1 × 2.84 GHz Kryo 485 Gold Prime, 3 × 2.42 GHz Kryo 485 Gold and 4 × 1.78 GHz Kryo 485 Silver) in order to obtain practical, and not simulated, results [19].
3. This system utilizes TensorFlow Lite as our machine learning framework (as opposed to our own framework's implementation) since standard TensorFlow is so widely used, is compatible with our existing federated blockchain learning framework and TensorFlow Lite converts TensorFlow models for use on IoT devices. Therefore, any existing system using Tensorflow will be able to receive the full benefit of our system with next to no overhead.

1.2. Related Works

Federated learning using blockchains has had an exponential surge in interest in recent years, especially for designing privacy enhancing and trustworthy AI, resulting in systems such as that by Yang et al. [20], which proposed a federated blockchain method using the PBFT consensus protocol. It can resist 33% of malicious users with a drastically lower energy cost than PoW but requires "Authorized" edge servers. Additionally, they use Multi-Krum as the federated aggregation scheme in order to achieve Byzantine-resilience. However, to the best of our knowledge, this proposed scheme has not been implemented on real hardware and utilizes cluster computing on the fog layer, requiring edge servers with significantly more power than standard IoT devices have to validate the blocks in the chain.

Islam et al. [21], on the other hand, utilizes drones as a method to ensure connection between devices running on the edge (pure mist computing) and utilizes differential privacy alongside federated blockchain learning to further enhance privacy. However, each entity in the system must register before being allowed to participate, which can cause issues if the system is required to grow and shrink dynamically.

1.3. Organization

The rest of this paper is organized as follows: Section 2 introduces federated learning and the blockchain as well as how we utilize these two different theories into a cohesive system. Section 3 analyzes the computational complexity and latency. Section 4 presents the results of our system running on a physical Android smartphone against the CIFAR-10 dataset in order to obtain results that demonstrate the real-world ability of our system. Finally, Section 5 discusses our findings and the direction we believe our future work will take.

2. Materials and Methods

2.1. Federated Learning

In the standard case, neural networks aim to approximate a function by minimizing the prediction loss with respect to the network's parameters:

$$\min_{\omega \in \mathbb{R}^d} \mathcal{T}(\omega) = \ell(\boldsymbol{x}, \boldsymbol{y}, \omega) \tag{1}$$

where ℓ is a chosen loss function, $\boldsymbol{x}, \boldsymbol{y}$ are the training (input and desired output) vectors and ω is the network's parameters.

However, with federated learning there are many networks training (potentially simultaneously) to form a global network, which may not have any training data of its own, and is an aggregation of these participating networks. Therefore, the *FedAvg* algorithm [22] (Equation (2)) is used to obtain the global weights for the global network, which may then be used as the initial network for the next round of federated learning.

$$\mathcal{F}(\omega) \triangleq \frac{1}{|\chi|} \sum_{i=1}^{N} |\chi_i| \cdot \mathcal{T}_i(\omega_i) \tag{2}$$

where for N participating networks, $|\chi_i|$ is the number of examples seen by network i (for $i \in \{1 \ldots N\}$), $\mathcal{T}_i(\omega_i)$ is the trained weights of network i as defined in Equation (1) and $|\chi| = \sum_{i=1}^{N} |\chi_i|$ is the total amount of data seen by all participants. Note that only the number of examples seen by each network is sent to the blockchain, no data are ever stored.

2.2. Decentralization with Blockchain

In order to replace vanilla federated learning's requirement for a centralized server we propose using a blockchain to tweak the paradigm to use a decentralized distributed ledger. This changes federated learning's architecture from the cloud/fog to the edge, where every device is independent and autonomous. The system will work even with only one node, and, in the event that all nodes go down, the system can recover fully since each node contains a copy of the accepted blockchain. This may not have been possible with a central server. The following details our design regarding the fundamental components of the blockchain.

2.2.1. Block

Our block format closely mirrors Bitcoin's format but with two major changes: the target formula and the federated components. The target is used to decide when the block has been mined via proof of work (PoW), which we chose over alternative, more green (both environmentally and chronologically) consensus mechanisms, such as proof of stake (PoS) [23–25]. Unfortunately, the downside of PoW is the energy cost; a miner who performs PoW must continually guess, in a deterministic manner, a hash that is less than the target, since the target is in big endian hexadecimal format we refer to it as a hash with more leading zeros than the target.

The criticism stems from the high computational cost of computing the hash which provides no computational benefit, other than the required effect of making it infeasible for a malicious node to pervert the system. However, as the blockchain is used to calculate the global update, the mining process provides some benefit, albeit tangentially. Furthermore, in an IoT system, this cost is somewhat negated as there are likely to be many interconnected devices and as such the problem can be split across them, much like mining pools.

Moreover, since the blockchain is primarily used as a trust mechanism for federated learning, the mining target difficulty can remain lower, reducing the computational cost; this additionally increases the rate of block addition to the chain which in turn reduces the time between each global network update. Therefore, lower-powered devices will have enough computing resources to generate hashes competitively, whilst still providing the same protection. As a result, we decided on a block rate of approximately one every 1.5 min. This rate is long enough for multiple local updates from different sources to be included in a block prior to the block being added to the chain, without being so long that the global update is outdated or a local device that misses out on the federated update grows stale.

PoS, on the other hand, would not be as suitable since it relies heavily on transactions, has no mining step, would give too much power to larger contributors, whose networks are likely to be less diverse. It also promotes coin hoarding, which negates the side benefit of blockchain rewards; this is what incentivizes institutions to utilize their spare computing power.

2.2.2. Mining

Mining of local updates via PoW requires storing the target in the block; however, the target size has the same number of bits as the hash. Therefore, much like bitcoin, we encode the target in 4 bytes:

$$\textbf{\textit{Target}} \overset{hex}{=} 0\text{x}\overbrace{\phi_1\phi_2}^{\Phi}\overbrace{\theta_1\theta_2\theta_3\theta_4\theta_5\theta_6}^{\Theta} \triangleq \Theta * 2^{8*(\Phi-4)} \tag{3}$$

where the first byte ($\phi_1\phi_2$) is an exponential scale and the lower three bytes contribute the linear scale. As with Bitcoin, we scale the exponential by 8, as there are 8 bits in a byte, simplifying the bit manipulation calculations. However, unlike Bitcoin, we scale the exponent by 4 (whereas Bitcoin scales by 3) in order to generate target values at the lower end of the spectrum since we use a lower block mining rate than Bitcoin.

2.2.3. Cryptography

We opted to use Keccak-256 for our cryptographic hashes instead of SHA256 and RIPEMD160, which Bitcoin uses. Keccak-256 is stronger compared to both and is used by Ethereum, which is a distributed state machine as opposed to a distributed ledger. Since we plan on adopting some of Ethereum's changes to the plain distributed ledger, such as smart contracts, keccak-256 is a more natural fit. However, we continue to use double hashing as Bitcoin does.

2.2.4. Networking

Using a peer-to-peer (P2P) network for our communication layer provides an essential benefit; whilst a single node will still be functional, the benefits of federated learning would be severely reduced. P2P networks are ideally suited to handle two vital situations: first, each node must be able to request a copy of the blockchain, including a list of addresses of other nodes which will receive the address of the connecting node and, secondly, the ability to broadcast information to all nodes accessing the chain, even those that may not be directly connected to this node.

By using a pair of UDP sockets (Algorithm 1), we can parallelize the communication and allow the communication to be distributed amongst different (local) devices. A hospital, for example, may have many IoMT devices but none with networking capabilities, just Bluetooth; they could therefore connect all IoMT devices to a single, network-capable IoT device, which would handle the networking and correct forwarding, much like Network Address Translation (NAT) with regards to WiFi routers. Consequently, any IoT device may participate, with the only requirement being a connection to a networking node, potentially via Bluetooth or even hardwired to a communication module, at some point downstream. Moreover, if a collection of IoT devices train as one unit, only one device needs to connect to the outbound UDP connection with all devices gaining the benefits.

2.3. *IoT Federated Learning*

All devices contain an instance of a, potentially pre-trained, TensorFlow lite model and train on incoming data, private to each participant. After a number of epochs, each participant may submit their network's weights and the number of different examples seen to the blockchain (Figure 2). The chain then contains the globally aggregated model that can then be used as the new initial network for the next round of local training.

Algorithm 1 UDP Pair Communication.

```
procedure INBOUND
    loop
        msg ← incomingMsg
        if ValidateIsNewestChain(msg) then
            chain ← msg
        else if msg ∉ addresses then
            addresses.append(msg)
            Outbound(Inbound.Address, msg)
        end if
    end loop
end procedure

procedure OUTBOUND(msg, addr = NULL)
    if msg == JoinNetwork then
        Broadcast(Join + Inbound.Address)
    else if addr == NULL then
        Broadcast(msg)
    else
        SendDirectMsg(msg, addr)
    end if
end procedure
```

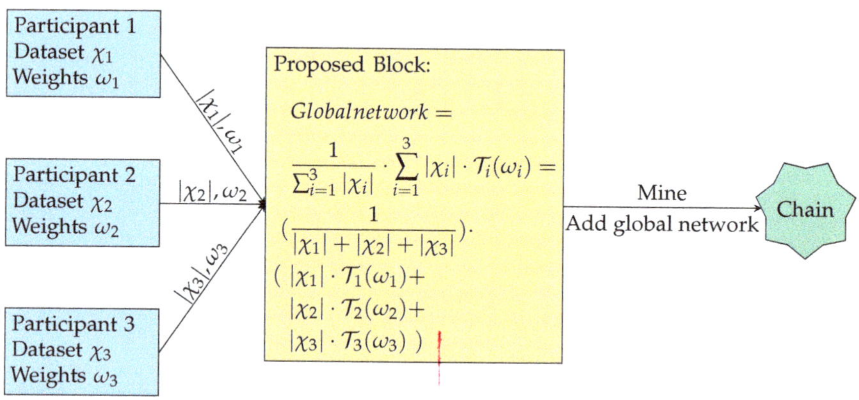

Figure 2. Example of three clients contributing to the blockchain. Each participant trains a local network for a set amount of epochs over their observed dataset χ_i resulting in the network's weight set ω_i. The number of examples seen $|\chi_i|$ and the weight set are passed to the proposed block to be aggregated into the global model using Equation (2). Once the proposed block is mined, it is added to the chain with the new global network in the block; this may be used by anyone with access to the chain and used for the next iteration of local training. Note $(x \cup y \cup z) \subseteq all\_possible\_data$.

3. System Complexity

In this section, we analyze the complexity and latency of the proposed system, including actual timings taken from the system running in debug mode on a physical Pixel 4 Android smartphone running Android 13 with a Qualcomm Snapdragon 855 Octa-core CPU (1 × 2.84 GHz Kryo 485 Gold Prime, 3 × 2.42 GHz Kryo 485 Gold and 4 × 1.78 GHz Kryo 485 Silver) to give real, practical values to fairly evaluate the system.

The first step of the system is the local training; using ω to denote the number of CPU cycles to execute one step of training (i.e., applying backpropagation to one input) then the complexity of an epoch of local training is defined as

$$\Theta_{local\_training} = \max_{\mathcal{D}_k \in \mathcal{D}} \left(\frac{\beta_{\mathcal{D}_k} \cdot \omega}{\Omega_{\mathcal{D}_k}} \right) \quad (4)$$

where $\beta_{\mathcal{D}_k}$ is the batch size of the input data used in each epoch, $\mathcal{D}_k$ is a participating device (from the set of all devices $\mathcal{D}$) and $\Omega_{\mathcal{D}_k}$ is the CPU frequency of the participating device $\mathcal{D}_k$. In our experiments (averaging over 144 epochs) the latency of $\Theta_{local\_training} = 23.68$ s.

Each participant then sends their model to the blockchain for validation, which consists of the PoW consensus algorithm followed by the aggregation of all models into a global model (using Equation (2)). Therefore the complexity of the global aggregation is as follows:

$$\Theta_{global\_aggregation} = \frac{\rho + \omega_{agg}}{\Omega_{\mathcal{V}}} \quad (5)$$

where ρ is the number of CPU cycles to execute the PoW algorithm and $\omega_{agg} = (N+1) \cdot \Xi + (N-1) \cdot \sigma$ is the number of CPU cycles to execute the aggregation of N models (where Ξ denotes scalar matrix multiplication and σ denotes matrix addition) and $\Omega_{\mathcal{V}}$ is the CPU frequency of the validator. In our experiments, consisting of four models, the latency of $\Theta_{global\_aggregation} = 106.83$ milliseconds which is a negligible cost owing to the large target for the PoW algorithm. However, in general, $\rho >> \omega_{agg}$ unless there are a large number of participating devices.

Running the entire system on our single device (sequentially, with no concurrency or optimizations applied) yielded a complete run time of 19.79 min consisting of 4 models training for 50 epochs each with global aggregation occurring every 25 epochs (i.e., twice in the complete run).

4. Results

To test our system, we used a simple convolutional network comprising two convolutional and max pooling layers, a final convolutional layer and two dense layers, to classify the CIFAR-10 dataset. Ordinarily, the model would be first pre-trained using TensorFlow before being converted to the TensorFlow lite format. However, we wished to train almost entirely on-device and so only ran for one epoch (against 100% of the CIFAR-10 training data) so as to have a starting point over a network with completely randomized weights; although we also pre-trained another instance of the network with 100 epochs to compare the effects of pre-training. The results of non-federated learning on these two networks are shown in Table 1 with the loss and accuracy after training further on the device (against only 25% of CIFAR-10's training data) for 50, 100 and 150 epochs. These results imply that the pre-training caused the model to overfit and is likely only useful if the data the device observes will differ from the pre-training data.

Table 1. Loss and accuracy of a neural network against CIFAR-10 test data trained and evaluated on an Android phone. Each network was pre-trained on a laptop for the specific number of epochs on 100% of the CIFAR-10 training dataset and then trained further on the Android phone for the specified number of epochs against 25% of the CIFAR-10 training data.

Pre-Trained Epochs	On-Device Trained Epochs	Final Loss	Final Accuracy
1	50	1.43	48.04%
	100	1.65	49.43%
	150	3.08	47.97%
100	50	1.73	48.02%
	100	3.43	46.94%
	150	4.42	46.67%

Next, we trained multiple instances of the single epoch pre-trained network via federated learning. Globally updating all participating networks after either 25 or 50 epochs and training for 50, 100 and 150 epochs, we tested the federated setting for 2, 4 and 8 participating networks. All participants were trained on an even split of 25% of CIFAR-10's training data such that no 2 networks saw the same example (an example of the case with 8 participating networks is shown in Figure 3. The results in Table 2 show that the accuracy is very similar to the non-federated context with the main difference within each configuration being loss increasing with more training despite the accuracy remaining virtually constant.

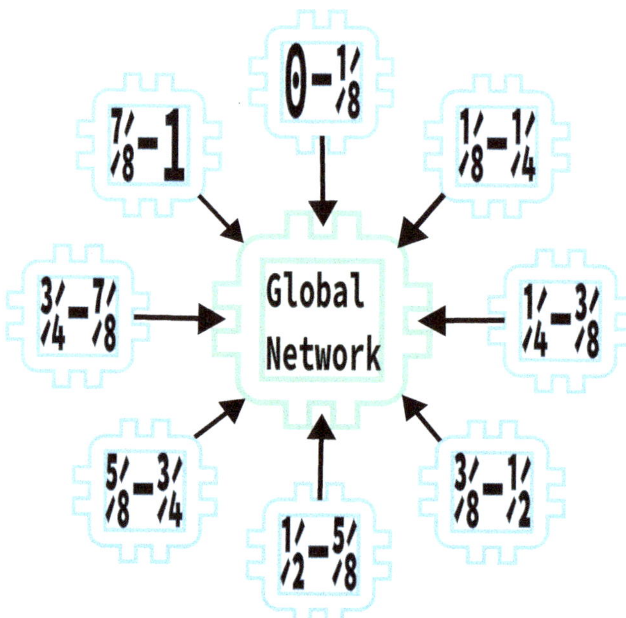

Figure 3. Federated Training of 8 models with an even split of the dataset. Given a dataset, χ, each local network, $i \in \{1 \ldots N\}$, trains on an even split of the dataset proportional to the number of networks such that for $\chi = \{\chi_1, \chi_2 \cdots \chi_k\}$, network i is trained on $\{\chi_{(i-1)k} \cdots \chi_{ik}\}$. These networks are then aggregated into a global network that performs as if it had been trained on χ despite not actually receiving any data, thereby preserving privacy.

Table 2. Loss and accuracy of a neural network against CIFAR-10 test data trained via federation and evaluated on an Android phone. Each participating network was trained on an even split of 25% of the CIFAR-10 training dataset with no participant seeing the same data.

Epochs per Global Update	Number of Participating Networks	Total Epochs	Final Loss	Final Accuracy
25	2	50	1.43	48.04%
		100	1.66	49.44%
		150	3.07	48.20%
50	2	50	1.43	48.03%
		100	1.66	49.19%
		150	3.05	47.93%

Table 2. *Cont.*

Epochs per Global Update	Number of Participating Networks	Total Epochs	Final Loss	Final Accuracy
25	4	50 100 150	1.43 1.64 3.06	48.04% 49.62% 48.12%
50	4	50 100 150	1.43 1.64 3.08	48.04% 49.65% 48.10%
25	8	50 100 150	1.43 1.65 3.08	48.04% 49.46% 48.28%
50	8	50 100 150	1.43 1.65 3.03	48.18% 49.33% 47.97%

5. Discussion

A new IoT-based, federated, decentralized and distributed learning approach is proposed with the aim of allowing learning on the edge by training multiple networks on multiple IoT devices and aggregating them into a global network that has seen no individual examples but is generalized for the specific task. We have produced a fully customizable Android application that allows on-device training of an arbitrary neural network via TensorFlow lite, allowing for existing networks to become federated and usable on an Android device in an extremely secure and privacy-enhancing manner.

However, there are a few components that we would like to address in future work: Smart contracts would be invaluable for automating tasks and sharing processing capabilities. Additionally, allowing the previous network iteration to supervise the next in a bootstrapping manner, forming a truly autonomous system. Finally, we wish to add homomorphic encryption to our system to further enhance the distributed manner of the system and to provide a novel, additional layer of privacy enhancement for both the machine learning models and the associated data.

Author Contributions: All authors contributed equally to this article. All authors have read and agreed to the published version of the manuscript.

Funding: Research was funded by the Engineering and Physical Sciences Research Council (EP/R513052/1).

Data Availability Statement: The data supporting this study's results can be found at https://github.com/SagaraBattousai/falcie/tree/0.1.0-AndroidResults (accessed on 3 April 2023) which is a git tag of the state of the project at the point it was used to generate our results.

Acknowledgments: We also acknowledge the invaluable inputs from Frank Po Wen Lo during this study.

Conflicts of Interest: The authors declare no conflict of interest.

Abbreviations

The following abbreviations are used in this manuscript:

MDPI	Multidisciplinary Digital Publishing Institute
IoT	Internet of things
IoMT	Internet of medical things
iid	Independent and identical distribution
P2P	Peer-to-peer
PoW	Proof of Work

References

1. Ishaq, M.; Afzal, M.H.; Tahir, S.; Ullah, K. A Compact Study of Recent Trends of Challenges and Opportunities in Integrating Internet of Things (IoT) and Cloud Computing. In Proceedings of the 2021 International Conference on Computing, Electronic and Electrical Engineering (ICE Cube), Quetta, Pakistan, 26–27 October 2021; pp. 1–4. [CrossRef]
2. Ghubaish, A.; Salman, T.; Zolanvari, M.; Unal, D.; Al-Ali, A.; Jain, R. Recent Advances in the Internet-of-Medical-Things (IoMT) Systems Security. *IEEE Internet Things J.* **2021**, *8*, 8707–8718. [CrossRef]
3. López Escobar, J.J.; Díaz Redondo, R.P.; Gil-Castiñeira, F. In-depth analysis and open challenges of Mist Computing. *J. Cloud Comput.* **2022**, *11*, 81. [CrossRef]
4. Chung, E.; Fowers, J.; Ovtcharov, K.; Papamichael, M.; Caulfield, A.; Massengill, T.; Liu, M.; Ghandi, M.; Lo, D.; Reinhardt, S.; et al. Serving DNNs in Real Time at Datacenter Scale with Project Brainwave. *IEEE Micro.* **2018**, *38*, 8–20. [CrossRef]
5. Fowers, J.; Ovtcharov, K.; Papamichael, M.; Massengill, T.; Liu, M.; Lo, D.; Alkalay, S.; Haselman, M.; Adams, L.; Ghandi, M.; et al. A Configurable Cloud-Scale DNN Processor for Real-Time AI. In Proceedings of the 2018 ACM/IEEE 45th Annual International Symposium on Computer Architecture (ISCA), Los Angeles, CA, USA, 1–6 June 2018.
6. Afroj, A.; Sahar, Q.; Naiyar, I.; Khalid, R. Fog, Edge and Pervasive Computing in Intelligent Internet of Things Driven Applications in Healthcare: Challenges, Limitations and Future Use. In *Fog, Edge, and Pervasive Computing in Intelligent IoT Driven Applications*; IEEE: Piscataway, NY, USA, 2021; pp. 1–26. [CrossRef]
7. Giang, N.K.; Lea, R.; Blackstock, M.; Leung, V.C.M. Fog at the Edge: Experiences Building an Edge Computing Platform. In Proceedings of the 2018 IEEE International Conference on Edge Computing (EDGE), San Francisco, CA, USA, 2–7 July 2018; pp. 9–16. [CrossRef]
8. Saeed, A.; Salim, F.D.; Ozcelebi, T.; Lukkien, J. Federated Self-Supervised Learning of Multisensor Representations for Embedded Intelligence. *IEEE Internet Things J.* **2021**, *8*, 1030–1040. [CrossRef]
9. Li, H.; Shou, G.; Hu, Y.; Guo, Z. Mobile Edge Computing: Progress and Challenges. In Proceedings of the 2016 4th IEEE International Conference on Mobile Cloud Computing, Services, and Engineering (MobileCloud), Oxford, UK, 29 March–1 April 2016; pp. 83–84. [CrossRef]
10. Abdellatif, A.A.; Mohamed, A.; Chiasserini, C. Automated class-based compression for real-time epileptic seizure detection. In Proceedings of the 2018 Wireless Telecommunications Symposium (WTS), Phoenix, AZ, USA, 17–20 April 2018; pp. 1–6. [CrossRef]
11. Emam, A.; Abdellatif, A.A.; Mohamed, A.; Harras, K.A. EdgeHealth: An Energy-Efficient Edge-based Remote mHealth Monitoring System. In Proceedings of the 2019 IEEE Wireless Communications and Networking Conference (WCNC), Marrakesh, Morocco, 15–18 April 2019; pp. 1–7. [CrossRef]
12. Chen, J.; Ran, X. Deep Learning with Edge Computing: A Review. *Proc. IEEE* **2019**, *107*, 1655–1674. [CrossRef]
13. TensorFlow For Mobile & IoT Overview. Available online: https://www.tensorflow.org/lite (accessed on 29 December 2020).
14. Labrèche, G.; Evans, D.; Marszk, D.; Mladenov, T.; Shiradhonkar, V.; Soto, T.; Zelenevskiy, V. OPS-SAT Spacecraft Autonomy with TensorFlow Lite, Unsupervised Learning, and Online Machine Learning. In Proceedings of the 2022 IEEE Aerospace Conference (AERO), Big Sky, MT, USA, 5–12 March 2022, pp. 1–17. [CrossRef]
15. Warden, P.; Situnayake, D. *TinyML*; O'Reilly Media, Inc.: Sebastopol, CA, USA, 2019; ISBN 978-1-492-05204-3
16. Rodriguez, A.; Li, W.; Dai, J.; Zhang, F.; Gong, J.; Yu, C. Intel Processors for Deep Learning Training. 2017. Available online: https://software.intel.com/content/www/us/en/develop/articles/intel-processors-for-deep-learning-training.html (accessed on 7 January 2021).
17. Hong, W.; Meng, J.; Yuan, J. Distributed Composite Quantization. In *Proceedings of the AAAI Conference on Artificial Intelligence*, New Orleans, LA, USA, 2–7 February 2018; AAAI Press: Palo Alto, CA, USA, 2018; Volume 32. [CrossRef]
18. Calo, J.; Lo, B. IoT Federated Blockchain Learning at the Edge. *arXiv* **2023**, arXiv:2304.03006 .
19. Krizhevsky, A.; Hinton, G. *Learning Multiple Layers of Features from Tiny Images*; Technical Report; University of Toronto: Toronto, ON, Canada, 2009. Available online: https://www.cs.toronto.edu/~kriz/learning-features-2009-TR.pdf (accessed on 25 February 2023).
20. Yang, Z.; Shi, Y.; Zhou, Y.; Wang, Z.; Yang, K. Trustworthy federated learning via blockchain. *IEEE Internet Things J.* **2022**, *10*, 92–109. [CrossRef]
21. Islam, A.; Amin, A.A.; Shin, S.Y. FBI: A Federated Learning-Based Blockchain-Embedded Data Accumulation Scheme Using Drones for Internet of Things. *IEEE Wirel. Commun. Lett.* **2022**, *11*, 972–976. [CrossRef]
22. McMahan, H.B.; Moore, E.; Ramage, D.; Hampson, S.; Arcas, B.A.y. Communication-Efficient Learning of Deep Networks from Decentralized Data. In Proceedings of the 20th International Conference on Artificial Intelligence and Statistics (AISTATS), Ft. Lauderdale, FL, USA, 20–22 April 2017; Volume 54.
23. Nair, P.R.; Dorai, D.R. Evaluation of Performance and Security of Proof of Work and Proof of Stake using Blockchain. In Proceedings of the 2021 Third International Conference on Intelligent Communication Technologies and Virtual Mobile Networks (ICICV), Tirunelveli, India, 4–6 February 2021; pp. 279–283. [CrossRef]

24. Chicaiza, S.A.Y.; Chafla, C.N.S.; Álvarez, L.F.E.; Matute, P.F.I.; Rodriguez, R.D. Analysis of information security in the PoW (Proof of Work) and PoS (Proof of Stake) blockchain protocols as an alternative for handling confidential information in the public finance Ecuadorian sector. In Proceedings of the 2021 16th Iberian Conference on Information Systems and Technologies (CISTI), Chaves, Portugal, 23–26 June 2021; pp. 1–5. [CrossRef]
25. Masseport, S.; Darties, B.; Giroudeau, R.; Lartigau, J. Proof of Experience: Empowering Proof of Work protocol with miner previous work. In Proceedings of the 2020 2nd Conference on Blockchain Research & Applications for Innovative Networks and Services (BRAINS), Paris, France, 28–30 September 2020; pp. 57–58. [CrossRef]

information

MDPI

Article

A Blockchain-Inspired Attribute-Based Zero-Trust Access Control Model for IoT

Samia Masood Awan [1], Muhammad Ajmal Azad [2,*], Junaid Arshad [2], Urooj Waheed [3] and Tahir Sharif [4]

1 Department of Computer Science, NED University of Engineering & Technology, Karachi 74200, Pakistan
2 School of Computing and Digital Technology, Birmingham City University, Birmingham B4 7BD, UK
3 Department of Computer Science, DHA Suffa University, Karachi 74200, Pakistan
4 College of Science and Engineering, University of Derby, Derby DE22 1GB, UK
* Correspondence: muhammadajmal.azad@bcu.ac.uk

Abstract: The connected or smart environment is the integration of smart devices (sensors, IoT devices, or actuator) into the Internet of Things (IoT) paradigm, in which a large number of devices are connected, monitoring the physical environment and processes and transmitting into the centralized database for advanced analytics and analysis. This integrated and connected setup allows greater levels of automation of smart systems than is possible with just the Internet. While delivering services to the different processes and application within connected smart systems, these IoT devices perform an impeccably large number of device-to-device communications that allow them to access the selected subsets of device information and data. The sensitive and private nature of these data renders the smart infrastructure vulnerable to copious attacks which threat agents exploit for cyberattacks which not only affect critical services but probably bring threat to people's lives. Hence, advanced measures need to be taken for securing smart environments, such as dynamic access control, advanced network screening, and monitoring behavioural anomalies. In this paper, we have discussed the essential cyberthreats and vulnerabilities in smart environments and proposed ZAIB (Zero-Trust and ABAC for IoT using Blockchain), a novel secure framework that monitors and facilitates device-to-device communications with different levels of access-controlled mechanisms based on environmental parameters and device behaviour. It is protected by zero-trust architecture and provides dynamic behavioural analysis of IoT devices by calculating device trust levels for each request. ZAIB enforces variable policies specifically generated for each scenario by using attribute-based access control (ABAC). We have used blockchain to ensure anonymous device and user registrations and immutable activity logs. All the attributes, trust level histories, and data generated by IoT devices are protected using IPFS. Finally, a security evaluation shows that ZAIB satisfies the needs of active defence and end-to-end security enforcement of data, users, and services involved in a smart grid network.

Keywords: smart cities; cyber security; Internet of Things; cyber-physical systems; zero-trust; ABAC; blockchain; IPFS

Citation: Awan, S.M.; Azad, M.A.; Arshad, J.; Waheed, U.; Sharif, T. A Blockchain-Inspired Attribute-Based Zero-Trust Access Control Model for IoT. *Information* **2023**, *14*, 129. https://doi.org/10.3390/info14020129

Academic Editors: Spyros Panagiotakis and Evangelos K. Markakis

Received: 12 January 2023
Revised: 9 February 2023
Accepted: 10 February 2023
Published: 16 February 2023

1. Introduction

The Industrial Revolution, benefiting from advancements within artificial intelligence, 5G, the Internet of Things, and blockchain technology has introduced a massive surge in technology inclusion, expansion, innovation, and research. Such paradigm shifts have highlighted the need for machine-to-machine and machine-to-human interactions where a huge amount of data transfer occurs during the process of communication devices setting up an Internet of Things network [1]. Alongside the need to transfer data at high speeds with low latency, the security of such systems is crucial due to applications dealing with sensitive user data or critical national infrastructure [2].

These types of massive communication handling require fool-proof security because whether the data come from home users or the data are being dealt with by any commercial company, such as smart industries or smart grid stations, a security breach can risk multiple human lives or the unavailability of resources offered by critical cyber-physical systems [2]. Smart grids have long been a crucial component of energy networks, incorporating a variety of instruments, such as IoT devices, sensors and linked gadgets, that monitor and analyse the physical processes. It has aided in the optimization of energy production, distribution, consumption, and storage. In 2007, the sophisticated attack on Iran's nuclear power plant disturbed the distribution and development of the country's nuclear energy resources [3,4]. On 25 December 2015, in the midst of a civil war, an electrical power grid in Ivano-Frankivsk was hit by the cyber attacker that left 80,000 people without electricity and affected many critical services [5]. Hence, security is the most crucial aspect of these cyber-physical systems nowadays. With the increase in the number of technologies coming out, the security risks associated with them are also increasing exponentially. It is impossible for security systems to achieve 100% efficiency, and even military-grade technologies are somewhat vulnerable when they are attached to the Internet [6–8].

Hackers have several ways to compromise systems if traditional boundary security measures are deployed. Detecting an intrusion in such a setup becomes increasingly challenging if an attacker successfully breaches that parameter layer of defence. In contrast to these trust-based systems, authentication provides a way to present the credentials that the user or machine is the legitimate user of the network. The traditional authentication and authorization system might not be directly deployed in the IoT network because of resource limitations and the dynamic nature of the network. The network requires a dynamic policy-based system that enforces policies in real-time considering the user's constraints as well as the dynamic nature of the network [9]. Within this setup, a zero-trust (ZT) model applies some kind of policy decision to authorise every action of a user or device. Every attempt to access data or resources is verified by the organisation, hence common modern attacks make it very difficult for intruders to impersonate or masquerade as an authenticated device or authorized user. ZT promotes a host-based monitoring approach where every host or owner device gets to set the criteria required to access it. Fine-grained data access control allows the host to dictate the intended audience and makes sure that the data cannot be accessed by any undesired user. Hence, ZT opens up new ways for security-enabled collaboration opportunities between organizations. On the implementation side, only regular updates of new technologies and methodologies with the pace of research and innovation, such as ZT, ABAC, and blockchain, can support a system to become less vulnerable against intruder attacks.

With the changing environment of networks and the new ways of communication among devices, a lot of effort is required to manage network security in real-time in a dynamic environment. The best practices of cybersecurity are becoming obsolete with the passage of time and new approaches are coming into the realization stage with every passing day. The issues associated with network security are no longer general, and the same policy and standards for each network cannot be replicated for every network. Therefore, aggressive encounter measures are required that not only support the network as a gatekeeper but also secure the system from malicious activities [10]. Access and authentication policies should be uniform at one end but must also be dynamic to reduce the vulnerabilities within the network in real-time. The Internet of Things deals with machine–machine and machine–human interactions over the Internet and blockchain is a distributed ledger primarily available for tamper-proof, hack-proof, and immutable recording of transactions into the ledger [11]. The combination of the IoT and blockchain-based networks somehow sorts out the problems associated with the domain. Similarly, access control generally implemented through MAC address, IP, and other tags is not sufficient. A modern and evolved approach is required to deal with network security [12]. In this paper, we have introduced a method to make security as efficient as possible compared to conventional ways. We proposed a system that ensures the security of IoT devices and users through the use of emerging concepts of zero-trust architecture, attribute-based access control, blockchain, and IPFS. The system will be used to sustain a network and communication as efficiently as possible to reduce real-time attacks through the implementation of real-time monitoring, dynamic policy generation mechanisms, and interminable monitoring of the various aspects of network security and communication.

Approach

A novel approach is required to deal with the above-mentioned challenges, the advanced technologies and techniques will play a vital role to get this job done. For example, the use of blockchain technology can secure the data and allow only recognized participants to join the network [13]. Similarly, a dynamic policy mechanism is required to create, authenticate, and recognize participants (IoT devices) in the non-trusted environment [14]. Each node/participant must authenticate first, before interacting with the system or the participants associated with the network. The non-trusted environment will reduce the chances of hackers exploiting the network by masquerading, man-in-the-middle attacks, and brute force attacks, which are the most commonly used techniques to attack the network [15]. The creation of blockchain wallets for each participant (each IoT device) at the time of registration of the new device will help to recognize and record the device details better in an automated way using smart contract technology, and IPFS secures all the information about the devices and the data generated by these IoT devices for any further pre-processing [16]. The ZTA is associated in such a way that the entire system has multiple divisions, and each division has its categories and priorities. Each division may be called a "Zone", and must have a PEP that enforces all policies within the network zone, and also routes all device communication requests through to the PDP where decisions are made to accept or reject the request, and when accepted it will create encrypted channels to entertain the interactions [17]. These PDPs are further connected to the PE that generates dynamic access policies. Due to the nature of IoT, devices are multivariant; therefore, the policy-making must be designed as per the variable attributes of the object and the subject of a certain request, hence the attributed-based access control should be introduced on top of ZTA [15].

The rest of the paper is organized as follows. Section 2 discusses the background of zero-trust architecture and the challenges for designing such architecture for IoT networks. Section 3 discusses related work in this domain. Section 4 provides the system architecture of zero-trust-based access control. Section 5 defines policies and Section 6 defines device attributes and management. Section 8 evaluates the approach and Section 9 concludes the paper.

2. Background

Zero-trust architecture (ZTA) is a practical implementation of the concept; trust is nothing but a vulnerability when it comes to network security [18,19]. The notion of zero trust is centred on network segmentation into "Microcores and Perimeters" (MCAP). Instead of having a trusted domain built by a perimeter around the network, ZT suggests everything, and everyone is untrusted even within the network perimeters [17]. Hence, it promotes the "never trust and always verify" principle even within the enterprise network.

ZTA divides the entire network into microcores and sets perimeters around each core. ZTA implementation includes a policy engine (PE) which generates access control policies, a policy administrator (PA) that evaluates a request to access an enterprise resource by applying the policies generated by PE, and a policy enforcement point (PEP) that enforces these policies by accepting or rejecting the received request as per the decision made by the PA. The core components of ZTA are shown in Figure 1.

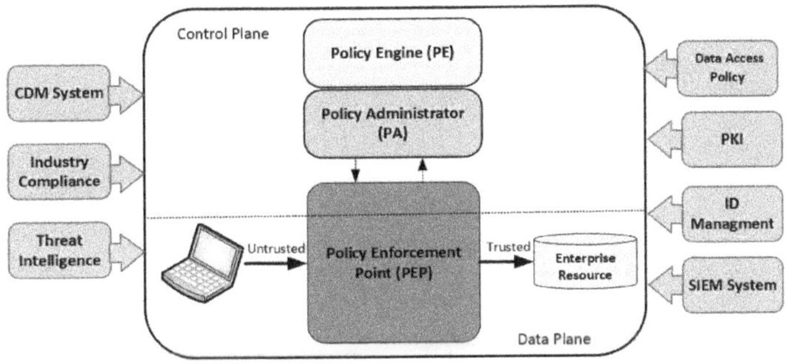

Figure 1. Core zero trust logical components.

Setting perimeter-based security is neither efficient nor possible for IoT networks as they are distributed in nature. Therefore, ZT is the perfect solution for all IoT security problems [1]. ZT provides a complete scheme for guaranteeing users access across amalgam infrastructures and networks through smartphones, computers, cloud applications, and other IoT devices [11].

2.1. Challenges

Network security, which deals with the high-volume data traffic from multiple sources, such as millions of IoT devices, is the top priority for any enterprise. The conventional ways of securing networks, such as firewalls, access control, etc., are becoming obsolete day by day and new innovative methods are required to deal with network topology, IoT device management, and overall the entire process of end-to-end communication, data storage devices, and data management [6]. The challenges presented in IoT networks are real-time monitoring, data handling, data storage, access management, trust-based or trust-less criteria to deal with the participant nodes and their behaviour, accessibility, roles, modes, and parameters and factors compromising security in real-time monitoring and management [2].

Some challenges that need to be addressed to enhance system security are as follows:

- **Authentication** refers to the verification of user credentials. Robust authentication mechanisms are required to identify valid users from ill-intended impersonators who try to gain illegal access to IoT devices and their data. Therefore, all the users and IoT devices should be registered, and their baseline behaviours need to be analyzed for instant detection of any behaviour anomalies such as impersonation or masquerade attacks due to the illicit use of valid user credentials.
- **Authorization** is another key aspect of IoT networks. The device owners have comprehensive rights and complete authority over the data generated by it, hence the type of access (read/write) to any device and its data can be granted or revoked based on the criteria set by the owner. Successful implementation of generic access control policies that evolve dynamically based on the current scenarios is one of the key challenges for huge IoT infrastructures.
- **Confidentiality** can be defined as the protection of system resources against unauthorized access. The degree of authorization required to access devices and data in an IoT network needs to be set intelligently in order to maintain the confidentiality of the classified information. Smart cities have every aspect of human life being connected and controlled with IoT devices. A data breach may result in life-threatening situations as sensitive information, such as the daily schedules or healthcare records of citizens, needs to remain concealed for their safety and well-being.
- **Privacy** means having full control or decision-making authority on how the user's data can be collected and used. Users have the right to protect their personal information, such as their daily schedules and medical and financial records, from being revealed

without their consent. Hence upholding the privacy of data generated by handheld gadgets, surveillance devices, or home IoT networks is one of the most important goals for any smart city infrastructure. The proposed framework needs to address all of these challenges and provide efficient solutions for them.

2.2. Attribute-Based Access Control

In attribute-based access control (ABAC), any access request is approved based on the attributes of the subject and the object. The identities, roles, functions, and other complex features of a subject are all posited as its attributes [17]. The attributes of the requester (subject) and the attributes of the requested (object) are both combined to form an access control policy to fulfil the security demand of the object's owner. Specified access policies are set by the object owners to determine whether the subject has appropriate privileges for the requested access based on these attributes [20]. Figure 2 shows the policy creation logic and components of ABAC.

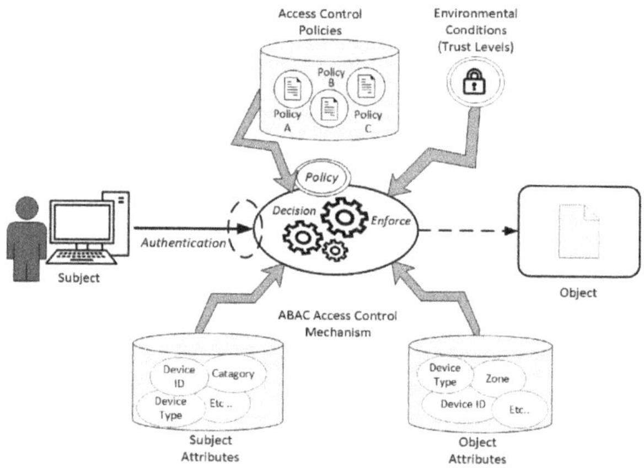

Figure 2. ABAC logical components.

Traditional access control models, such as discretionary access control (DAC), mandatory access control (MAC), and role-based access control (RBAC), are completely centralized. Based on the distributed, decentralized, and dynamic architectures of IoT networks, attribute-based access control (ABAC) is regarded as the most suitable approach for IoT scenarios. ABAC provides the strong dynamicity, scalability, and flexibility required for IoT environments to implement fine-grained control over the access requests for every device [21].

2.3. Interplanetary File System

Another vital component of our proposed system is the InterPlanetary File System (IPFS). It is a file system where data are stored in the form of uniquely identifiable blocks by multiple peers following a distributed approach [22]. The IPFS maintains all versions of a file as separate individual blocks and cryptographic hashes are assigned to each block as a unique identifier, which means no two different blocks in the system can have the same cryptographic hash. While searching for some content, it is located and accessed by its assigned hash value [16]. In a huge IoT infrastructure with millions of devices, the amount of data generated by these devices cannot be stored on-chain and large off-chain storage reserves are required. IPFS provides a distributed and secure solution to all storage issues as it has such a massive data storage capacity [23].

3. Related Works

With the vastly used cloud applications and IoT networks, traditional network security approaches such as building a wall between trusted and untrusted devices and the trusted local networks do not work anymore. The need for secure and smart access control where no trusted networks or devices exist has been fulfilled by ZTA. Various variants of ZT have been proposed and implemented by researchers to satisfy their network's unique security demands. In [24], Pedro Assunacao discussed a ZTA that eliminates static credentials, applies multi-factor authentication, and maintains a log of all devices and network traffic.

In [25], authors suggested context-based ZT access control to overcome the challenge of a secure and heterogeneous Moodle application. This framework is an application of a model that provides access control to an e-Learning platform called Moodle. Through the implementation of the zero-trust model, a positive performance on the webserver has been seen. However, to evaluate the non-functional performance of the zero-trust model, additional tests need to be carried out.

With the scalability of IoT networks, there is a need for trust management controls that could safeguard the systems against malicious attacks. To overcome this, a centralized validation mechanism is required. The authors in [11] have presented an IoT-based zero-trust model which enables a novel hierarchical mining concept. According to the authors, IoT infrastructure is a zero-trust model, and cannot be trusted. To overcome this, a blockchain-based middleware, Amatista, has been introduced. The mining platform integrates distributed validation authorities for the IoT at different levels of trust. Firstly, context-based mining has been introduced. Secondly, publish/subscribe provisioning of data has been introduced. Amatista has been evaluated using IoT sensors and edge networks, and it has been concluded that the system can handle not only the infrastructure but the transactions as well.

In [19], authors have suggested a similar mechanism in the context of smart cities. The authors have applied the network classification to extend the idea of zero trust. As per ZTA, the framework divides the system into separate MCAPs for web access, mobile access, and database, and a blockchain node is attached to each MCAP for request authentication. The IoTs are also divided into eight different categories based on their risk analysis calculated on three factors, i.e network capability, risk score, and data risk. All the MCAPs are connected by a segmented gateway. Every time an access request is generated, the blockchain node attached to the targeted MCAP verifies it using a smart contract, and access is granted if the request is considered genuine and is verified. Although the system has no implementation in the real-life IoT network, the authors claim to address multiple security concerns including the risk-based MCAPs for IoT devices. However, the IoT data transferred over the network are not secured.

Chen and Qiao in [1] implemented a smart healthcare system for 5G networks where they divided their entire system into four dimensions, i.e., object, subject, environment, and behaviour. Fine-grained access policies are being defined by using machine learning and deep learning that use real-time threat and trust levels generated for all IoMT devices and users. In Fabric-IoT [21], the authors have implemented hyper-ledger Fabric-based access control for IoT using smart gateways where every IoT device sends its encrypted data to a URL. The framework comprises three smart contracts: one creates, modifies, and deletes policies, the second one assigns URLs to IoT devices, and the third one enforces access control. Nevertheless, to achieve better performance, the system's scalability needs to be improved.

In [26], the researchers implemented attribute-based access control on IoT sensor data by using a rule-based proactive engine which helps to generate new rules and policies, monitors the environment, and helps the PDP decide on what to do in case of any sudden changes by creating a behaviour baseline saving all the previous transactions in the PIP database. However, details on how these policies will be implemented need further description. In [27], a secure IoT system using ABAC is implemented by IPFS and smart contracts. All transactions generated by IoT devices along with all the policies are saved in the IPFS database in the form of hashed blocks. When a user sends a request to access any IoT device's data, the PDP requests attributes from PIP and policy rules from PAP, matches the two, and decides whether to grant this access or deny it.

In [28], to preserve the privacy of e-health data authors have proposed using blockchain with non-interactive zero-knowledge proof-based key authentication to manage the device authentication process for millions of medical things joining the network. While the process ensures that no unauthorized device joins the e-health network, it does not deploy any dynamic mechanism to identify intrusions caused by device compromise after they have completed their authentication process.

The authors in [29,30] have presented a blockchain-based data provenance system which uses the Merkel chain blockchain property to maintain a chain of custody for data. While the system maintains complete data access logs, it does not provide any trust management to handle the dynamics of machine-to-machine communication. Analysing a baseline machine behaviour can help to revoke access whenever a machine misbehaves. Through a complete analysis of related works as mentioned in Table 1, we discovered that most of the already proposed architectures either lack proper dynamic policy generation, which allows the systems to automatically create new policies for previously unseen situations, or they lack a completely decentralized architecture where any IoT device can get its request instantly processed by any available node without any delay or a centralized authority being involved in the process [31]. Even if we use secure and anonymous device authentication using zero-knowledge proofs, it does not guarantee that the device will remain uncompromised [32]. In this paper, we have proposed an architecture that is truly decentralized, completely dynamic, and will process every request based on the current environment and the behaviour of the IoT device instead of old-school role-based or identity-based authentication for processing access requests. The Internet of Things (IoT) has introduced a smart lifestyle. Home appliances, power plants, vehicles, and healthcare, we are aiming to automate the entire metropolitan infrastructure [33]. With all our systems connected with the IoT, the most crucial concern is if we can trust the current IoT authentication and access control infrastructure after connecting it to millions of "things" [34]. The current systems are not trustworthy enough to have our lives depend on them. The following design is inspired by ZTA and helps to curtail the mistrust of the current systems.

Table 1. Analysis of existing approaches to zero trust for IoT.

Paper ID	IoT Domain	Utilized Techniques	Contribution	Limitations
[1]	IoT in Healthcare	5G, Zero Trust, Attribute-based Access Control	The system uses trust assessment and risk level of objects to dynamically grant access based on attributes and performs traffic monitoring, load matching, access control, and auditing by using ML and DL.	Specific to healthcare scenario and therefore focused on access to resources rather than communication requests.
[11]	Hierarchical Management in IoT	Blockchain	Introduced a novel hierarchical mining the concept of using twotier miners for contextbased validation.	Other hierarchies of validation should be introduced as consensus on twotiers is time expensive. Authors should also include more specialized smart contracts for IoT.
[19]	Securing IoT devices using Zero Trust and blockchain	Zero Trust, Blockchain	The proposed framework divides the system into separate MCAPs and it uses risk factors to categorize IoT devices into different zones. Blockchain nodes are attached to each MCAP for request authentication.	All data is to be stored in blockchain where transaction per second rate is very slow and the management server is a single centralized server that defies the decentralized nature of the proposed model.
[21]	Fabric-IoT: A blockchain-based Access Control System in IoT	Hyper-Ledger Fabric, ABAC	Using smart gateways, Fabric IoT uses a hyper-ledger-based approach to implement ABAC.	Scalability is the biggest limitation for fabric-IoT along with minimal support for IoT application integration.

Table 1. *Cont.*

Paper ID	IoT Domain	Utilized Techniques	Contribution	Limitations
[25]	Context-based Access Control and Trust Scores in Zero Trust Campus Networks	Zero Trust	Secures the Moodle Application for university-wide open and heterogeneous research network using zero trust	It lacks policies for the policy engine as well as for trust score metrics.
[35]	FairAccess: a new Blockchain-based access control framework for the Internet of Things	Bitcoin-based Blockchain and OrBAC	Secures the IoT devices by using identity-based and permission-based access control policies	The approach does not analyse the dynamic IoT device behaviours and hence is not ideal for evolving scenarios of machine-to-machine communications.
[36]	CES Blocks—A Novel Attribute-Based Access Control Scheme Using Blockchain for IoT	Consortium Blockchain, ABAC	Secures the IoT devices using ABAC and records all attributes and requests as blockchain transactions by using a simple hash and signature protocol	It lacks creation of new policies along with calculation for device trust scores.
[26]	Context-aware and Attribute-based Access Control Applying Proactive Computing to IoT System	ABAC access control, rule-based proactive engine	Implemented ABAC on IoT sensor data using rule-based Proactive Engine which helps to generate new rules and policies, monitors the environment, and helps the PDP to decide what to do in case of any sudden changes by creating a behaviour baseline, saving all the previous transactions in the PIP database.	The paper only discusses the data received from IoT sensors and actions to be initiated based on this data but does not mention how users' access requests will be entertained.
[27]	IoT architecture based on ABAC smart contract	ABAC, IPFS	A secure IoT system using ABAC is implemented by using IPFS and smart contracts. All transactions generated by IoT devices are saved in the IPFS database along with all the policies in the form of hashed blocks based on which it is decided whether to grant access or deny it.	Static ABAC policies do not consider environmental or behavioural attributes while granting any access control request.
[37]	Securing Home IoT Environments with Attribute-Based Access Control	ABAC access control, NIST NGAC	The proposed framework suggests securing IT devices by using ABAC policies by defining attributes for Subject, Object, and Network.	Uses a set of predefined policies and no new smart dynamic policies can be made by the system at run-time to counter a new undefined scenario.
[38]	BlockShare: A Blockchain-Empowered System for Privacy-Preserving Verifiable Data Sharing	Blockchain, Zero-Knowledge Proof	Uses a newly defined data structure to store all e-health records for sharing.	While the approach emphasises anonymous data sharing, it does not consider access control and hence is not suitable for D2D communication.
[28]	Preserving Privacy in Mobile Health Systems Using Non-Interactive Zero-Knowledge Proof	Blockchain, ABE, Non-Interactive Zero-Knowledge Proof, IPFS	An authentication scheme that is lightweight enough to run on e-Health devices with minimal resources to provide a secure device authentication mechanism.	It does not detect a compromised device once it has completed the secure device authentication process.

4. System Architecture

The framework contains a consortium blockchain network and a network of IoT devices whose attributes serve as policy components for policy creation and implementation processes. A block architecture of the proposed ZAIB system is shown in Figure 3.

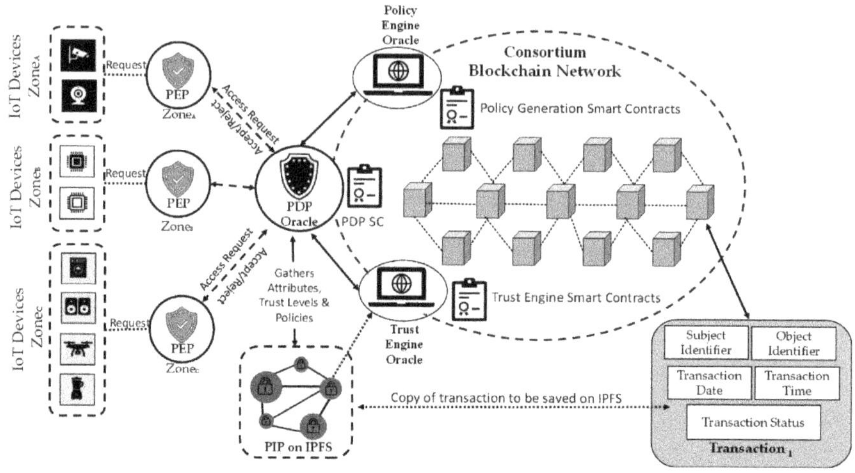

Figure 3. Architecture of the ZAIB framework.

A blockchain component is added to ZAIB for facilitating different IoT devices to communicate freely, securely, and anonymously on the network. To ensure device and data security, ABAC access control mechanisms are being implemented through smart contracts for device communication management. The policy engine (PE) oracle receives requests to make new policies and triggers the policy engine smart contract (PESC) to make new access policies for ABAC. To save IoT device attributes for ABAC, the data generated by them, all the policies generated by the policy engine (PE), and trust-level histories for device behaviour analysis, we are using IPFS. IPFS stores large-sized files easily, hence small-block-size issues are resolved. IPFS provides secure storage due to automatic resource mapping and hashed data; it is also connected with smart contracts, hence the authenticity of all information stored on the IPFS can be checked by comparing it with the transactions stored on the blockchain ledger. To implement zero-trust architecture, a trust engine oracle triggers the trust calculation smart contract which calculates the trust level of different devices considering several factors from their behaviour history stored in the ledger. Lastly, PDP smart contracts approve or reject IoT device requests for device-to-device communication.

4.1. Device Registration on Blockchain

Blockchain is a vital component of the system as it provides anonymous and secure D2D communication using smart contracts and its immutable distributed ledger [39]. Cryptographic key pairs provide the security feature in blockchain wallets. On registration of every new IoT device, an account is assigned to it to call contracts or initiate transactions. The system architecture of registration of new IoT devices is shown in Figure 4. The device's authentication and transaction anonymity are ensured due to the blockchain wallet [40]. The PBFT consensus algorithm is selected, as the frequency of requests is very high, and consensus needs to be reached very quickly. All the device attributes are saved in PIP which is implemented as an IPFS storage, and device management smart contracts are installed on the device.

Whenever an IoT device generates a communication request, the PEP acts as a gateway to pass it along to the PDP oracle, which triggers the PDP smart contract, hence recording this request as a transaction on the ledger. The PDP smart contract checks if any policies regarding such a request exist and decides to accept or reject the request based on these policies and the decision is recorded as a transaction. If it is found that such a request does not exist, it generates a request to create a new policy and triggers the PESC; this transaction is also recorded on the chain. Whenever a request is processed, the trust level smart contract is triggered and the new trust value for the IoT devices is saved both as a transaction on the chain and also in the PIP. A hashed link of all the data, trust levels, and policies stored in a block on the IPFS-based PIP is also stored in the chain which later on can be used for data validation.

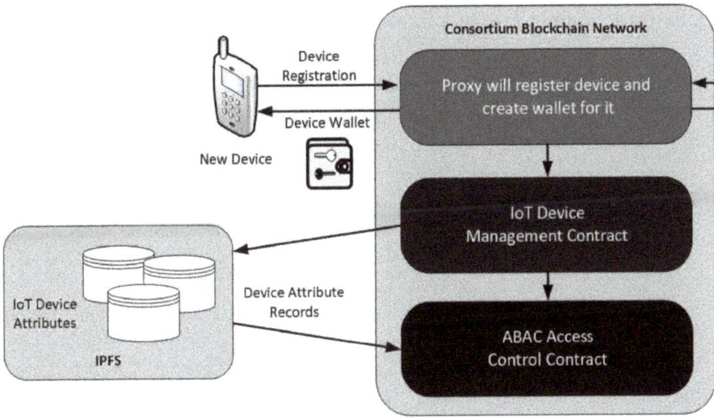

Figure 4. New IoT device registrations.

4.2. Hashed Storage of IoT Data Using IPFS

In our proposed system, the IPFS is responsible for storing the attributes of all connected IoT devices, smart contracts, access policies, trust level history for all the connected devices, and the data generated by our IoT devices in a very secure manner. Data generated by IoT devices, even the audio, video, and images can be encrypted and stored in blocks [30]. At any instance, the authenticity of the policies or trust levels stored in the IPFS can be checked by comparing the IPFS hashed blocks with the on-chain transaction to ensure that data or policies on IPFS have never been tampered with or corrupted. Figure 5 presents how data is stored in our blockchain system.

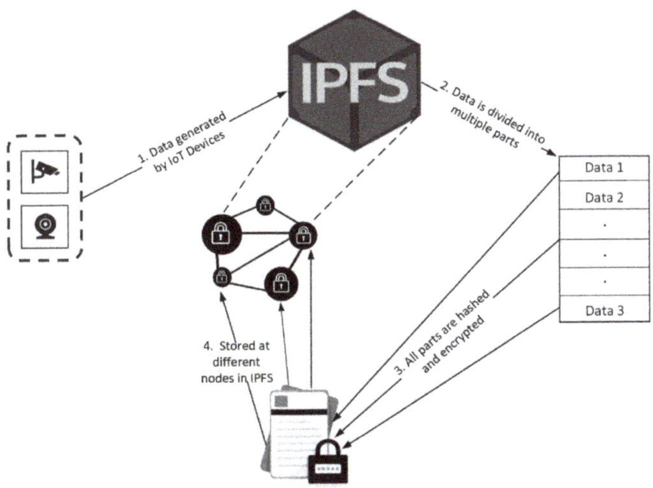

Figure 5. Storage of IoT data.

4.3. Zero-Trust Architecture

Since there are no trusted devices, trusted systems, or trusted users, all access and device-to-device communication requests need to be monitored and granted only when they are tested as valid access requests [15]. The integration of ZT in the IoT network and all its connected devices to provide complete and utmost security requires the key features of ZTA, such as micro-core, perimeters, and trust calculations, to be added to the infrastructure [17].

4.3.1. Zone Division

To implement ZTA, our entire IoT network is divided into different micro-cores, called "Zones". On any network, IoT devices can be categorized into zones based on their physical location, device categories, and priorities [41]. For example, if ZT is applied to a smart home, various similar category devices can be grouped to form different zones, for example, all kitchen appliances, which may include microwaves, refrigerators, coffee makers, juicers, blenders, etc., can be assigned a separate zone. Similarly, a home surveillance and security devices zone may contain all the cameras, smoke detectors, and smart locks. When applied to a huge smart edifice such as a smart city, separate zones can be identified in every division of the metropolitan infrastructure [42].

Each zone has its own policy enforcement point (PEP) that receives every communication request from all the devices and routes it to the connected policy decision point (PDP) which makes all policy decisions and accepts or rejects them based on the policies defined by the policy engine (PE). If a request is accepted, the PEP creates an encrypted channel to facilitate IoT device interactions.

4.3.2. Policy Enforcement

Each policy decision point oracle (PDPO) has multiple PEPs attached to it. To make decisions for all requests submitted by the PEPs, the PDPO reads the policies and device attributes from the PIP and the current trust levels of each device from the TE. It makes continuous dynamic decisions to accept or reject the request by putting the acquired data depicting the run-time status of the system, the network, and its involved devices into the policy. If no suitable policy is found to review the present request, it demands PEO to generate a new policy for the current scenario as shown in Figure 6.

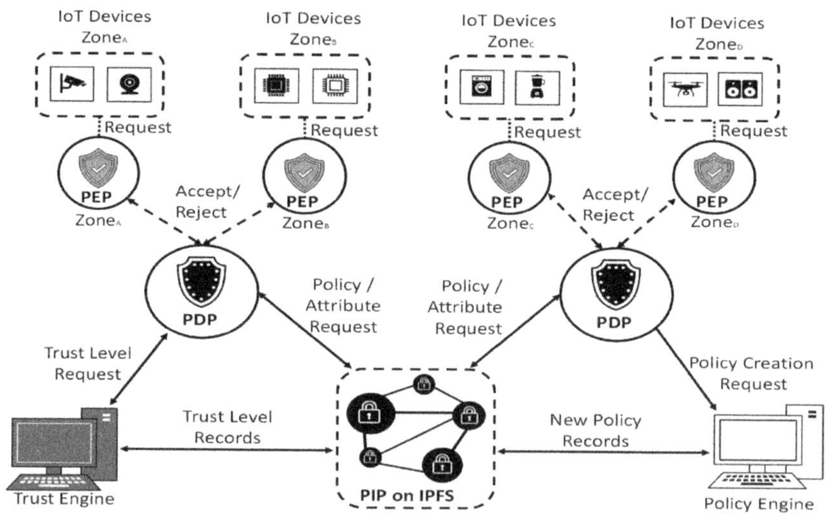

Figure 6. Zero-trust architecture for IoT devices.

The policy engine (PEO) generates new policies when the PESC is triggered, these policies are based on the set of policy frameworks provided by the network administrator dynamically. A detailed description of the policy creation process of these policies and the basic guidelines to be followed during this policy generation is shown in Section 5.

4.4. Trust Engine

One very important component of the system is the trust engine (TE) that calculates trust levels for IoT devices on the network [25]. The TEO is connected to the PDPO to provide updated trust levels of subject (S) and object (O) IoT devices for policy evaluation as shown in Figure 7.

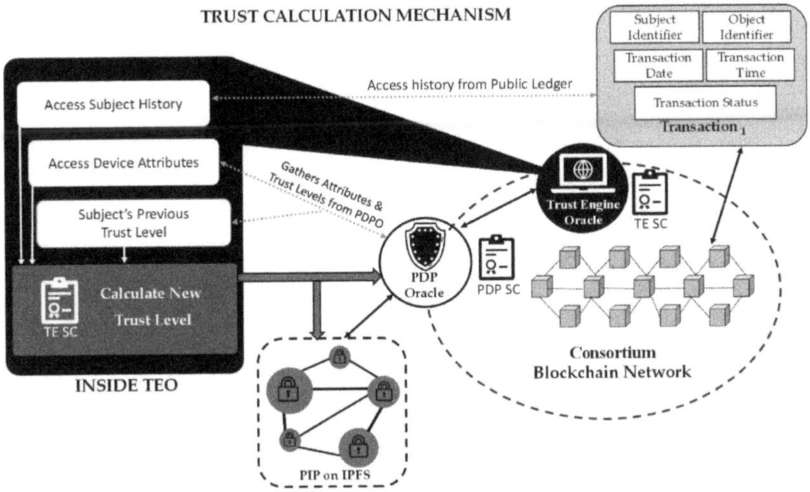

Figure 7. Trust calculation mechanism.

Possible Inputs for Trust Score Calculations

The most important feature for trust calculation is behaviour analysis. This behaviour analysis is carried out by the TESC by accessing the request history of devices from the PIP [43]. The device access history helps to determine the baseline behaviour for each device which is then saved periodically in IPFS storage [1]. A new trust score is generated for each device by comparing its current behaviour with its baseline behaviour [25]. A drastic change in behaviour results in a decrease in its trust level, whereas a persistent behaviour increases the trust level of the device, as shown in Figure 8.

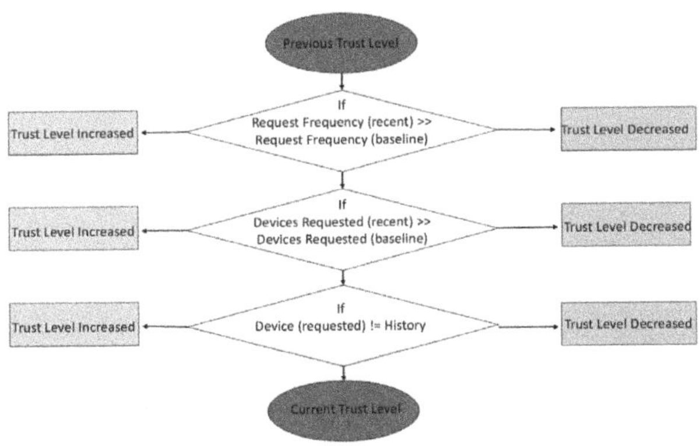

Figure 8. Suggested trust level calculations.

4.5. Access Control Model for Device-to-Device Communication

The decision-making policies for granting or denying a certain device-to-device communication request are generated using the ABAC model based on the features discussed below.

4.5.1. Attributes of IoT Network

Due to the multivariate nature of possible IoT interactions, attributes play a vital role in the decision-making process. Policies are designed based on these variable attributes of both the object and subject of a certain interaction request [26]. This section defines and describes the attributes for object and subject IoTs, the basic entities for ABAC. The next phases define how these attributes inhabit the policy information point (PIP) and how the policy engine (PE) uses these attributes to create new policies.

Every ABAC request must have a subject (S) that initiates the request, an object (O) that is the device the subject wants to commence communication with, a nature (N) which represents the nature of communication, and an environment (E) which represents the network at the time of generation of the request [37]. Based on these factors, the ABAC request format can be represented as

$$\textbf{\textit{IoT Access Request}} = < Subject(S), Object(O), AccessType(A), Environment(E) >$$

A received request can be allowed or denied based on the ABAC policies for the combination of attributes of the subject (S), object (O), the access type (A) of the communication request, and the environment (E). We described various key attributes that can be examined by ABAC policies.

Device Attributes: An IoT device can be a subject if it initiates the data request from another device or it wants to transmit data to another device. The object is the device the subject wants to communicate with [44]. Conventional device attributes are

- $DeviceIdentifier$: =The unique blockchain wallet id assigned to every IoT device.
- $DeviceType$: =The devices can be categorized into different types, such as smart TV, cameras, drones, sensory devices, and smart vehicles.
- $DeviceAge$: =The number of days since the device was first registered on the IoT network.
- $DevicePriority$: =Devices may be assigned different priority levels depending on the sensitive nature of their data and the security clearance level required to access them.
- $DeviceTrustLevel$: =Device trust level is to be calculated by the TESC based on the device's previous behaviour and its request pattern on the IoT Network.
- $DeviceCategory$: =Devices can be categorized as entertainment, healthcare, controllers, surveillance, monitoring, diagnostic, etc. A certain category of devices can be allowed to communicate with each other or with devices from other categories.
- $DeviceZone$: =Every device is assigned a zone or group once it registers on the network. Before a certain age and trust level is achieved, a device can only communicate with the devices in its zone.
- $DeviceLocation$: =The physical location of the device can also be stored as some of the policies can depend on the proximity of the devices.
- $DeviceStatus$: =Once a write connection is established with a monitoring device, the object device enters a locked status for all other write requests.
- $Networkidentifier$: =For some devices that might need the network identifier, this attribute will be a combination of sub-fields such as IP address and subnet mask.

Subject (S) is an IoT device requesting to initiate communication with an IoT device object (O), both IoT devices will have all these attributes assigned to them. Based on the values assigned to these attributes of subject (S) and object (O), different policies will be generated to accept or reject a generated communication request.

Access Type (A): The operations permitted on an IoT device are accessing the data from the devices or sending new control messages to the devices. The sending of new control messages can also be called a "Write" operation while accessing the device data can be termed as a "read" operation [45]. The nature of access demanded by a user is mentioned in this access type (A) field. The attributes for Access_Type are:

1. read: =data size
2. read_all: =data size

3. write: =size of message
4. write_all: =size of message

Environment (E): The environment for any communication is the network itself; these are the external parameters for both the subject (S) and the object (O). An example of environmental attributes that need to be recorded are date and network time, assuming that the standard synchronization policies such as network time protocol (NTP) are in place [44].

4.5.2. Attribute-Based Access Control Policy Model

Numerous kinds of data can be generated by IoT devices. For example, cameras capture videos, a microphone captures sound, and sensors capture humidity, temperature, and light. All of these features are very important and if someone with malicious intent gets access, it might even end up putting lives at risk [21]. To ensure that only trusted devices can communicate with IoT devices on the network, ABAC access policies are implemented. The establishment of a connection between two devices is shown in Figure 9.

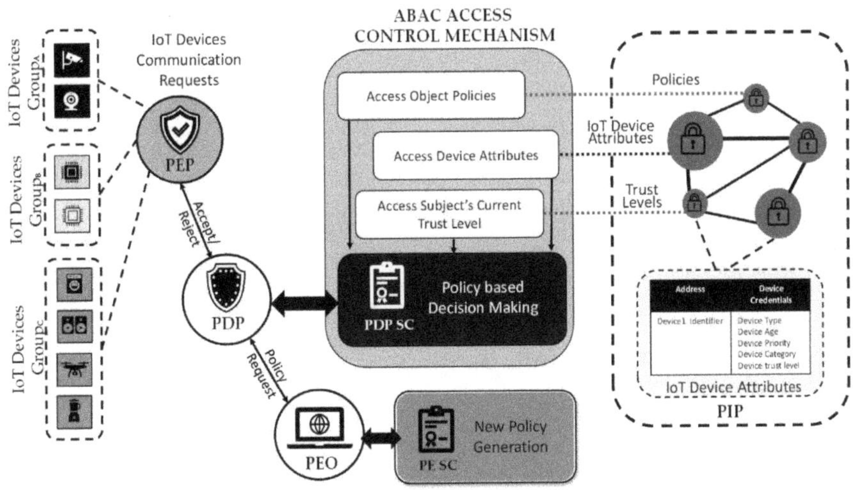

Figure 9. ABAC access policy implementation in ZAIB.

A brief description is also shown in steps 1 to 5.

1. Subject (S) requests to initiate communication with object (O).
2. The request is received by the smart gateway.
3. The request is forwarded to PDP.
4. PDP requests PIP for attributes of both subject (S) and object (O).
5. Based on device type, category, priority, and current trust levels (provided by the trust engine), the policy engine decides to accept/reject the request.
6. The PDP enforces the decision made by PE and, if access is granted, establishes a secure encrypted channel for safe D2D communication.

Based on the steps discussed previously, the ABAC device access policies can be defined as mentioned in Algorithm 1:

Algorithm 1 An algorithm for policy

Require: $\boldsymbol{Policy} = Subject_{attributes}, Object_{attributes},$
Require: $Subject_{attributes}$ = Device Identifier, Device Type, Device Age, Device Priority, Device Category, Device Zone.
Require: $Object_{attributes}$ = Device Identifier, Device Type, Device Status, Device Priority, Device Category, Device Zone.
Require: $Environment_{attributes}$ = Date, Time
Require: $TrustLevels$ = Subject Trust Level, Object Trust Level, Network Trust Level
  if $Permission == 0$ **then**
    $AccessGranted$
  else if $Permission == 0$ **then**
    $AccessDenied$

5. Policy Creation Framework

To generate automated policies, some ground rules have been set that help the PE in its policy-making activities. The policies required for the efficient and secure operation of an IoT infrastructure can be characterized as follows:

5.1. Device Acceptance Policies

Whenever a new device joins the network, it needs to be characterized and the current devices on the network need to be protected from it until it is verified as a trusted and safe device [46]. These policies should be independent of the specific device features for them to be applied to all kinds of devices. After running the diagnostics for the security state of the device and registration of the device, it will take some time to communicate with a few devices slowly and gradually as it ages and the baseline behaviour remains consistent to build up the trust level, only then can the new device request to communicate with highly trusted devices. To secure the network against a new device, some generic policies should be defined. A few sample protective policies for limiting access are provided in this section.

Sample Policy 1: *A new IoT device cannot request communications with more than a certain number of devices in a specific acceptance time.*

The new IoT device is the subject (S), another IoT device is the object (O), and the generation of object access requests by the subject is the desired action. The environment time and the registration time of the device help to calculate the age of the device on our network. A specific time interval is set as device acceptance time during which a new device can only access a limited number of devices. This policy ensures that the new device does not try to access and communicate with all devices on the network before it gains a certain trust, and that its security status is checked. The attribute fields critical for the mentioned access policy are

- *Subject: ⟨ Device Identifier, Device Age ⟩*
- *Object: ⟨ Device Type, Device Identifier ⟩*
- *Environment: ⟨ Date, Time ⟩*

Now, such a policy is very generic and essentially captures the security essence.

Sample Policy 2: *a new IoT can only communicate with devices in its zone until it reaches a specific age.*

The new IoT device is the subject, another IoT device is the object, and the generation of object access requests by the subject is the desired action. The purpose behind setting such a policy is to ensure that the new device does not try to send broadcast messages to devices across all different zones. This policy also helps in the development of a baseline behaviour of the devices and helps limit access to all zones until a certain age and trust level is achieved by this new device. The attribute fields critical for the mentioned access policy are

- *Subject ⟨ Device Identifier, Device Age, Device Zone ⟩*
- *Object ⟨ Device Identifier, Device Zone ⟩*
- *Environment ⟨ Date, Time ⟩*

5.2. Device Access Policies

Access to every IoT device cannot be granted to every other IoT device if it generates a request. The access limitation policies make sure that only valid requests get accepted while all other requests get rejected. Attributes such as device type, device category, and device priority of both subject and object are considered while creating these access policies [37]. A device access request is accepted only when the subject has a certain priority, and trust level that matches the object, and the device category allows the kind of access type of the generated request. Some of the sample access policies are defined as

Sample Policy 3: *An IoT device can only communicate with another IoT device if it matches the priority combined with the trust level required to access that device.*

Here an IoT device is the subject, another IoT device is the object, and the acceptance of object access requests by the subject is the desired action. This policy allows access to an IoT device if and only if the subject has a combined value of its priority and current trust level greater than or equal to that of the object device. As the write access is granted to a monitoring device, the device status of the object is set to Lock. The attribute fields critical for the mentioned access policy are

- *Subject: ⟨ Device Identifier, Device Priority, Device Trust Level ⟩*
- *Object: ⟨ Device Identifier, Device Priority, Device Trust Level ⟩*
- *Access_Type: ⟨ read ⟩*
- *Environment: ⟨ Date, Time ⟩*

Sample Policy 4: *only monitoring type devices can not send control data to any other device.*

Here, one IoT device is the subject, another IoT device is the object, and transmission of a control message is the desired action. We do not want any device to be able to change the settings of another IoT device unless it is an authorized and trusted monitoring device. The attribute fields critical for the mentioned access policy are

- *Subject: ⟨ Device Identifier, Device Priority, Device Trust Level ⟩*
- *Object: ⟨ Device Identifier, Device Status, Device Priority, Device Trust Level ⟩*
- *Access_Type: ⟨ write⟩*
- *Environment: ⟨Date, Time⟩*

Sample Policy 5: *an IoT device can receive control data from only one monitoring device at a certain instance of time.*

Here, an IoT device is the subject, another IoT device is the object, and transmission of a control message is the desired action. While a monitoring device has established a connection with an IoT device and is sending some control instructions, hence changing the other device's settings, no other device should be allowed write access to such an object. We do not want multiple devices to be able to change the settings of another IoT device simultaneously. Hence, the device status of the object is checked whenever a written request is received, and the request is set as pending until the device is set free and its status is turned back to unlock. The attribute fields critical for the mentioned access policy are

- *Subject: ⟨ Device Identifier, Device Priority, Device Trust Level ⟩*
- *Object: ⟨ Device Identifier, Device Status, Device Priority, Device Trust Level ⟩*
- *Access_Type: ⟨ write ⟩*
- *Environment: ⟨ Date, Time ⟩*

5.3. Device Access Limitation Policies

Countermeasures need to be taken to avoid the possibility of any flooding attack. No device should be allowed to access all the devices available on the network simultaneously. In ZTA, the traffic is monitored and zones are mentioned; hence, a rogue device that tries to initiate broadcast requests is sent to a quarantine zone where the device is reset, hence setting its age back to zero and a full scan of the device's security status is performed to detect the cause of such malicious activities. To achieve this goal, the trust level of a device is decreased as it initiates any broadcast request. To enforce a guaranteed rejection of random access requests, some rules can be set to create regional broadcast boundaries as follows:

Sample Policy 6: *A monitoring device can send control data to multiple IoT devices at a certain instance of time if they all belong to the same zone.*

Here a monitoring IoT device is the subject, a set of multiple IoT devices in a certain zone is the object, and transmission of a control message is the desired action. Monitoring devices can establish multiple concurrent write connections with other IoT devices and send some control instructions if and only if they all belong to the same zone. We do not want a change of control settings for multiple devices in multiple zones to occur simultaneously. Hence, the device zone of the object is checked whenever a "write-all" request is received. The attribute fields critical for the mentioned access policy are

- *Subject: ⟨ Device Identifier, Device Priority, Device Trust Level ⟩*
- *Object: ⟨ Device Identifier, Device Status, Device Priority, Device Trust Level, Device Zone ⟩*
- *Access_Type: ⟨ write_all ⟩*
- *Environment: ⟨ Date, Time ⟩*

Sample Policy 7: *Only a controlling/monitoring device can initiate connections to all devices in a zone simultaneously.*

Here, a monitoring IoT device is the subject, a set of all the IoT devices in a zone is the object, and transmission of a control message is the desired action. To ensure the security of our IoT network, only supervising devices are allowed to have concurrent access to all devices in a certain zone. This guarantees that no rouge device is allowed access to multiple devices. If any non-controlling device initiates a "write" request or a "read_all" request, its trust level is depleted, and it is quarantined until a complete security clearance. The attribute fields critical for the mentioned access policy are

- *Subject: ⟨ Device Identifier, Device Status, Device Priority, Device Trust Level, Device Zone ⟩*
- *Access_Type: ⟨ read_all / write_all ⟩*
- *Environment: ⟨Date, Time⟩*

Sample Policy 8: *Broadcast messages cannot be sent across the network by any device.*

Here the subject IoT device tries to transmit a write_all control message to all the devices connected across the IoT network. To safeguard our IoT network from flooding attacks, broadcasting messages across the entire network is strictly prohibited. This behaviour is considered malicious and such a device is quarantined instantly. The attribute fields critical for the mentioned access policy are

- *Subject: ⟨ Device Identifier, Device Priority, Device Trust Level ⟩*
- *Access_Type: ⟨ write_all ⟩*
- *Environment: ⟨ Date, Time ⟩*

Table 2 summarises the aforementioned sample policies. Numerous other rules may be developed in the same way to handle various events depending on the requirements of the system.

Table 2. Description of all sample ABAC policies.

Policy	Description
Policy 1	A new IoT device cannot request communications with more than a certain number of devices in a specific acceptance time.
Policy 2	A new IoT can only communicate with devices in its zone until it reaches a specific age.
Policy 3	Any IoT device can only communicate with another IoT device if it matches the priority combined with the trust level required to access that device.
Policy 4	Only monitoring-type devices can send control data to any other device.
Policy 5	An IoT device can receive control data from only one monitoring device at a certain instance of time.
Policy 6	A monitoring device can send control data to multiple IoT devices at a certain instance of time if they all belong to the same zone.
Policy 7	Only a controlling/monitoring device can initiate connections to all devices in a zone simultaneously.
Policy 8	Broadcast messages cannot be sent across the network by any device.

6. Attribute Management Framework

The attribute management framework is a fundamental component of our system, which extracts and stores all the required attributes of every IoT device connected to the network by working persistently with the policy information point (PIP). The attribute management framework consists of several modules responsible for the compilation and preservation of the attributes of the overall system. In this section, we have discussed the modules that will be useful in extracting the attributes from the respective entities.

6.1. Device Attribute Management

ZAIB's entire ABAC mechanism works on device attributes, hence obtaining and maintaining a proper and updated storage of these attributes is one of the most important aspects of the ZAIB framework. ZAIB requires every IoT device to get registered as soon as it joins the network. To extract device attributes, a device fingerprinting mechanism will vigorously fingerprint different devices and record them. These attributes will be associated with the device wallet ID assigned to each device and hence will be stored in PIP's device database maintained on the chain for active devices and stored on the IPFS for all non-active devices that ever joined the network. The basic fingerprinting techniques defined in [47,48] represent different ways of identifying different device-related attributes, for example in [49] the authors suggest that a TCP port scan reveals enough information to help classify an IoT device. None of the mentioned approaches can individually satisfy our needs but a combination of a few approaches depending on the numerous network-level header fields, payload classification, and other cyber-physical features of devices can help to identify a device successfully.

6.2. User Attribute Management

Our smart citizens and network administrators will be the ones controlling our cyber city's one or more IoT devices. The trust engine determines the extent of access assigned to each user based on the device attributes, user attributes, and additional behaviour attributes such as the device and user's access histories and trust levels. We will be counting on the fingerprinting mechanism to distinguish between data and control packets.

6.3. Network Traffic Attribute Management

Each network device will request the PEP to access any other device on the network, which, when processed according to the policies, will result in being granted or denied. Hence, each transaction will be recorded in the public ledger. The network traffic attributes comprise the packet header fields and the traffic flow statistics. For extracting the packet header fields, we will use the packet capture module and extract the necessary fields. Within the same module, we will incorporate scripts that will appropriately record the necessary flow metrics and record them in a flow monitor module.

7. ZAIB Workflow and Scenario

To explain the working of ZAIB, a complete system workflow is presented as a common use case example of access request of an IoT device, such as a weight sensor, to another IoT device, such as an industrial conveyer belt motor

7.1. ZAIB Workflow

Every new user or device needs to be registered on the network to gain a blockchain wallet with public and private key pairs. This makes the entire communication anonymous and hence completely secured. Since all the communications are encrypted, this further improves the security. The entire workflow of the system is defined in the steps below:

1. After registration, a new device becomes a part of the IoT network and it can request to access any device on the network.
2. Once the request is made, it is received by the PEP from where it is forwarded to the PDPO.
3. The PDPO collects the attributes and trust levels from the PIP and requests the PIP to check if any policy regarding the access of the object by the subject exists.
4. If the policy exists, the PDP SC is triggered that implements the policy and accepts or rejects the request.
5. If such a policy is not found, a request for policy generation is sent to the PEO.
6. After receiving the request, the PEO triggers PE SC that generates the policy based on the role of the subject, its trust level, the type and category of the device, along with the trust level, type, and category required to access the object.
7. Once the policy is generated, the PEP enforces it.
8. If access is allowed, PEP generates an encrypted channel to facilitate secure communication between subject and object. If it is denied, the PEP informs the subject about the request rejection.
9. Every transaction is recorded in the PIP as it is used for determining device trust level and identifying behaviour anomalies.
10. The request and the decision taken on that request are both stored in the distributed ledger as transactions, creating an immutable history of all device activities on the IoT network. Any alteration in PIP can easily be detected by matching its records with the ledger transactions.
11. The TE SC is triggered every time a transaction is accepted or denied and it updates the device trust level based on this new transaction and the device's previous behaviour.

An overall workflow of the system is shown in Figure 10.

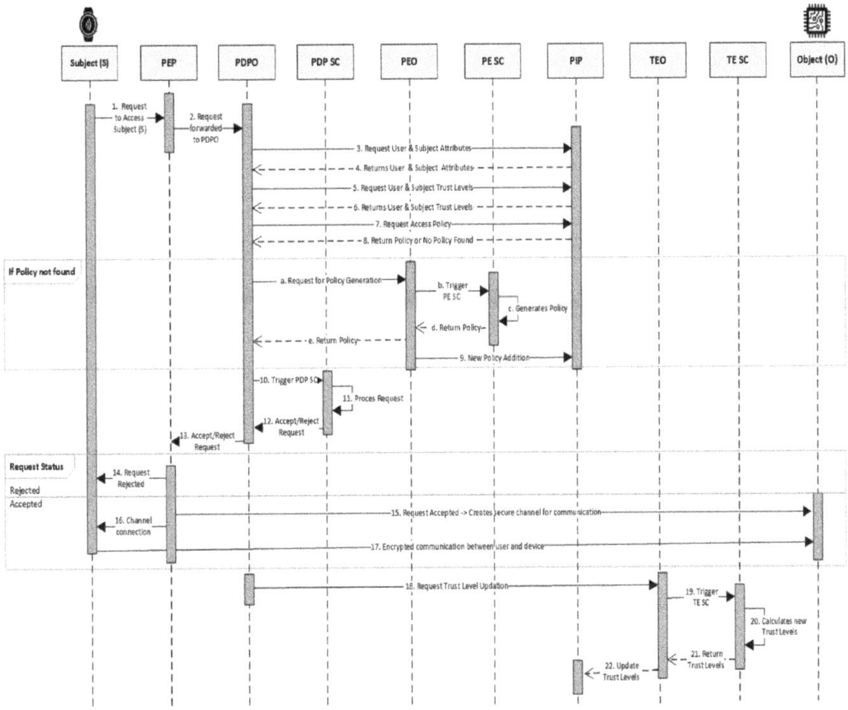

Figure 10. Overall workflow of ZAIB.

7.2. Working Scenario

As a "proof of concept" ZAIB, the proposed framework is a generic access control framework that can be applied to several IoT application systems, including those for smart homes, transportation, smart industries, and health. The following scenario is taken into consideration as a typical use case to exemplify the user experience and practicality of the provided framework. In this scenario, the object is a smart security camera set outside the main door while the subject requesting a picture and approval is a door lock, which will unlock automatically whenever the house owners arrive at the door. The smart security camera can be built using a Raspberry Pi 2 board with its dedicated camera while the lock will also consist of a Raspberry Pi 2 that will unlock whenever the facial image received from the camera matches one of the owner's images recorded in the SD card. Both the Raspberry Pi will be connected to WiFi to provide remote access. The door lock might take several different actions depending on the privileges as stated by the policies, whereas the camera serves as a resource whose access requests need to be managed. In the current scenario, the camera will take a snapshot and save it on the Raspberry Pi SD card. As per rules set by the access control policies, the lock will request to access the screenshot of the person at the door to compare it to its presorted data and give the requester remote access using a request transaction through our PEP. Hence, the access request will consist of the following four major components:

* Subject: the smart door lock;
* Object: the smart security camera at the door;
* Access_Type: request to take an image and read it;
* Environment: current date, current time.

An initial implementation of the proposed protocol is demonstrated in Figure 11.

After registration, both devices become a part of the IoT network. The subject (*Lock*) now requests to access the snapshot captured by the object (*Camera*). The request is received by the PEP of the smart home zone from where it is forwarded to the PDPO. The PDPO collects the attributes and trust levels from the PIP and requests the PIP to check if any policy regarding the access of the object (*Camera*) by the subject (*Lock*) exists. As the policy exists, the PDP SC has triggered and implements the policy and accepts this request. If the policy does not exist, a policy generation request is sent to the PEO. The PEP enforces the policy, but generates an encrypted channel to facilitate secure communication between subject and object. The PIP records this transaction as it will be used later on for determining device trust level and identifying behaviour anomalies. The request and the decision taken on that request are both stored in the distributed ledger as transactions, creating an immutable history of all device activities on the IoT network.

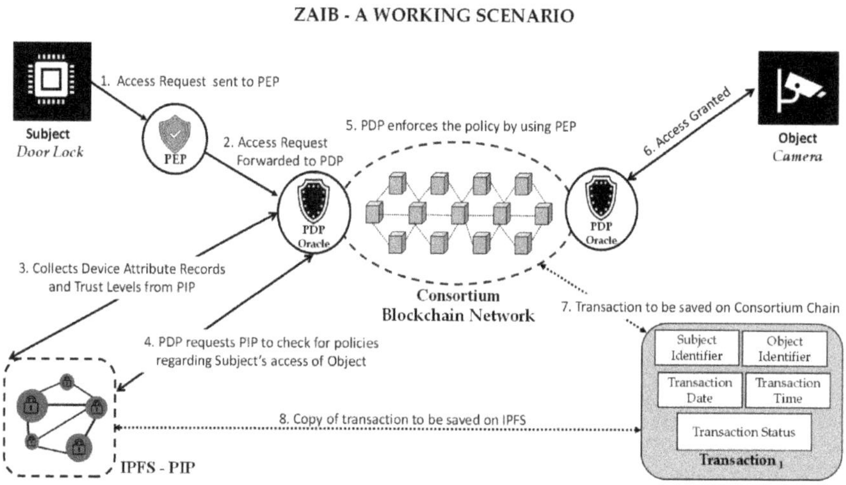

Figure 11. A working scenario of ZAIB implementation.

8. Evaluation

The distributed and vastly scattered nature of IoT devices, actuators, and sensors work flawlessly devoid of any human interventions. The standard centralized access control mechanisms cannot secure extremely distributed massive IOT infrastructures, such as smart cities where administration, traffic, business, hospitals, and citizens, i.e., all stakeholders, need everything to be connected to the Internet [50]. Millions of IoT devices and sensors will come together to form smart networks for smart transmission grids, smart security, smart healthcare, smart roads, and smart cars. A vulnerability in security will not only end in financial losses but also risks the lives of thousands of citizens [19]. The proposed framework provides fail-safe security as it not only ensures the authentication and authorization of users and devices but also maintains data privacy and confidentiality. This section discusses how these security requirements have been met by ZAIB. The analysis of all proposed sample ABAC policies is shown in Table 3.

8.1. Device Authentication

ZAIB ensures device and user authentication by using blockchain wallets. Public key cryptography certifies the device credentials, whereas the device security check is conducted periodically to have the latest update on the device status. After making sure that only certified devices join the network, device behaviour history is also secured which helps identify any anomalies in device behaviour and immediately decreases its trust level while initiating a security check to verify if the device was tampered with in any way. To prevent a malicious device from misbehaving and infecting other devices, it is quarantined immediately and is only allowed to rejoin the network and communicate with other devices when they have been restarted and have obtained security clearance. Therefore, ZAIB offers protection against attacks such as impersonation or masquerade attacks.

Table 3. Analysis of all proposed sample ABAC policies.

Policy	Authentication	Authorization	Confidentiality	Privacy
Policy 1	✓	✓	✓	✓
Policy 2	✓	✓	✓	✓
Policy 3	✓	✓	✓	✓
Policy 4	✓	✓	**X**	**X**
Policy 5	**X**	✓	✓	**X**
Policy 6	✓	✓	**X**	**X**
Policy 7	✓	✓	✓	✓
Policy 8	**X**	**X**	✓	✓

8.2. Authorization

ZAIB provides a dynamic access authorization by implementing attribute-based access control in a massive IoT infrastructure. Monitoring dynamic parameters like the latest trust and risk levels of IoT devices along with static device attributes such as the priority, category, and device network for every respective request ensures that access will not be granted to any device with behaviour anomalies. Hence, the proposed scheme provides comprehensive dynamic access control compared to DAC, MAC, RBAC, or even static ABAC.

8.3. Confidentiality

ZAIB provides the protection of system resources against unauthorized access. The degree of authorization required to access devices and data in ZAIB is set intelligently by using public-key authorization with smart wallets and continuous behaviour analysis of all IoT devices to ensure security against identity theft or masquerade attacks to maintain the confidentiality of classified information.

8.4. Privacy

Privacy means having full control or decision-making authority on how the user's data can be collected and used. ZAIB ensures that no unauthorized person or device can gain access to devices owned by an individual or data generated by them, hence ensuring the basic right to privacy for every individual.

To evaluate the effectiveness of the suggested policies, let us assume a scenario like a smart home. With all the home appliances and security equipment attached to the Internet and being monitored by central controlling devices, the proposed policies will make sure that no unauthorized perpetrator gets access to your devices. In case an attacker gets access to your home network and tries to plant a new IoT device of their own, policies 1 and 2 limit their communication requests and also limit their access to only a certain zone, hence reducing the attacker's access radius to a minimum area. Policies 3 and 4 stop the subject from communicating with any object device on the home network by applying the priority, age, trust level, and device type filters, allowing communication if the above-mentioned attributes are matched. Here, priority, age, and trust level are all dynamic attributes that change with time and hence provide improved security as devices once deemed eligible for communication will be denied the same privilege if any behaviour anomalies are detected and their trust levels are reduced. This dynamic behaviour analysis ensures security even if any device on the home network is compromised by an intruder. Policies 5 and 6 dictate the behaviour of monitoring devices and keep check on their commanding areas by limiting them to particular zones. Hence, if a malicious device is registered in the kitchen, it can not access security devices or devices in other zones such as the bedrooms or lounges. Policies 7 and 8 ensure that only a monitoring device can connect to multiple devices simultaneously and broadcasting any message on the network is prohibited, hence the chances of an intruder infecting all devices on networks are very slim and almost down to zero. Therefore, the discussion suggests that the proposed system provides complete security, authentication, data privacy, and confidentiality to all users utilizing IoT devices registered on the network.

8.5. Computational and Storage Tradeoff

We have identified a trade-off between computational and spatial complexity for the realistic implementation of this system. Keeping all device attribute data on-chain will result in processing all access requests more quickly, but the larger ledger size will raise computation costs. However, if an off-chain data repository is maintained, the smaller ledger size will result in greater costs and delays while reducing computing complexity.

9. Conclusions

This paper addresses the security challenges for large IoT-based infrastructures such as smart cities and provides a zero-trust and ABAC-based dynamic solution for confronting these challenges. Issues such as user privacy, device authentication, and authorization have been resolved by modelling a framework that provides a completely secure device-to-device communication mechanism by not only considering the existing security levels of the network but also considering the behaviour anomalies of the devices to instantaneously detect any kind of intrusion or device misconduct by using device trust levels. The framework is modelled by dividing IoT networks into various zero-trust zones and an ABAC framework that specifies subject, object, and network attributes. Several policies were defined based on these attributes to enforce reliable and secure machine-to-machine communication which is being recorded by the immutable distributed ledger for advanced accountability. We have also discussed an attribute management framework that captures and calculates the important device attributes required for implementing the ABAC policies. In future, some still-relevant major problems in this field, such as computation and space overheads, need to be addressed to find the optimal equilibrium for the best system performance.

Author Contributions: All authors contributed equally to this article. All authors have read and agreed to the published version of the manuscript.

Funding: This research received no external funding.

Data Availability Statement: Not applicable.

Conflicts of Interest: The authors do not have financial or professional conflicting interest.

Abbreviations

The following abbreviations are used in this manuscript:

ABAC	Attribute-Based Access Control
D2D	Device-to-Device
DAC	Discretionary Access Control
IoT	Internet of Things
IPFS	Interplanetary File System
MAC	Mandatory Access Control
MCAP	Microcore And Perimeter
PA	Policy Administrator
PAP	Policy Administration Point
PDP	Policy Decision Point
PDPO	Policy Decision Point Oracle

PE	Policy Engine
PEO	Policy Engine Oracle
PEP	Policy Enforcement Point
PIP	Policy Information Point
RBAC	Role-Based Access Control
SC	Smart Contract
TE	Trust Engine
TEO	Trust Engine Oracle
ZAIB	The name of the proposed architecture (ZTA and ABAC for IoT using Blockchain)
ZT	Zero Trust
ZTA	Zero-Trust Architecture

References

1. Chen, B.; Qiao, S.; Zhao, J.; Liu, D.; Shi, X.; Lyu, M.; Chen, H.; Lu, H.; Zhai, Y. A Security Awareness and Protection System for 5G Smart Healthcare Based on Zero-Trust Architecture. *IEEE Internet Things J.* **2021**, *8*, 10248–10263. [CrossRef] [PubMed]
2. Syed, A.S.; Sierra-Sosa, D.; Kumar, A.; Elmaghraby, A. IoT in Smart Cities: A Survey of Technologies, Practices and Challenges. *Smart Cities* **2021**, *4*, 24. [CrossRef]
3. What Is Stuxnet? 1999. Available online: https://www.trellix.com/en-us/security-awareness/ransomware/what-is-stuxnet.html (accessed on 30 December 2022).
4. U.S. Institute of Peace. Israeli Sabotage of Iran's Nuclear Program. 2021. Available online: https://iranprimer.usip.org/blog/2021/apr/12/israeli-sabotage-iran%E2%80%99s-nuclear-program (accessed on 12 April 2021).
5. Zetter, K. Inside the Cunning, Unprecedented Hack of Ukraine's Power Grid, Published in Wired. 2010. Available online: https://www.wired.com/2016/03/inside-cunning-unprecedented-hack-ukraines-power-grid/ (accessed on 30 October 2010).
6. Razzaq, M.A.; Gill, S.H.; Qureshi, M.A.; Ullah, S. Security issues in the Internet of Things (IoT): A comprehensive study. *Int. J. Adv. Comput. Sci. Appl.* **2017**, *8*, 383. [CrossRef]
7. Arshad, J.; Azad, M.A.; Abdeltaif, M.M.; Salah, K. An intrusion detection framework for energy constrained IoT devices. *Mech. Syst. Signal Process.* **2020**, *136*, 106436. [CrossRef]
8. Arshad, J.; Azad, M.A.; Mahmoud Abdellatif, M.; Ur Rehman, M.H.; Salah, K. COLIDE: A collaborative intrusion detection framework for Internet of Things. *IET Netw.* **2019**, *8*, 3–14.
9. Trilles, S.; Calia, A.; Belmonte, Ó.; Torres-Sospedra, J.; Montoliu, R.; Huerta, J. Deployment of an open sensorized platform in a smart city context. *Future Gener. Comput. Syst.* **2017**, *76*, 221–233. [CrossRef]
10. Pacheco, J.; Hariri, S. Anomaly behavior analysis for IoT sensors. *Trans. Emerg. Telecommun. Technol.* **2018**, *29*, e3188. [CrossRef]
11. Samaniego, M.; Deters, R. Zero-trust hierarchical management in IoT. In Proceedings of the 2018 IEEE International Congress on Internet of Things (ICIOT), San Francisco, CA, USA, 2–7 July 2018; pp. 88–95. [CrossRef]
12. Bruno, E.; Gallier, R.; Gabillon, A. Enforcing access controls in IoT networks. In *Proceedings of the International Conference on Future Data and Security Engineering*; Springer: Berlin/Heidelberg, Germany, 2019; pp. 429–445. [CrossRef]
13. Zimmer, B. LISA: A Practical Zero Trust Architecture. In *Proceedings of the Enigma 2018 (Enigma 2018)*; USENIX Association: Santa Clara, CA, USA, 2018.
14. Alramadhan, M.; Sha, K. An overview of access control mechanisms for internet of things. In Proceedings of the 2017 26th International Conference on Computer Communication and Networks (ICCCN), Vancouver, BC, Canada, 31 July–3 August 2017; pp. 1–6. [CrossRef]
15. Kindervag, J.; *Build Security into Your Network's DNA: The Zero Trust Network Architecture*; Forrester Research Inc.: Cambridge, MA, USA, 2010; pp. 1–26.
16. Muralidharan, S.; Ko, H. An InterPlanetary file system (IPFS) based IoT framework. In Proceedings of the 2019 IEEE International Conference on Consumer Electronics (ICCE), Las Vegas, NV, USA, 11–13 January 2019; pp. 1–2.
17. Rose, S.W.; Borchert, O.; Mitchell, S.; Connelly, S. Zero Trust Architecture. 2020. Available online: https://www.nist.gov/publications/zero-trust-architecture (accessed on 1 February 2023). [CrossRef]
18. Babiker Mohamed, M.; Matthew Alofe, O.; Ajmal Azad, M.; Singh Lallie, H.; Fatema, K.; Sharif, T. A comprehensive survey on secure software-defined network for the Internet of Things. *Trans. Emerg. Telecommun. Technol.* **2022**, *33*, e4391. [CrossRef]
19. Dhar, S.; Bose, I. Securing IoT Devices Using Zero Trust and Blockchain. *J. Organ. Comput. Electron. Commer.* **2020**, 1–17. [CrossRef]
20. Zhang, Y.; Li, B.; Liu, B.; Wu, J.; Wang, Y.; Yang, X. An attribute-based collaborative access control scheme using blockchain for IoT devices. *Electronics* **2020**, *9*, 285. [CrossRef]
21. Liu, H.; Han, D.; Li, D. Fabric-IoT: A blockchain-based access control system in IoT. *IEEE Access* **2020**, *8*, 18207–18218. [CrossRef]
22. Benet, J. Ipfs-content addressed, versioned, p2p file system. *arXiv* **2014**, arXiv:1407.3561.
23. Naz, M.; Al-zahrani, F.A.; Khalid, R.; Javaid, N.; Qamar, A.M.; Afzal, M.K.; Shafiq, M. A secure data sharing platform using blockchain and interplanetary file system. *Sustainability* **2019**, *11*, 7054. [CrossRef]
24. Assunção, P. A Zero Trust Approach to Network Security. In Proceedings of the Digital Privacy and Security Conference 2019, Miami, FL, USA, 15–17 May 2019.

25. Lukaseder, T.; Halter, M.; Kargl, F. Context-based Access Control and Trust Scores in Zero Trust Campus Networks. In *Sicherheit 2020*; Gesellschaft für Informatik e.V.: Bonn, Germany, 2020. [CrossRef]
26. Picard, N.; Colin, J.N.; Zampunieris, D. Context-aware and attribute-based access control applying proactive computing to IoT system. In Proceedings of the 3rd International Conference on Internet of Things, Big Data and Security (IoTBDS 2018). SCITEPRESS, Madeira, Portugal, 19–21 March 2018; pp. 333–339. [CrossRef]
27. Zhang, X.; Jiang, X. IoT architecture based on ABAC smart contract. In Proceedings of the 2020 3rd International Conference on Advanced Electronic Materials, Computers and Software Engineering (AEMCSE), Shenzhen, China, 24–26 April 2020; pp. 122–128. [CrossRef]
28. Tomaz, A.E.B.; Do Nascimento, J.C.; Hafid, A.S.; De Souza, J.N. Preserving Privacy in Mobile Health Systems Using Non-Interactive Zero-Knowledge Proof and Blockchain. *IEEE Access* **2020**, *8*, 204441–204458. [CrossRef]
29. Ruan, P.; Anh Dinh, T.T.; Lin, Q.; Zhang, M.; Chen, G.; Chin Ooi, B. Revealing Every Story of Data in Blockchain Systems. *SIGMOD Rec.* **2020**, *49*, 70–77. [CrossRef]
30. Ruan, P.; Chen, G.; Dinh, T.T.A.; Lin, Q.; Ooi, B.C.; Zhang, M. Fine-Grained, Secure and Efficient Data Provenance on Blockchain Systems. *Proc. VLDB Endow.* **2019**, *12*, 975–988. [CrossRef]
31. Ferraiolo, D.; Chandramouli, R.; Hu, V.; Kuhn, R. A comparison of attribute based access control (ABAC) standards for data service applications. *NIST Spec. Publ.* **2016**, *800*, 178.
32. Arnold, R.; Longley, D. Zero-knowledge proofs do not solve the privacy-trust problem of attribute-based credentials: What if alice is evil? *IEEE Commun. Stand. Mag.* **2019**, *3*, 26–31. [CrossRef]
33. Arasteh, H.; Hosseinnezhad, V.; Loia, V.; Tommasetti, A.; Troisi, O.; Shafie-khah, M.; Siano, P. Iot-based smart cities: A survey. In Proceedings of the 2016 IEEE 16th International Conference on Environment and Electrical Engineering (EEEIC), Florence, Italy, 7–10 June 2016; pp. 1–6. [CrossRef]
34. Waheed, U.; Khan, M.S.A.; Awan, S.M.; Khan, M.A.; Mansoor, Y. Decentralized Approach to Secure IoT based Networks using Blockchain Technology. 3C Tecnología_Glosas de innovación aplicadas a la pyme (2019). Available online: https://dialnet.unirioja.es/servlet/articulo?codigo=6933920 (accessed on 13 January 2023).
35. Ouaddah, A.; Abou Elkalam, A.; Ait Ouahman, A. FairAccess: A new Blockchain-based access control framework for the Internet of Things. *Secur. Commun. Netw.* **2016**, *9*, 5943–5964. [CrossRef]
36. Durga, R.; Poovammal, E.; Ramana, K.; Jhaveri, R.H.; Singh, S.; Yoon, B. CES Blocks—A Novel Chaotic Encryption Schemes-Based Blockchain System for an IoT Environment. *IEEE Access* **2022**, *10*, 11354–11371. [CrossRef]
37. Bezawada, B.; Haefner, K.; Ray, I. Securing home IoT environments with attribute-based access control. In Proceedings of the Third ACM Workshop on Attribute-Based Access Control, Tempe, AZ, USA, 21 March 2018; pp. 43–53.
38. Peng, Z.; Xu, J.; Hu, H.; Chen, L.; Kong, H. BlockShare: A Blockchain empowered system for privacy-preserving verifiable data sharing. *Bull. IEEE Comput. Soc. Tech. Comm. Data Eng.* **2022**, *1*, 14–24.
39. Alevizos, L.; Ta, V.T.; Eiza, M.H. Augmenting Zero Trust Architecture to Endpoints Using Blockchain: A Systematic Review. *arXiv* **2021**, arXiv:2104.00460.
40. Ding, S.; Cao, J.; Li, C.; Fan, K.; Li, H. A novel attribute-based access control scheme using blockchain for IoT. *IEEE Access* **2019**, *7*, 38431–38441. [CrossRef]
41. Yan, X.; Wang, H. Survey on Zero-Trust Network Security. In *Proceedings of the International Conference on Artificial Intelligence and Security*; Springer: Berlin/Heidelberg, Germany, 2020; pp. 50–60. [CrossRef]
42. Weerapanpisit, P.; Trilles, S.; Huerta, J.; Painho, M. A Decentralized Location-Based Reputation Management System in the IoT Using Blockchain. *IEEE Internet Things J.* **2022**, *9*, 15100–15115. [CrossRef]
43. Bernabe, J.B.; Ramos, J.L.H.; Gomez, A.F.S. TACIoT: Multidimensional trust-aware access control system for the Internet of Things. *Soft Comput.* **2016**, *20*, 1763–1779. [CrossRef]
44. Xu, R.; Chen, Y.; Blasch, E.; Chen, G. Blendcac: A smart contract enabled decentralized capability-based access control mechanism for the iot. *Computers* **2018**, *7*, 39. [CrossRef]
45. Cruz-Piris, L.; Rivera, D.; Marsa-Maestre, I.; De La Hoz, E.; Velasco, J.R. Access control mechanism for IoT environments based on modelling communication procedures as resources. *Sensors* **2018**, *18*, 917. [CrossRef]
46. Eidle, D.; Ni, S.Y.; DeCusatis, C.; Sager, A. Autonomic security for zero trust networks. In Proceedings of the 2017 IEEE 8th Annual Ubiquitous Computing, Electronics and Mobile Communication Conference (UEMCON), New York, NY, USA, 19–21 October 2017; pp. 288–293. [CrossRef]
47. François, J.; Abdelnur, H.; Festor, O. Automated behavioral fingerprinting. In *Proceedings of the International Workshop on Recent Advances in Intrusion Detection*; Springer: Berlin/Heidelberg, Germany, 2009; pp. 182–201. [CrossRef]
48. Radhakrishnan, S.V.; Uluagac, A.S.; Beyah, R. GTID: A technique for physical device and device type fingerprinting. *IEEE Trans. Dependable Secur. Comput.* **2014**, *12*, 519–532. [CrossRef]

49. Sivanathan, A.; Gharakheili, H.H.; Sivaraman, V. Can we classify an iot device using tcp port scan? In Proceedings of the 2018 IEEE International Conference on Information and Automation for Sustainability (ICIAfS), Colombo, Sri Lanka, 21–22 December 2018; pp. 1–4. [CrossRef]
50. Gabillon, A.; Gallier, R.; Bruno, E. Access controls for IoT networks. *SN Comput. Sci.* **2020**, *1*, 1–13. [CrossRef]

information

Article

Betraying Blockchain: Accountability, Transparency and Document Standards for Non-Fungible Tokens (NFTs)

Kristin Cornelius

Center for Information as Evidence, University of Los Angeles, Los Angeles, CA 90095, USA; krisbcorn@ucla.edu

Abstract: Transparency and accountability are important aspects to any technological endeavor and are popular topics of research as many everyday items have become 'smart' and interact with user data on a regular basis. Recent technologies such as blockchain tout these traits through the design of their infrastructure and their ability as recordkeeping mechanisms. This project analyzes and compares records produced by non-fungible tokens (NFTs), an increasingly popular blockchain application for recording and trading digital assets, and compares them to 'document standards,' an interdisciplinary method of contract law, diplomatics, document/interface theory, and evidentiary proof, to see if they live up to the bar that has been set by a body of literature concerned with authentic documents. Through a close reading of the current policies on transparency (i.e., CCPA, GDPR), compliance and recordkeeping (i.e., FCPA, SOX, UETA), and the consideration of blockchain records as user-facing interfaces, this study draws the conclusion that without an effort to design these records with these various concerns in mind and from the perspectives of all three stakeholders (Users, Firms, and Regulators), any transparency will only be illusory and could serve the opposite purpose for bad actors if not resolved.

Keywords: blockchain; records; non-fungible tokens; transparency

Citation: Cornelius, K. Betraying Blockchain: Accountability, Transparency and Document Standards for Non-Fungible Tokens (NFTs). *Information* **2021**, *12*, 358. https://doi.org/10.3390/info12090358

Academic Editors: Spyros Panagiotakis and Evangelos K. Markakis

Received: 30 June 2021
Accepted: 26 August 2021
Published: 31 August 2021

Publisher's Note: MDPI stays neutral with regard to jurisdictional claims in published maps and institutional affiliations.

1. Introduction

Broadly, transparency and accountability are current topics of interest. Researchers from several fields, regulators from various sectors, and the vast general public are searching for the best way forward to keep the major technology companies that shape our daily behavior accountable for the data they collect and sell. We know these practices have a manipulative effect, more profound than any other data capture/advertising business model of the past [1]. Developments in technology have allowed for a sophisticated Wizard of Oz-like presentation that hides how its algorithms interpret the data it collects and aggregates; however, concepts of transparency are thrown around as claims that the increasingly sophisticated technology can provide these qualities implicitly [2,3]. Some applications (i.e., Google, Facebook, Microsoft), for instance, make use of design capabilities to influence users' data privacy settings, a practice now being deemed "dark patterns" [1]. Yet, the 'smart' software embedded in many of our daily practices such as with pervasive computing and the internet of things (IoT) brings with it new opportunities for hidden contracts, automated data collection, and increasingly context algorithms. In response, new technological infrastructures such as blockchain applications explicitly tout anonymity, making use of decentralized computing systems that render tracing users nearly impossible. This has many benefits including encrypted records that are reportedly immutable, allowing for them to act as currency, notaries, authentication markers, even contracts [4], with the downside being that the anonymity offered can attract bad actors [5]. As blockchain proponents tout the transparency and accountability they say is built into its architecture, it is important that it is not considered a *solution* to the aforementioned issues without a rigorous critique of its methods for solving them.

Since its recent spike in use and popularity around 2016–2017, blockchain technology's uses have included cryptocurrencies, smart contracts, and non-fungible tokens (NFTs), increasingly among others. Broadly, as a blockchain creates a ledger of immutable records, it can be diced up into portions that can be given value as currency (e.g., Bitcoin), can be coded to execute automated transactions (e.g., Ethereum), and can be tied to assets as a record of ownership (e.g., non-fungible tokens or "NFTs"). Past work in this space has discussed the potential dangers of assuming some of these blockchain technologies can replace certain types of documents that have standards and practices which have been around for centuries. For instance, the issues with standard form contracts can be exacerbated in this blockchain environment [6,7], the freedom of contract principle can be exploited to allow for the cementing of unfair terms [8], and, claiming that blockchain produces immutable records when the technology has facilitated nefarious activity can cause all sorts of issues for regulators—especially when these records are used for evidence [5]. Other research has noted the inconsistencies with the technology and some of its claims [5,6,9]. A sample of this research suggests that to solve some of these transparency issues the roles of responsibility should be considered; should blockchain applications have fiduciary responsibilities, for example? What are the proper notation practices as NFTs are essentially documentation of ownership? [9]. And who is responsible for solving its environmental impact? There are still many unanswered questions that need to be addressed to figure out the possible issues with this new digital environment.

Of the examples of blockchain technology mentioned, NFTs have gained renown for their large purchase price, bringing in sometimes millions of dollars to own one. Recently an NFT was auctioned off at Christie's for $69 million that a represented digital artwork, for instance [10]. At the time of writing (Aug 2021), the total trading value of NFTs in the last 24 h is $3,656,194,242.58 [11]. NFTs are generally created with the ERC-721 standard written in the Solidity language and make use of public blockchain protocols and platforms such as Ethereum, EOS, Cardano, Flow, and Tron, among others [10–13]. They are called "non-fungible" since they do not exist in a one-to-one ratio with other assets (such as cryptocurrencies do). Essentially, NFTs make ownership of an asset verifiable and since each asset is unique, each NFT record has a unique value relevant to the value of the asset itself. While NFT applications make use of other cryptocurrencies and are tied to their markets in some respects [13], they are not currencies themselves, but rather more like records of ownership. The IoT environment is increasingly making use of decentralized computing systems and blockchain applications to tie real world items to digital records that can prove ownership and trading rights. Sales of assets, transfers of ownership, even rental or insurance agreements might have a NFT associated with it. However, these applications will need to make use of automated smart contracts that control the terms associated with the transaction, bringing about a host of issues that go back to how users have engaged with standard form contracts for decades [7]. Additionally, industry has yet to reckon with the compliance issues presented with an anonymous technology, and, along those lines, regulators will need a method that will aid in updating outdated regulatory schema so that it can address these issues.

This paper adds to past research on the critical application of document standards to blockchain applications, in this instance, a study of NFTs as records. By 'document standards' it is referring to the types of mechanisms that have been applied to records and documents over time to maintain their authenticity. This draws upon contract law, document theory, diplomatics, records management, evidence and the descriptions for compliance as outlined in regulatory schema. These standards have been developed in nuanced, iterative ways with details that describe markers which provide actual validity and thus viability with documents. This could include chain of custody, records life cycles, even interrogating the epistemological underpinnings of how documents function and how users engage with them. As a methodology, it is not only theoretical and qualitative, but also pragmatic since there are real issues that need to be solved in a regulatory sense. Moreover, if these blockchain records are not interrogated sufficiently, the consequences would be that they *exacerbate* the issues with records and their possible nefarious uses, so it is imperative that these studies take place now. And although blockchain technology is proving to contribute to environmental issues [14], when this aspect becomes more sustainable, it seems like this technology is here to stay and could transform business practices and records management across the space entirely.

While the conversation is being had around transparency for regular technology companies, blockchain technology has sat comfortably within this sphere due to the public nature of its records, the immutability of the information, and the supposed ease of which it is queried [10]. However, this paper argues that the transparency provided by blockchain technology largely serves either as a checkmark for firms or, at most, an easier information dump to sift through for regulators. The transparency that it provides does not necessarily translate to accountability. In fact, too much information can have the opposite effect if not curated properly [2,3].

This study examines current blockchain NFT implementations to test to assess their output to see if there is enough information to produce a viable record. NFTs were chosen since they blur the line between financial token and record of proof [9]. As the records produced from these applications act as both representative of value but also as something attached to another thing (being a digital file or real-world asset), the needs associated with the record are unique and worth rigorous analysis that could be generalized across the field. This space still needs much more work in terms of exploring the benefits of document standards for blockchain technology as a way of interrogating its actual levels of transparency. It should be made clear that this study is *not* conclusive in scope—rather, it simply examines the idea that the use of these document standards as points of measurement for various audiences or perspectives (i.e., Users, Firms, Regulators) could be helpful, and details a methodology to this purpose. It also calls for more research in these areas by highlighting where the work needs to be done. The results of this preliminary study show document standards are indeed necessary for this technology to grow and thrive in the way that is fair and transparent and in line with its motivation.

2. Materials and Methods

This section outlines the methods and materials used in this study of NFTs and provides a way of looking at blockchain records more generally. It builds upon the body of past research on the critical application of document and recordkeeping standards to blockchain applications, as well as appropriate document and records theories that have not yet been applied. This section defines the phrase 'document standards,' or the types of mechanisms that have been employed over centuries to records and documents to maintain their authenticity. This proposed method draws upon these fields that deal with documents, records, interfaces, and evidence to determine standards of measurement, rather than take for granted that an encrypted technological architecture will produce it automatically. These standards have been developed in nuanced, iterative ways over time and promote particular features that designate qualities such as validity, reliability, and authenticity to records and other types of documents [15,16].

This rest of this section details the terms that are operationalized and provides brief summaries of which document standards affect which stakeholder. It is intentionally hefty as a large part of this project is to provide a novel methodology, in addition to the case study, that can be used to study other NFT or blockchain instances in the future.

2.1. Terms Defined

This section explains the terms and concepts used throughout this study. Some of these terms are more self-evident than others. For instance, "blockchain" and "non-fungible tokens" are described in terms that other scholars and industry people have described several times [10,12,17]. Others, such as "document standards" are laid out in terms that have been aggregated from prior research. It is useful to define these terms so that this method is as apparent and replicable as possible. It is imperative that this space continues to get interrogated so that the research can help create the best possible uses for each instantiation of this technology.

2.1.1. "Blockchain"

A blockchain is a secure, distributed ledger comprised of a chain of record-type entities. These entities are cryptographically chained together and publicly replicated across each node of a decentralized computer network. There are generally two types of blockchain technology infrastructures—public and private ('permissioned'). Public blockchain infrastructure uses the transparency of multiple public copies of the ledger to ensure the accountability and accuracy of the entities. Since these are publicly copied onto each node of the peer-to-peer network, there is no centralized point of attack [18]. To fraudulently change records on the publicly distributed ledger, an attacker must control the majority of nodes in the network, a computationally expensive and improbable event. Some blockchain technology supporters promote variations of this technology such as smart contracts that aim to overturn centralized governance provided by third-party oversight with "immutable, unstoppable, and irrefutable computer code" that instantiates the "tamper-proof" records [19], which allows these 'contracts' the ability to "self-enforce" [20]. For a permissioned blockchain, the records are distributed within a closed network that requires permission; however, it still sports the features of consensus amongst multiple computing devices. Examples of public blockchains include Bitcoin, Ethereum, EOS, Cardano, Flow, and Tron.

2.1.2. "Non-Fungible Tokens" (NFTs)

An NFT is a unique cryptographic record linked to an asset, typically a piece of art, music, collectable, or another presumed valuable object [10,12]. These could be thought of as similar to trading cards for the digital age. Basically, NFT protocols provide an underlying distributed ledger for records, and combine it with transactions that make them exchangeable in a peer-to-peer network [12]. These records are called 'tokens' that can be bought, traded, or sold much like physical assets in some ways and radically different in others. Since they run on blockchain technology that claims to prove validity of the ownership of an asset, ideally all transactions associated with this relationship (i.e., between record and actual object) are recorded. NFTs are generally bought, sold, and traded from 'wallets' and can be explored, if public, on websites such as Blockchain.com, TokenView, and BTC.com.

There are a few notable features of NFTs that have been suggested to promote stability and consistency [12]. Figure 1 is a helpful diagram that shows the NFT process and the roles of each actor, reproduced from one of the first systemized studies of NFTs [12].

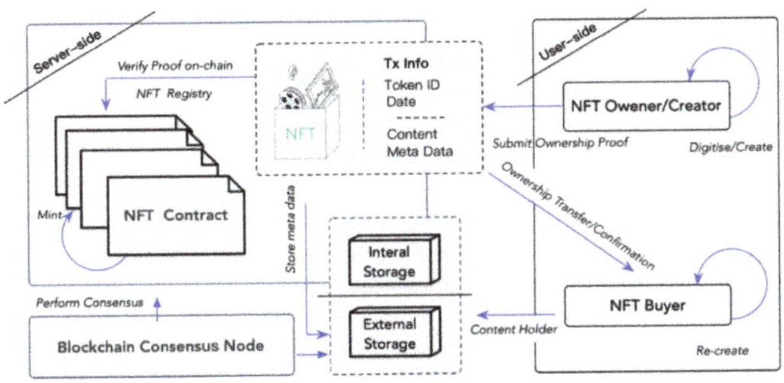

Figure 1. "Model of NFT Systems" (taken from Wang, Qin, et al. "Non-fungible token (NFT): Overview, evaluation, opportunities and challenges.").

Essentially this process consists of two roles: NFT owner and NFT buyer. An NFT owner digitizes the raw data from the transaction into the proper format, then stores it on a database external to the blockchain (or on a blockchain, but that is more costly). The owner then signs the transaction with a hash (or cryptographic signature made of a string of numbers) and sends the data to a smart contract. The smart contract processes the data, then mint or trades it on the blockchain as a transaction. Once it has been confirmed through mathematical consensus, the NFT is linked permanently to the unique hash identifier and the distributed blockchain record, where it cannot be changed.

2.1.3. "Document Standards"

This paper utilizes the phrase 'document standards,' which stands in for the multiple, interdisciplinary mechanisms that have been applied to records and documents over the centuries in order to maintain their authenticity. This draws upon document theory [21,22], diplomatics [12,13], interface theory [23,24], evidence [25–27], and regulatory schema that depend on records management, thus speaking to the interdisciplinarity of this method. These standards have been developed in nuanced, iterative ways over time with details that describe markers that provide actual validity and thus viability with documents and records. This includes how chain of custody and ownership is maintained, which information on users should be stored, even interrogating the epistemological underpinnings of how documents function and how users engage with them. This body of work was compiled organically from various fields, but they all share a common humanistic or sociological attention to the subject or user position that allows for a type of activism throughout their arguments. As a methodology, it is not only theoretical and qualitative, but also pragmatic since there are real issues that need to be solved in a practical and regulatory sense.

2.1.4. "Users"

This project separates off each of the stakeholders into general categories. The first of these categories is 'Users,' which includes any person that makes use of blockchain technology in any capacity. This may include people who are using a public blockchain administerd in a decentralized way, and also those who administer the technology in permissioned chains for their own purposes. It does not include the technology itself, for which there is a good argument to be made that there is a certain amount of agency that can be coded into the technology (for example, as a 'user' who perpetuates certain transactions on the blockchain such as smart contracts). Rather, this category is purporsely excluding technology from this category and looks to any real person whom did the coding at some point. This helps determine where the responsibilty lies when studying accountability and transparency issues in practice.

2.1.5. "Firms"

The category called 'Firms' is defined as any business entity that adminsters, facilitates, or incorporates a blockchain into its infrastructure. This includes public blockchain offerings that create the design for tokens which are 'mined' for consensus with the computing power across peer-to-peer devices. It also includes the permissioned blockchain administers whom incorpoarate blockchain databases into their current business environment. Lastly, it includes the intermediary companies that provide the services which facilitate blockchain applications; for instance, those that offer 'wallets' to allow for the buying, selling, and holding of tokens, or 'explorers' that allow for the reading of the cryptographic records of public systems. Examples of 'firms' include Ethereum (which provides the protocol and public blockchain), eToro (which is a platform that makes use of Ethereum protocols to create NFTs), and even Christie's Auction house (which is now selling NFTs).

2.1.6. "Regulators"

The third study looks at blockchain technology from the perspective of 'Regulators,' which includes any entity whose job it is to produce policies and practices that would protect those in the category 'users' from any malfeasance that originated in the 'firms listed in the second category. These institutions have the vast task of sorting through a technology that may have already outpaced current policy. This is especially important since some of the orignal motivations behind blockchain technology was to sidestep these exact regulatory bodies [19]. Current regulatory schemas are being developed for big tech companies across the board, including the General Data Protection Regulation (GDPR) by the EU and the California Consumer Privacy Act (CCPA) in the US, which push for more transparent disclosure practices and protections for patrons of these applications. In the financial space, regulatory agencies have already begun to tackle blockchain applications (e.g., Financial Crimes Enforcement Network [FinCen], Office of Foreign Assets Control [OFAC], Internal Revenue Service [IRS]). This study tries to be aware of the financial regualtory space as it is highly specialized and developing, yet stll applicable to NFTs in a similar fashion as cryptocurrencies and smart contracts. Moreover, it draws upon the new consumer protection regimes (e.g., GDPR and CCPA) to consider that other types of regulatory schema that no doubt will be applied to blockchain applications as well.

2.2. *Project Design*

This project is designed so that it provides a methodology for examining certain qualities of blockchain records that would provide transparency and accountability. It takes the definitions in the previous subsection and applies the category 'document standards' to 'blockchain technology' (specifically NFTs) to see if they are up to par and to imagine how they might be improved. It does so from the three perspectives of (1) Users, (2) Firms, and (3) Regulators. The following details the queries that were conducted for each perspective. The results of these queries are in the next section.

2.2.1. Queries for Users

- Copyright: How do you clearly and conspicuously "attach" terms and conditions to an NFT and ensure that those terms follow the NFT and bind subsequent owners? How are evolving copyright issues handled (public domain laws, estate changes, etc.)?
- Standard form contracts: Do these tokens present the same issues as other standard form contracts if this same genre of contract is used to lay out terms?
- Interface design: Do these records match user expectations and follow usability and design best practices? Are they designed with users in mind? Who is their audience?

2.2.2. Queries for Firms

- Jurisdiction: If a Firm makes use of a public blockchain to create their NFTs, do they streamline transactions or obscure necessary data (e.g., accd. to FCPA, SOX, UETA)?
- Liability: Do the current records protect from liability or increase risk?
- Compliance: Do the current NFT records aid or hinder other compliance or record-keeping efforts?

2.2.3. Queries for Regulators

- Investigation: Do the records produced by NFTs allow for the same procedures when investigating a bad actor?
- Evidence: Do the current records serve as viable evidence? What are the consequences if they do (accd. to the US Federal rules of Evidence) but do not have the required components of legitimate records (accd. to document standards)?
- Policy: How does blockchain aid or hinder the regulations set forth by KYC, OFAC, FASB, SOX, and GDPR, among others?

2.3. *Limitations*

Although effort was put into ensuring that this study is as sound and replicable as possible, it does have its limitations. This methodology follows the dominant research method from the past several decades, "interdisciplinarity," which has benefits in its application to the perspectives of multiple involved parties or stakeholders. The limits to this method, however, are that it has been accused of overlooking potentially contrasting assumptions in each discipline, promoting what was identified early on as "conceptual confusion" [28]. This often occurs through the motivation to find similarities between the disciplines that may not exist; but it can be remedied by noting explicit differences and contexts throughout the analysis, including between vastly different origins, history, objects of study, and interest [29]. While there is not enough space in this paper to give each of these fields the full treatment they deserve, the interdisciplinarity of this study is still useful as the fields utilized (that make up the category called "document standards") work toward the common goal of maintaining the qualities that determine whether a record is viable.

Additionally, some of the advantages of interdisciplinary research that this method hopes to benefit from include some logistical advantages such as a wider audience and other, loftier goals such as possibly more 'normative' conclusions, which might be more well-rounded and humane and consider "trade-offs" and "principles" (including ethical concerns) [30]. Also, writing for a larger, multi-disciplinary audience not only encourages more productive solutions that may not have occurred otherwise, but also provides a close textual study that is qualitative and holistic.

3. Results

This section describes the results of exploring the above questions, including the interpretation of the results and what it might mean for blockchain technology going forward. Below, the results of the application of document standards to the aforementioned queries from each of the three perspectives (i.e., Users, Firms, Regulators) is shown.

3.1. Description of Results

This section will describe in clear, succinct summaries how these queries were considered in this study. The document standards that were applied to each query are explained, as well as a brief acknowledgment of the context within which these standards were developed. The purpose of these results is not to prove that these queries *have* been answered by this study; rather, it is to show that these questions exist and that they *can* be answered.

3.1.1. Results for Users

In the case of NFTs, their purpose is to tie a record to prove the ownership of an asset, whether it be a digital file or a real-world item. While at first this may seem promising because it is a more immutable record due to its cryptographic qualities, this record can present very significant issues. First, most records of ownership that are used for this purpose generally need to have several qualities for them to be viable, including reliability, validity, and authenticity [12,13]. These qualities come from certain information being documented, demonstrating chain of custody and ownership, and most importantly to answer the queries for Users, the terms of this ownership over time. This last concern requires that the record makes use of a contract that will lay out the terms of copyright, use, transfer of ownership, etc. The concern here centers around two issues; (1) that the NFT record will begin to use *standard form contracts* for this purpose, which includes all the issues associated with this genre of contract [31–35], (2) that the immutability of the record will enforce these terms beyond even what normative contracts negotiate, not allowing for the flexibility that contracts require, and (3) that the design of the NFT record will further obscure any important information for the User with unfamiliar or deceptive design practices. These queries are answered within the history and context of contracts and design since they are a vital part of these records being viable for asset ownership.

As warned about in the literature around standard form contracts (e.g., Terms of Service), these 'zombie contracts' may appear as traditional contracts on the surface, but under deeper scrutiny, have "several distinct features that sit in very deep tension with contract [doctrine]" [31]. Contract law is generally viewed as a remedial "institution" whose function is to adjudicate any issues that arise between two individuals or entities after transactional activity as they arise [32]. Generally, contract law enables two or more self-governing parties to document shared goals. With *standard form* contracts, however, one party drafts the terms, relying on past types of similar contracts. In other words, notions of 'standardization' here are being defined and perpetuated by 'standard practice.' Without regulatory response to inhibit egregious terms, one party of these contracts is at a vast advantage in the arrangement and, further, there are concerns that powerful players could take control of their governance, ensuring any regulations result in their favor. [33].

Recent studies confirm that standard form contracts reinforce certain inequities between classes—not only through unregulated clauses, but also in terms of who advocates for fairness. Because of a difference of perception and knowledge of contracts amongst users, which varies according to socio-economic class, "elite customers" are less likely to advocate for the fairness of the agreement and are only motivated to stop egregiousness when it threatens them personally, further solidifying the distance between the "haves" and "have nots" [31,34]. This both reinforces a general distrust in the legal system for the lower classes as well as reactionary costs for the Firm such as evasion measures (e.g., piracy, hacking, misuse of services) that might come about consequently. Since there is a great amount of anonymity with blockchain, contracts used in an NFT situation where owners and traders/buyers are not identified could exacerbate some of these power imbalances with more immutable, boilerplate terms—an even stronger win for the more powerful entity with no path toward restitution other than those that are illegal and put other users at risk.

However, there are some areas where blockchain technology could help with securing contract terms. One of these areas is with unilateral modification clauses. These common clauses allow for a service provider to change the other terms of the contract at will, essentially making the other promises in the contract "completely illusory" and at the determination of only one party [35]. In other words, for a user, the concept of a contract document as a stable entity is disrupted by the mere fact that it could change at any time without their knowledge of this change. As Preston and McCann (2011) ask: "If the service provider can change the contract at will, why bother to call it a contract at all?" [35] If the records provide more concrete documentation, it could potentially provide a chain of records that prevents this clause from being enforced, which is significantly better than the Terms of Service agreements that change at will without a record of the past terms. There are normative benefits to contracts that have the flexibility to change terms as the relationship between the parties develop, and blockchain-made smart contracts have been critiqued from this standpoint prior [6]. These issues are not unfixable if thought is put into them before the issues that are already present are cemented into the genre. There is a body of research that is dedicated to studying the regulation of the issues associated with these contracts and it should be considered if they become a part of the purpose and practice of the records that support NFTs.

An additional area of study that could help NFT records from the perspective of Users, especially since the result is user-facing, would be the area of interface and document theory. These disciplines work toward understanding the concepts of documents and interfaces, both of which would be useful in the process of understand what users expect from these records. If they remain somewhat exclusive and specific to the realm of blockchain output, the average user will not find them meaningful, and consequently not be as transparent as they purport to be. For instance, information and media theorists have expanded on earlier ideas but locates the 'informativeness' of a document (and thus and definition of information itself) in its materiality, institutional embeddedness, and historical contingency, and recognition of user subjectivity, rather than in a theory that assumes an 'intentional substance' of a phenomenon [21–24]. In other words, once all the social and political forces that configure documentary practices are considered, "the genie is out of the bottle: the informativeness of documents, when recognized as dependent on practices is also dependent on what shapes and configures them" [21]. This body of work provides an analysis of documentary agency and seeks to understand how even fundamental understandings of a document, text, or interface can be undermined by a design that is inaccessible and unfamiliar [23,24]. This way of thinking is not only beneficial for Users, but for all stakeholders as it considers the audience and context of the record, thereby making the outcome more useable, more transparent, and, ultimately, more accountable.

3.1.2. Results for Firms

The biggest concern for Firms in regards to blockhain technology is the issues that come about from compliance policy and regulation. Some US laws that affect blockchain applications include recordkeeping determinations of the Foreign Corrupt Practices Act of 1977 (FCPA), the Sarbanes-Oxely Act (SOX), and the Uniform Electronic Transaction Act (UETA) of 2000. Civil systems such as the E.U. have also sought to find ways to keep coroporations accountable with the information they store and disclose, including GDPR (and subsequently California's CCPA), for which eventually Firms from the blockchain space will also be held accountable. As a new sphere of corporate activity has surrounded blockchain technology, these new companies, which are often start-ups, are concerned with liability and compliance as they coud become major issues to their endeavors.

Produced in reaction to a string of several major corporate scandals such as Enron and WorldCom, SOX was the most comprehensive accounting reform enforcement since the FCPA from the late 1970s [36]. Both regulatory schema outlined accounting and bookkeeping requirements for companies, with FCPA handling transparency in dealing with foreign officials and SOX concentrating on the alteration and destruction of records. Both acts also endorse the need for more thourough recordkeeping requirements to prevent and prosecute white collar crime and fraud. Specifically, the FCPA requires companies to make accurate and complete records and devise methods for an internal system of accounting controls [31]. SOX expanded on these basic requests, asking for transparent and accurate corporate records (Title III, Section 302) and both internal and external auditing assessments. It also outlines specific actions regarding the destruction of or tampering with records involved in "official proceedings" (Title VIII also called the "Corporate Fraud Accountability Act of 2002"). Although blockchain records claim to be immutable, there are ways to code in acts of deletion with smart contracts or to fork the chain, which changes the outcome of the order of the records. If this happened with a public blockchain application, a bad actor could code in the deletion of the record, violating SOX and making the Firm liable for sanctions and fines.

On the other hand, the UETA laws of the early 2000s were meant to streamline commercial activity online across multiple jurisdictions [37], yet in doing so, lowered the recordkeeping requirements for digital transactions, including removing the retention of paper copies and relaxing the types of actions that signal agreement amongst parties. These laws ultimately benefit NFTs and blockchain technology as they lower the standard for compliance and do not require any extra disclosures on the part of whomever is managing the blockchain (the viewers and wallet companies, for instance). This only goes so far, however, as other more recent laws supercede these and require these types of companies to retain knowledge of their customers for money-laundering reasons (remember, this does not the blokchain itself, as no one is responsible exactly for those records).

Newer EU policies, including the General Data Protection Regulation (GDPR) that was enacted in May 2018, specifies, among other principles, an accountability mechanism referred to as "the right to know" that describes corporate responsibility to inform consumers about any collected personal data and the algorithms that affect a user's experience in this regard [38]. Clarity of information to the public is one of the major principles of GDPR and requires transparency in terms of the contracts that govern the way this information is handled. From these nerw regulations came other regulatory schema have been formed such as the California Consumer Protection Act (CCPA) that requires transparency and disclosuresfor technology companies and complicates this already complex regulatory landscape. It is difficult to know how these new regulatory paradigms will play out with blockchain technology since NFTs are built publically and are used for many different purposes and for various 'audiences.' And while the one of the limitations of this study is the simplification of these groups of interested parties, the study overall is useful in its ability to consider the liability that the firms whom facilitate these applications might be subjected to.

3.1.3. Results for Regulators

Regulatory bodies are concerned with how they might protect consumers and the public from corporate activity they deem harmful. Generally, this process of accountability involves either preventative measures such as compliance requirements or accountability measures after the fact that investigate and collect evidence for prosecution. With blockchain technology, this process can be hindered by the issues with authentic documents and how these documents serve as proof for investigative purposes. As mentioned in the last section, preventive measures and compliance do not always lead to easier culling for regulatory bodies or prosecuting entities. This section focuses on their perspective and the difficulties blockchain technology presents when building a case against a Firm, which looks to the rules and uses of evidence to determine how effective these records are in this realm.

Records have long been associated with manifestations of evidence, even if evidence can take many other forms and instantiations [26,27]. The history and uses of evidence suggest that understandings of what comprises evidence are dependent upon the paradigm within which evidence will be acquired, assessed, and introduced, not just simply how they support an argument logically. For example, even if a record is produced that seemingly claims to prove something over something else objectively, the record itself must prove that it is an authentic document before it can be used in the case. Thus, the circumstances around the record and its creation, history, uses, etc. must also be rigorously considered. This project makes the case that considering legal conceptions of evidence from a more holistic, philosophical perspective such as in the research of legal scholar John Henry Wigmore (1863–1943), is a more useful route than simply relying on static definitions such as those laid out in legal doctrine.

Wigmore's work argued that "facts are evidence insofar as they play a role in a teleologically directed argument" [25]. In other words, his work found that the process of using evidence to prove a case required viewing it as three distinct layers of information: (1) as a proposition (hypothesis); (2) as specific elements of law that need to be satisfied; and (3) as material evidence and facts that make up the narrative of the case. This view of evidence fixes the "worn out legal system" of the nineteenth century that relied mostly on numerical systems and had "no understanding of the living process of belief." This paradigm, while over a century old, is useful to consider as it brings back the nature of human judgement and essentially what is overlooked if blockchain technology stands in for legitimate records.

Wigmore distinguishes the uses of evidence into two categories: (1) the analysis that details the informal logic of reasoning and argumentation, which he called Proof, and (2) the "rules of procedure," which he called Admissibility. Proof consists of the practice concerned with "the ratiocinative process of continuous persuasion." Admissibility consists of the procedural rules developed by the law and based on "litigious experience and tradition, to guard the [jury] against erroneous persuasion." These distinctions can help us consider the "probative force" of a piece of evidence, which describes its tendency to support or to negate the first piece of information, the proposition or hypothesis. A record in this paradigm could have two outcomes once it is interpreted in a trial: Proof or non-Proof. Currently, the Federal Rules of Evidence govern the process of evidence admissibility and discretion in trial court. A difference in Wigmore's conception of evidence is that he makes distinctions between other aspects Proof, including argumentation, inference, and probative value, rather than superficially codifying them into procedure doctrines [25]. This project confirms the usefulness of these early distinctions by Wigmore, providing a rigorous method of logical inquiry that benefits current evidentiary paradigms in which these aspects might be discounted.

A more holistic and humanistic understanding of evidence is especially important since blockchain records could give the false sense that somehow, they implicitly have the qualities needed for Proof—the issue being that unless the technology and infrastructure of the blockchain being considered is scrutinized properly and understood thoroughly, these qualities are not ensured. In other words, if an investigator or regulator or even a jury or judge misunderstand the technology behind a blockchain record and think that somehow it purports some type of reliable information because of its cryptographic features or architectural design, these records will be *more* dangerous than had they not been made of blockchain. The technology could provide a misleading testament to validity when actually they could *aid* in nefarious activity.

3.2. Interpretation of Results

The results (Table 1) can mean only one thing for blockchain environments—more work is needed before they accomplish the transparency it claims to provide. Below a screenshotted is provided of two recent NFT records taken from the blockchain explorer Adapools.com. These records show the information that is typically displayed on a blockchain record. It is also useful to see a real example of a NFT record to analyze how it might be improved to address the issues just mentioned. Newly proposed methods [39] that take apart the components of a document could consider the components as they currently exist and reconstruct the design of the record to include appropriate context and information from each of the three perspectives. The rest of this section details how a few important aspects of this example NFT record could be improved.

Table 1. Description of Results.

Perspective/Stakeholder	Area of Concern	Applicable Document Standards	Examples
User	• Design • Usability • Ownership	• UX/Human Factors Principles • Contract Literature (esp. standard-form) • Copyright Literature	• Clear coherent design that considers the needs of user ownership over time, including: (1) "Name" (2) "Publisher" (3) "Collection" (4) "Artist" • Fair contract terms • Appropriate contextual elements that utilize familiar design conventions • No deceptive design that perpetuates inequities
Firm	• Assets • Recordkeeping • Compliance • Risk Assessment and Reduction	• Recordkeeping standards (i.e., validity/reliability) • Asset document conventions and standards • Electronic contract standards and compliance (e.g., UETA, FCPA, AML, SOX) • Transparency requirements such as data disclosures (e.g., GDPR, CCPA)	• Storing customer information for AML and preventing deletion of records for SOX • Current asset record requirements: (1) Description (2) Location (3) Procurement (4) Life Cycle (5) History (6) Depreciation value (7) Insurance (8) Maintenance (9) Ownership distinctions (10) Barcode or serial number (11) Warranty information

Table 1. *Cont.*

Perspective/Stakeholder	Area of Concern	Applicable Document Standards	Examples
Regulator	• Evidence • Protection • Transparency	• Evidentiary requirements • Discrete regulatory schema that are effective	• Viable evidentiary components, including: (1) Parallel contracts in other mediums (2) Searches, subpoenas (3) Payment history between parties (4) Communication between parties (e.g., phone calls, email) (5) Witness accounts (6) Blockchain environment-specific conditions, (e.g., authentication protocols, automated features, triggers, oracles) (7) Personal computer evidence (acquired through search warrants) (8) Wallet managing services, along with ISPs, phone records

For Users, the two most important considerations are attached contract terms and usability. From Figure 2a,b it is apparent that the information is neither contextualized nor understandable in terms of the normative conception of asset records of ownership. Further, although the cryptographic hash promises to make the record unique (which is does in a technological sense), it is difficult to decipher the difference between the digital objects when similar records are inquired about as you can see in Figure 2b. In terms of the contract issues, this record is not sophisticated enough to have attached terms; however, one could imagine what that might look like in this instance. Some of the initial, basic information for a contract arrangement is present, yet not clear—the two parties could be the "Name," "Collection," "Artist," or "Publisher." It is not apparent from the information shown what the relationship is between these entities. Moreover, if the basic ownership information is not clear, then the projected trajectory of their relationship is even more unclear. In a contract situation, this information would be clarified upfront in a Preamble, for instance. As an example of how much contracts could be skewed in this environment, consider the effort that generally goes into the conventions and allowances of very nuanced language choices in a non-blockchain produced contract [40].

Additionally, the anonymity of these entities allows for the record to exist publicly and be authenticated on a public blockchain. While this is important for 'transparency' in this respect, if a contract situation is entered, which in the case of a NFT it seems would be quite common, then it could preclude clear and conspicuous terms from also benefiting from this transparency. If there is a power imbalance (or information asymmetry) between the two contracting parties, this might be especially concerning [34].

For Firms, the current design of NFTs could confuse some of the requirements and responsibility of a corporate entity. As Firms are concerned with liability, an ambiguous set of metadata might not satisfy the information needed for FCPA. As the anonymity can obscure the location information, the rules for foreign transaction are also obscured. This might cause the company to be at risk for compliance issues unless this information has been obtained and stored, which would then reduce the anonymity that allows for the transparency that produces immutable records. This is one area that needs much more focused attention—sorting out personal information, privacy, and compliance while negotiating the needs of a decentralized architecture and maintain the integrity it claims to promote.

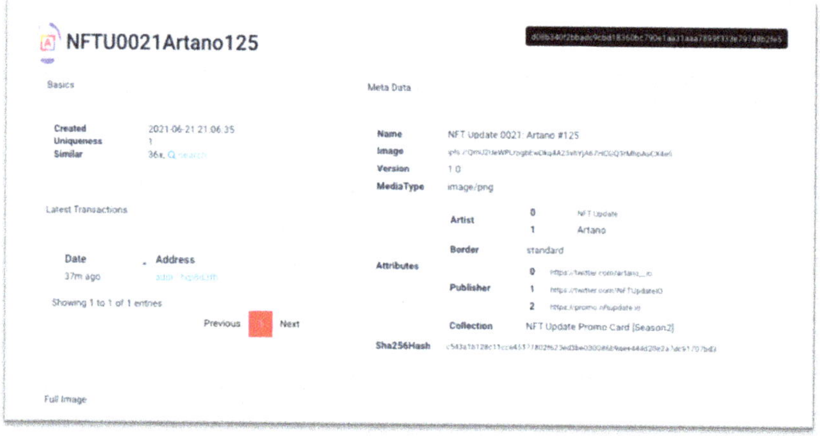

(a)

(b)

Figure 2. This is a figure. Schemes follow another format. If there are multiple panels, they should be listed as: (**a**) An example of a public NFT record; (**b**) A search from the "Similar" link on Figure 1. Source: Adapools.org.

Moreover, it could be argued possibly that blockchain records have simply been streamlined for easy transaction under the UETA laws. Yet, if nefarious activity such as money laundering took place—for instance, if someone 'bought' an asset with tainted currency and makes use of the anonymity that it provides, the transaction could be a liability for the company. Like foreign transactional requirements, if the company does not retain the information about the parties behind the NFT, the laws that require financial institutions to have this data such as Know-Your-Customer (KYC) laws. If a participant in a NFT transaction is sanctioned by the Office of Foreign Assets Control (OFAC), this could be a major violation of the rules set forth by FinCen or other agencies. Since NFTs are records of ownership, it makes sense to look at how these types of records have been considered previously. This includes how they are situated, criticized, used as evidence, retain their authenticity, and more. For example, is important to know what information has been required on these types of records. Then, when concerning NFTs, it should be negotiated how this information will be displayed or retained with privacy concerns. For instance, blockchain technology in general might have trouble complying with new disclosure laws (i.e., GDPR and CCPA) and something like 'privacy by design' could be necessary. Making use of UX and interface design theories (as outlined in the Users' section) could also aid firms in this task [41].

Additionally, it is in the interest of a firm to correctly manage their assets so that they can accurately document the value of their company and avoid unnecessary risk. This is where considering previous document standards are useful. Asset records, to be useful, must contain certain information according to the financial and business world. This includes: (1) a description of the asset, (2) the exact location of the asset (what about link rot? is it hybrid of local/cloud etc.), (3) procurement details including purchase dates and price history, (4) life expectancy, (5) depreciation value, (6) insurance and compliance details, (7) maintenance history, including repairs and downtime, (8) the owner of the asset versus the user of the asset, (9) the barcode or serial number, (10) warranty information, and the list goes on. It benefits the business to have all this information since a lot of these transactions are *financial* and money laundering is a real concern. Anti-money laundering (AML) rules require that a company know their customer (KYC) and have certain data on them or else they become liable. In the video game world, it would be worth considering who would be responsible in this case—the video game creators or the group that maintains the NFT protocol? In the blockchain world, this needs to be negotiated with the movement that is motivated to be both transparent *and* anonymous. Obviously, including all that information on an NFT record would be a task that needs some thinking in terms of design.

For Regulators, the main concern in terms of blockchain technology is how it aids or hinders their goal of keeping Firms accountable and providing the proper scrutiny to stop the disreputable uses of its features. This includes the rigorous examination of blockchain records as evidence, including not accepting that the technology itself creates validity or reliability in a diplomatic sense [15,16]. As Wigmore points out, the creation of an argument using evidence such as those used in litigation, does not depend solely on the qualities of the records. These arguments are contextual and if all parties involved do not understand the records fully, which is nearly impossible with NFT records in their current form (Figure 2a), then the case will not be fully examined appropriately. This could have major consequences down the road if the precedent is set that the records *are* viable in their current form. The qualities they purport to have will never actually be developed. For instance, in previous work, a criminal investigator from the U.S. Internal Revenue Service (IRS) provided a list of all the information needed for prosecution of a blockchain case that involved smart contracts [7]. This list includes:

(1) A parallel paper contract with agreements from both parties (public records
(2) Searches, subpoenas to record keeping authorities)
(3) A history of payments between A and B (subpoenas to corresponding financial institutions)
(4) Email, Phone, other types of communication (subpoenas or search warrants)
(5) Undercover, if necessary (to reveal and confirm any allegations of fraud on the part of the various parties)
(6) Witness accounts from the inception of the agreement through the present (interviews)
(7) Conditions on the blockchain that would reveal payment/non-payment
(8) Their corresponding truth/falseness in real life
(9) Authentication protocols for both users and their conditions (are the transactions automated or triggered by one party or the other, and for what reason?)
(10) Personal computer evidence (acquired through search warrants or consent searches)
(11) Wallet managing services, along with ISPs, phone records and other third-party record keepers would be acquired through subpoena or court order.

The list is quite thorough, but it gives a sense of what could be required to build a case against a Firm. While it is not practical to assume or even hope that a Firm will be required to provide this information since one of the major features of blockchain application is the benefits of an anonymous system, if Regulators are serious about preventing or prosecuting bad actors, then some thought is due in this area. Perhaps there is a way to keep the information in a secured location so that the Firm can comply with regulatory schema—however, this would require that it be kept in a centralized system, which ultimately would undermine the protection that a public chain offers, and which patrons currently enjoy.

4. Discussion

One of the most important aspects of the issues highlighted by this discussion is that the main feautre of a public blockchain—its transparency, which allows it to perpetuate claims of accountability—does not *ensure* that this accountability takes place without other considerations. While private information is generally protected, this anonymity also allows for the easier facilitation of nefarious activity such as contracts with unfair terms created by asymmetric power arrangements, ambigous and unfamiliar design, compliance issues, money-laundering issues, and more. Thinking through some of these issues premptively and prior to these practices becoming normalized could help begin a real, sustainable future for blockchain. Future research on this topic could help negotiate privacy concerns and anonymity benefits with transparency claims, for instance. Pinpointing responsible parties and maintenance issues, deletion concerns, and record metadata requirements is a daunting task with distributed infrastructure in some respects, but is also a worthwhile endeavor so as to make sure that the features of the technology are not abused.

There can be a few conclusions drawn from this preliminary study. The first is that the current implementations of blockchain technology need a bit of brainstorming and creative work in terms of its records and the audiences for which they are intended. Simplified conclusions from this study are as follows:

- *Users*: Blockchain records for Users could be developed with standard form contract issues in mind, as well as with usability and interface theory conceptualizations of how users engage with the technology and the records it produces. This might include designing the records so that they include context and familiarity, which would be useful in holding the parties of a contract accountable and providing those with the information they need to litigate a more powerful party. It could also normalize some of these design conventions so that the average person could understand their information, creating best practices that could iteratively be made better moving forward. *This is especially important for IoT applications as they are embedded into the fabric of everyday life, proliferating contracts for each instance of each transaction.*
- *Firms*: Blockchain records for Firms could be improved with some research that focuses on the information that should be required for companies to retain on their own records. Also, it should be considered how this would contribute to compliance or a lack thereof. Negotiating privacy concerns with the information that the company needs to be compliant is one of the biggest concerns for Firms at the moment. Reducing risk and liability for this new type of company encourages innovation in this space *as well as* sustainability. More users will begin to make use of blockchain applications as trust in the companies increases, which will help streamline some of the compliance requirements into more practical applications.
- *Regulators*: Blockchain records present several quagmires for Regulators. Since the decentralization of the blockchain ledger offers protection for Users, it is difficult to negotiate this protection with the necessary information needed to keep the Firms accountable. One possible way forward is to create personal data stores for consumers that could be accessed in an investigation [41]. There are still issues with this idea such as the inability to access this data due to multiple entities being involved, yet it could be studied as a model to prompt the sorting out of what information needs to be on the blockchain and who is responsible for its retention. Making nuanced and effective regulatory schemes is a concern being tackled for all technology applications. Along these lines, new disclosure laws such as GDPR and CCPA can be satisfied as effort is put forth in this space.

Ultimately the main conclusion that can be drawn from this study, and why it is named 'Betraying Blockchain ... ,' reiterates that although it seems like the technology associated with these records proves a sense of viability in its immutability or transparency, those qualities should not be taken for granted as there is still much work to do in this space. And while blockchain, smart contracts, and NFTs are seemingly solving some of the issues present in the currently popular centralized data-collection application model, they cannot solve them adequately unless the context and situation of the application is considered. That includes looking at how this new technology replaces the past functions of records and how well it lives up to solving the problems associated with that practice. The IoT future will include applications that are involved in many aspects of our lives and involve sensitive data, so simply switching to a different type of infrastructure should not be thought of as a cure-all. Hopefully this work will free up interested parties to make use of more interdisciplinary research that moves from just cryptography to document and record theory and practices. The technology is exciting, but the future that awaits is dependent on us making use of and building upon the work that has come before us. This will then, in turn, support the goals of those utilizing the qualities that the technology should provide as the records can live up to the standards that have been developed over many centuries; only then can blockchain technology can be considered a 'development' in its truest sense.

Funding: This research received no external funding.

Institutional Review Board Statement: Not applicable.

Informed Consent Statement: Not applicable.

Data Availability Statement: Not applicable.

Conflicts of Interest: The authors declare no conflict of interest.

References

1. Deceived by Design. Forbruker Rådet. Norwegian Consumer Council. 2018. Available online: https://fil.forbrukerradet.no/wp-content/uploads/2018/06/2018-06-27-deceived-by-designfinal.pdf (accessed on 4 August 2021).
2. Koivisto, I. The anatomy of transparency: The concept and its multifarious implications. EUI MWP, Cadmus, European University Institute Research Repository. 2016. Available online: http://hdl.handle.net/1814/41166 (accessed on 4 August 2021).
3. Obar, J.A. Sunlight alone is not a disinfectant: Consent and the futility of opening big data black boxes (without assistance). *Big Data Soc.* **2020**, *7*, 1–5. [CrossRef]
4. De Filippi, P.; Wright, A. *Blockchain and the Law: The Rule of Code*; Harvard University Press: Cambridge, MA, USA, 2018.
5. Cornelius, K.B. Smart Contracts as Evidence: Trust, records, and the future of decentralized transactions. In *International Handbook of Internet Research*; Springer: Berlin/Heidelberg, Germany, 2018.
6. Levy, Smart Contracts Levy; Karen, C. Book-Smart, Not Street-Smart: Blockchain-Based Smart Contracts and The Social Workings of Law. *Engag. Sci. Technol. Soc.* **2017**, *3*, 1–15.
7. Cornelius, K.B. Standard Form Contracts in a Smart Contract Future. *Internet Policy Rev.* **2018**, *7*, 2. [CrossRef]
8. Cornelius, K.B. Smart Contracts and the Freedom of Contract Doctrine. *J. Internet Law* **2018**, *22*, 3–11.
9. DuPont, Q. Blockchain Identities: Notational Technologies for Control and Management of Abstracted Entities. *Metaphilosophy* **2017**, *48*, 634–653. [CrossRef]
10. Doan, A.; Johnson, R.; Rasmussen, M.; Snyder, C.L.; Sterling, J.; Yeargin, D.G. NFTs: Key US Legal Considerations for an Emerging Asset Class. *Jones Day*. 14 May 2021. Available online: https://www.jdsupra.com/legalnews/nfts-key-u-s-legal-considerations-for-6366844/; https://www.coingecko.com/en/nft (accessed on 4 August 2021).
11. Wang, Q.; Li, R.; Wang, Q.; Chen, S. Non-fungible token (NFT): Overview, evaluation, opportunities and challenges. *arXiv* **2021**, arXiv:2105.07447.
12. Ante, L. *The Non-Fungible Token (NFT) Market and Its Relationship with Bitcoin and Ethereum*; Blockchain Research Lab: Hamburg, Germany, 2021.
13. Imbault, F.; Swiatek, M.; de Beaufort, R.; Plana, R. The green blockchain: Managing decentralized energy production and consumption. In Proceedings of the 2017 IEEE International Conference on Environment and Electrical Engineering and 2017 IEEE Industrial and Commercial Power Systems Europe (EEEIC/I&CPS Europe), Milan, Italy, 6–9 June 2017; pp. 1–5. [CrossRef]
14. Duranti, L. Diplomatics: New Uses for an Old Science. *Arch. Assoc. Can. Arch.* **1989**, *28*, 7–27.
15. Duranti, L. Reliability and Authenticity: The Concepts and their Implications. *Arch. Assoc. Can. Arch.* **1994**, *39*, 5–10.
16. Suciu, G.; Nădrag, C.; Istrate, C.; Vulpe, A.; Ditu, M.C.; Subea, O. Comparative analysis of distributed ledger technologies. In Proceedings of the 2018 Global Wireless Summit (GWS), Chiang Rai, Thailand, 25–28 November 2018; IEEE: Manhattan, NY, USA, 2018.

17. Wall, L. *Smart Contracts' in a Complex World*; Federal Reserve Bank of Atlanta: Atlanta, GA, USA, 2016. Available online: https://theihs.org/blog/5-advantages-of-interdisciplinary-research/; https://www.frbatlanta.org/cenfis/publications/Notesfromthevault/1607 (accessed on 4 August 2021).
18. Mills, D.C.; Wang, K.; Malone, B.; Ravi, A.; Marquardt, J.; Badev, A.I.; Brezinski, T.; Fahy, L.; Liao, K.; Kargenian, V.; et al. *Distributed Ledger Technology in Payments, Clearing, and Settlement*; Finance and Economics Discussion Series 2016-095; Board of Governors of the Federal Reserve System: Washington, DC, USA, 2016. [CrossRef]
19. Szabo, N. Smart Contracts: Building Blocks for Digital Markets. *Alamut*. 1996. Available online: http://www.alamut.com/subj/economics/nick_szabo/smartContracts.html (accessed on 4 August 2021).
20. Frohmann, B. Revisiting "What is a Document?". *J. Doc.* **2009**, *65*, 291–303. [CrossRef]
21. Buckland, M. Document Theory: An Introduction. 2013. Available online: People.ischool.berkeley.edu/~{}buckland/zadardoctheory.pdf (accessed on 4 August 2021).
22. Drucker, J. Entity to Event: From Literal, Mechanistic Materiality to Probabilistic Materiality. *Parallax* **2009**, *15*, 7–17. [CrossRef]
23. Drucker, J. Humanities Approaches to Interface Theory. *Cult. Mach.* **2011**, *12*. Available online: http://svr91.edns1.com/~{}culturem/index.php/cm/article/download/434/462 (accessed on 4 August 2021).
24. Anderson, T.; Twining, W. *Analysis of Evidence: How to Do Things with Facts Based on Wigmore's Science of Judicial Proof (Law in Context)*; Northwestern University Press: Evanston, IL, USA, 1991.
25. Furner, J. Conceptual Analysis: A Method for Understanding Information as Evidence and Evidence as Information. *Arch. Sci.* **2004**, *4*, 233–265. [CrossRef]
26. Yeo, G. Concepts of Record (1): Evidence, Information, and Persistent Representations. *Am. Arch.* **2007**, *70*, 315–343. [CrossRef]
27. Benson, T.C. Five Arguments Against Interdisciplinary Studies. *Issues Integr. Stud.* **1982**, *1*, 38–48.
28. Vosskamp, W. From Specialization to the Dialogue Between the Disciplines. *Oakl. Univ. Issues Integr. Stud.* **1986**, *4*, 17–36. Available online: https://our.oakland.edu/handle/10323/4017 (accessed on 4 August 2021).
29. Glod, B. *The Significant Advantages of Interdisciplinary Research*; Institute for Humane Studies: Menlo Park, CA, USA, 2016.
30. Leib, E.J.; Eigen, Z.J. Consumer Form Contracting in the Age of Mechanical Reproduction: The Unread and the Undead. *Univ. Ill. Law Rev.* **2017**, 65–108. Available online: https://ir.lawnet.fordham.edu/faculty_scholarship/883/ (accessed on 4 August 2021).
31. Griffin, R.C. Standard Form Contracts. *N. Carol. Cent. Law J.* **1978**, *9*, 158.
32. Kim, N. *Wrap Contracts: Foundations and Ramifications*; Oxford University Press: Oxford, UK, 2013.
33. Ventuini, J.; Louzada, L.; Maciel, M.; Zingales, N.; Stylianou, K.; Belli, L.; Magrani, E. Terms of Service and Human Rights: An Analysis of Online Platform Contracts. Revan. 2016. Available online: http://internetgovernance.fgv.br/sites/internetgovernance.fgv.br/files/publicacoes/terms_of_services_06_12_2016.pdf (accessed on 4 August 2021).
34. Preston, C.; McCann, E.W. Unwrapping Shrinkwraps, Clickwraps, and Browsewraps: How the Law Went Wrong from Horse Traders to the Law of the Horse. *BYU J. Public Law* **2011**, *26*, 1.
35. Demming, S.H. The Potent and Broad-Ranging Implications of the Accounting and Record-keeping Provisions of the Foreign Corrupt Practices Act. *J. Law Criminol.* **2006**, *96*, 465.
36. Robert, A. Wittie and Jane K. Winn. Electronic Records and Signatures under the Federal E-SIGN Legislation and the UETA. *Bus. Lawyer* **2001**, *56*, 293–340.
37. Werbach, K.; Cornell, N. Contracts Ex Machina. *Duke Law J.* **2017**, *67*, 313–382. Available online: https://scholarship.law.duke.edu/dlj/vol67/iss2/2 (accessed on 4 August 2021).
38. Gustavo, A., Jr. Blockchain and CCPA. *St. Clara High Technol. Law J.* **2021**, *37*, 231.
39. Hutton, C. *Language, Meaning and the Law*; Edinburgh University Press: Edinburgh, UK, 2009.
40. Norval, C.; Kristin, B. Cornelius, Jennifer Cobbe, and Jatinder Singh. Disclosure by Design: Document engineering for meaningful data disclosures. 2021; Unpublished work.
41. Janssen, H.; Cobbe, J.; Norval, C.; Singh, J. Decentralised Data Processing: Personal Data Stores and the GDPR. *Int. Data Priv. Law* **2020**, *10*, 356–384. [CrossRef]

information MDPI

Article

A New Data-Preprocessing-Related Taxonomy of Sensors for IoT Applications

Paul D. Rosero-Montalvo [1,*,†], Vivian F. López-Batista [2,†] and Diego H. Peluffo-Ordóñez [3,4,†]

1 Computer Science Department, IT University of Copenhagen, 2300 Copenhagen, Denmark
2 Department of Computer Science and Automatics, University of Salamanca, 37008 Salamanca, Spain; vivian@usal.es
3 Morocco and SDAS Researh Group, Modeling, Simulation and Data Analysis (MSDA) Research Program, Mohammed VI Polytechnic University, Ben Guerir 43150, Morocco; diego.peluffo@sdas-group.com or diego.peluffo@aunar.edu.co
4 Faculty of Engineering, Corporación Universitaria Autónoma de Nariño, Pasto 520001, Colombia
* Correspondence: puldavid87@gmail.com
† These authors contributed equally to this work.

Abstract: IoT devices play a fundamental role in the machine learning (ML) application pipeline, as they collect rich data for model training using sensors. However, this process can be affected by uncontrollable variables that introduce errors into the data, resulting in a higher computational cost to eliminate them. Thus, selecting the most suitable algorithm for this pre-processing step on-device can reduce ML model complexity and unnecessary bandwidth usage for cloud processing. Therefore, this work presents a new sensor taxonomy with which to deploy data pre-processing on an IoT device by using a specific filter for each data type that the system handles. We define statistical and functional performance metrics to perform filter selection. Experimental results show that the Butterworth filter is a suitable solution for invariant sampling rates, while the Savi–Golay and medium filters are appropriate choices for variable sampling rates.

Keywords: Internet of Things; sensor; machine learning; computational intelligence; data analytics; data pre-processing

Citation: Rosero-Montalvo, P.; López-Batista, V.; Peluffo-Ordóñez, D. A New Data-Preprocessing-Related Taxonomy of Sensors for IoT Applications. *Information* **2022**, *13*, 241. https://doi.org/10.3390/info13050241

Academic Editors: Spyros Panagiotakis and Evangelos K. Markakis

Received: 11 April 2022
Accepted: 5 May 2022
Published: 9 May 2022

1. Introduction

Internet of Things (IoT) technology allows electronic devices to be deployed in indoor and outdoor environments to collect data [1]. Commonly, these IoT devices consist of a microcontroller, sensors, a battery, and wireless communicationelectronic devices to be deployed in indoor and outdoor environments to collect data. IoT devices can be installed in harsh scenarios due to their flexible development [2]. Nowadays, about 22 billion IoT devices are uploading data to the cloud. Every year, this number increases exponentially to continue collecting data through a wide variety of sensors. These data are used to train machine learning (ML) models, powerful tools that can find hidden knowledge in data that describes a phenomenon or human behavior [3]. However, constantly uploading data to the cloud causes bottlenecks in the communication channel, and in some cases, the stored data are not processed for a specific purpose [4]. Hence, cloud computing servers have to delete data periodically to avoid storage overload. Consequently, data quality is essential to reduce the complexity of the ML model, and it is necessary to send only relevant data to be processed. Therefore, after the data gathering process, a data pre-processing step is required to eliminate errors, which means both stages are part of the ML pipeline [5]. There are several repositories in different areas where researchers and developers can obtain databases to deploy and test ML models. They assume that the data are cleaned before being put into the repository. Nevertheless, this is not the case for IoT environments, where sensors gather data in situ because they describe specific parameters such as environmental conditions, gas concentration, and location. In conclusion, data gathering and pre-processing are obligatory stages of building an ML application in IoT environments.

The data collection stage in IoT environments needs to handle uncontrolled conditions such as environmental changes, construction failures of microcontrollers and sensors that cause poor calibration, and vibrations in their working environment, among others [6]. Therefore, model inference can reduce performance, resulting in the use of complex models when describing a phenomenon or human behavior [7]. The pre-processing stage produces reliable, accurate, repeatable, and error-free data [8]. Thus, the electrical signal obtained by the sensors should be acquired with an adequate sampling rating and proper tuning of analog-to-digital converters [9]. On the software side, digital filters are applied when data have been stored on servers. However, sensors have different data collecting procedures, such as digital-analog converters, communication ports, and pulse trains [10]. Therefore, the cloud can not apply a standard filtering process to all the features stored. Additionally, this data flow over communication channels increases security concerns and decreases user confidence in the system. Therefore, new computational paradigms propose decentralized computing where some ML stages run closer to the user, which means performing data-preprocessing locally [11]. It is worth pointing out that these algorithms can be deployed on IoT devices due to the microcontrollers' increasing computing capacity, which will not affect battery consumption [12]. Additionally, sensors vendors are working to give the IoT developer robust libraries to improve sensor management [13]. However, sensor data need to be pre-processed before sending it to the cloud [14].

Data filtering removes noise by comparing each signal component to the rest and eliminating the unusual ones. The most relevant criteria and their principal algorithms are infinite impulse response (IIR) with the approximations Butterworth, Bessel, and Chebyshev; finite impulse response (FIR) with the windows Hamming, Tayler, Barlett, and Blackman; and smoothing filters with the algorithms: mean, average, Gaussian, and Savi–Golay [14]. For more information about digital filter design, we suggest following these works [15,16]. These filter criteria depend on the sampling rate at which the IoT device is configured, the collection procedure of each sensor, and the application. However, previous sensor taxonomies focus on hardware characteristics without considering their primary purpose of collecting data. In addition, data filtering criteria are applied for each IoT development, which consumes additional time for IoT researchers and developers.

It is necessary to define a new sensor taxonomy related to the data collection and pre-processing processes that fits the filtering criteria to be part of the whole ML pipeline [17]. Therefore, this work introduces a new sensor taxonomy oriented to pre-processing data on-device according to the type of sensor used in the IoT application. Consequently, we need to define how IoT devices collect data through sensors to determine the suitable filter for each case. Our summarized contributions are:

- We define a new sensor taxonomy related to data gathering and data pre-processing on-device.
- We determined that the main sensor characteristic for classification is sampling rate.
- We introduce a data filtering scheme using the most representative algorithms/models of infinite impulse response (IIR), finite impulse response (FIR), and smoothing filters by setting specific sampling rates for each sensor type.
- We compare data filtering criteria to select the suitable ones for the proposed taxonomy of sensors and ensure its usefulness in computationally constrained IoT environments.
- We performed tests on sensor data with statistical and functional metrics.

The main result of this work is defining the Butterworth filter as a suitable criterion for analog sensors with invariant sampling rates. Meanwhile, Savi–Golay fits analog sensors with varying sampling frequencies. The average filter is adequate with this signal in digital pulse train sensors. Savi–Golay and medium filters remove noise and preserve the main signal characteristics regarding communication protocol sensors.

The rest of the manuscript is structured as follows: Section 2 shows related works and signal filtering background. Then, Section 3 introduces the proposed sensor taxonomy. Next, the methodology is shown in Section 4. Results are presented in Section 5 with the statistical and sensor functionality metrics to define the filter algorithm we chose. Finally, Section 6 shows conclusions and future work.

2. Background

In this section, we present a summary of previous sensor taxonomy and data filtering works.

2.1. Early Studies Sensors

The increasing use of electronic devices in the industry has opened up opportunities to develop different types of sensors. Indeed, new technology trends such as the Internet of Things (IoT) allowed expanding the research areas where sensors are used. Therefore, new ways to describe/classify them are relevant, since they play a significant role in the data-gathering stage of the entire machine learning application pipeline. Thus, in the early stages of sensor development, works such as MacRuairi et al. [18] presented sensor requirements taxonomies to match specific sensors with real scenarios. Then, Fowler et al. [19] presented a survey related to the materials that sensors are made from. Following this classification scheme, works such as Tuukkanen et al. [20], Noel et al. [21], Cornacchia et al. [22], and Khanh et al. [23] presented sensors surveys for specific areas, such as piezoelectric sensors, health monitoring, wearable sensors, and intelligent agriculture, respectively. In recent years, Abdel Azeem et al. [17] have shown the fundamentals, challenges, opportunities, and taxonomy of sensors in IoT environments describing the needs and usages of each one. They also presented a wide array of previously proposed solutions, comparing them to each other and providing brief descriptions of the issues addressed by each category of that taxonomy. Finally, works such as Latifi et al. [24] and Anajeba et al. [25] presented early intuitions about improving the security of the communication channel in IoT environments.

In the ML application pipeline, Morrison et al. [26] present an innovative survey in sensor data collection and analytical systems. Additionally, Infanteena et al. [27] showed a survey on compressive data collection techniques for IoT devices and analyzed their features. Finally, in this research area, Tiboni et al. [28] described sensors and actuators in exoskeletons using the machine learning pipeline.

2.2. Data Pre-Processing

The most relevant works in this field started with Zhang et al. [14] presenting a relevant work about a data H_∞ filtering approach for wireless sensor networks (WSNs) in nonuniform sampling periods with optimization techniques. Then, Deepshukha et al. [29] designed a low-power digital FIR filter on FPGA for noise reduction in a WSN. Later, Bose et al. [2] presented an analysis of contemporary lossy compression algorithms using the signal characteristics of sensor data. At the same time, Safaei et al. [30] showed a novel approach to integrating time-series analysis, entropy, and random forest-based classification. For their part, Kowalski et al. [31] presented a review and comparison of smoothing algorithms for one-dimensional data noise reduction in specific sensors and environments. Timo et al. [12] presented outlier detection from non-smooth sensor data, as they worked in spatial discontinuities in the data, such as those arising from shadows in photovoltaic (PV) systems. Saad et al. [32] analyzed how quantization affects distributed graph filtering over both time-invariant and time-varying graphs. We bring insights into the quantization effects of the two most common graph filters: the finite impulse response (FIR) and auto-regressive moving average (ARMA) graph filters. In addition, we have proposed robust filter design strategies that minimize the quantization noise for time-invariant and time-varying networks.

Several works have delved into data pre-processing in IoT devices, but most introduced approaches for specific scenarios without a rationale for the selected filter criterion.

On the one hand, outlier detection is a complex task for determining if external causes have corrupted the data. Therefore, as mentioned in [33], the filtering process must be carried on before the outlier detection stage. On the other hand, filtering can avoid physical constraints by giving a clean dataset to implement different stages. Finally, the literature review allowed us to observe open challenges in data filtering, such as the lack of a sensor taxonomy related to the data acquisition process and the establishment of adequate sampling rates for each type of sensor.

3. Proposed Sensor Taxonomy

We propose classifying sensors into three groups considering the sampling rate and how sensors send information to the microcontroller. Figure 1 illustrates this taxonomy.

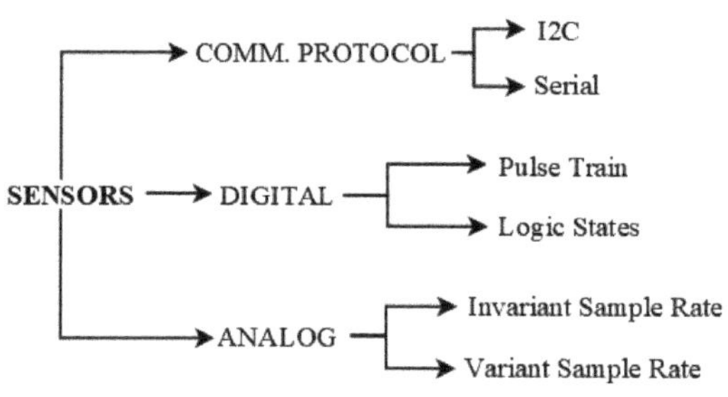

Figure 1. Proposed taxonomy of IoT sensors considering data processing characteristics.

3.1. Analog Sensors

These sensors mostly have passive elements and operational amplifiers for the hardware conditioning of the electrical signal and deliver it analogously to the microprocessor to convert it to digital form (analog–digital conversion) [34]. The ability to recreate the original signal is related to the resolution of the ADC, which is the number of bits that the microprocessor has for this process. Therefore, the sampling rate is the most relevant characteristic of the filter implementation criteria. Hence, we divide them into two categories:

- Invariant sampling rate: These sensors are developed for collecting signals continuously to detect changes in a main characteristic. For example, the processing of human electrical activity through electromyography (muscle), electrocardiogram (heart), electroencephalogram (EEG), or galvanic skin response (hands).
- Variant sampling rate: These sensors run a couple of times a day due to their applications. They do not have a specific sampling frequency because the system focuses on taking the same number of samples each time it is activated [6].

3.2. Digital Sensors

These sensors each contain a tiny microcontroller to perform the ADC process by themselves and send the data to the main microcontroller in two ways:

- Pulse train sensors: variate their pulse train frequency when the transducer detects that a physical magnitude such as temperature, humidity, or distance is changing. Therefore, capacitors are often used in this type of sensor.
- Logic states sensors: use only two logical values, 3.3 vs. or 5 vs., when detecting a physical magnitude, no matter their variations, and 0v when the sensor cannot catch the magnitude. Thus, for example, the human presence sensor can not give us more information about the phenomenon, just its presence.

3.3. Sensors by Communication Protocol

They are the most complex sensors because they have a microcontroller inside whose main objective is to obtain the best signal of the physical magnitude. These sensors also implement a communication protocol to connect sensors in series. Therefore, only a few pin connections are necessary to handle many sensors. Furthermore, these communication protocols define a master device (microcontroller) to coordinate the slave devices' (sensors) communication. Nowadays, sensor vendors, such as SparkFun, perform new socket connections to develop the electronic systems quickly.

- Serial communication: A sensor uses one pin to transmit messages and another pin to receive them. This protocol extensively adds wireless protocols to the IoT device, such as Bluetooth.
- I_2C: They have a new socket connection called Qwiic (Connect System uses 4-pin JST connectors to quickly interface development boards with sensors). This standard also allows connecting 127 sensors using just two pins. One is the clock rate, and the other is the transmitter line.

4. Methodology

The proposed methodology determines the sensors used and the data sampling required to implement filters. First, it is necessary to mention that the FIR and IIR filters are implemented only in the sensors with invariant sampling rates and the signal smoothing technique on the rest. However, the metrics used for both criteria are: signal-to-noise ratio (SNR), mean squared error (MSE), mean absolute error (MAE), root-mean-square error (RMSE), and R2 score. Figure 2 shows the mentioned process.

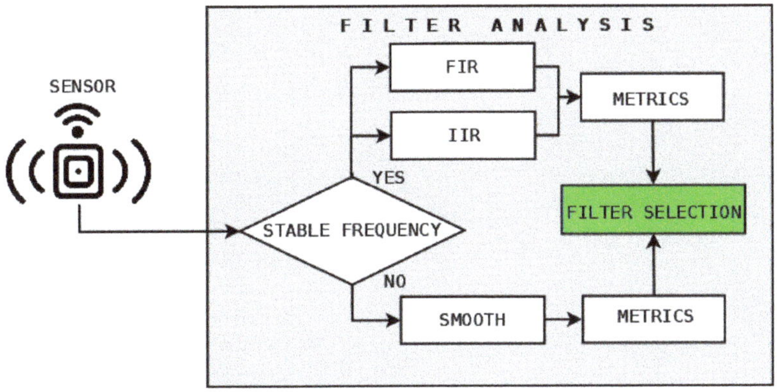

Figure 2. Sensor data and pre-preprocessing analysis.

4.1. Sensors' Characteristics

The most commonly used sensors were identified from the reviewed related works. As a result, the relevant research areas are smart farming, cities' environmental conditions analysis, and human illness. Therefore, sensors chosen regarding the proposed taxonomy were ECG Pulse Sensor (Bio-signal), Force Sensitive Resistor (FSR) (specific propose), Flex Sensor (Specific propose), Humidity and Temperature Sensor DHT-22 (pulse train), Gas Sensor MQ-135 (pulse train), and CO_2 sensor-SCD30 (I_2C/serial), UV sensor-VEML6075 (I_2C/serial). These sensors are from the same sensor vendor company SparkFun. We avoided using logical state sensors because they would not allow us to have data filtering criteria with only two values. Moreover, the sensors' communication protocol offers us the same ability to use I_2C and serial protocol. Table 1 shows the principal characteristics of each sensor used.

Table 1. Most commonly used sensors in IoT devices regarding the proposed taxonomy.

Sensor Type	Sensor	Characteristics
Bio-Signals	ECG (pulse sensor)	Detects changes in the volume of a blood vessel that occur when the heart pumps blood. To do so, they emit infrared, red or green light (550 nm) towards the body and measure the amount of reflected light with a photodiode or phototransistor. It has an operating voltage between 3.3 and 5 volts with a power consumption of 4 mA.
Specific Propose	Flexometer	Produces a variable resistance according to the degree to which it is bent. In this sense, the sensor converts the bending into different values of electrical resistance.
	Force	The force-sensing resistance sensor (also called FSR) varies its internal resistance when pressure is applied to its sensing area. As of this effect, the output voltage changes as well. Thus, the higher the pressure, the higher the output voltage.
Pulse train	Humidity and Temperature (DTH11)	This sensor sends a calibrated digital signal containing an 8-bit microcontroller. In addition, it contains two resistive sensors (NTC and humidity). It uses one-wire communication (pulse train).
	gas NOx (MQ135)	This air quality sensor detects gas concentration in various percentages. The output signal presents TTL voltage levels to be processed by a microcontroller.
Cx I2C	CO_2 (SCD 30)	This is a high quality non-dispersive infrared (NDIR) based CO_2 sensor capable of detecting from 400 to 10,000 ppm with an accuracy of $\pm$ (30 ppm + 3%).
	UV (VEML)	This sensor implements a simple photodiode to measure UVA (320–400 nm) and UVB (280–320 nm) radiation levels. With this data, it can read the intensity of these types of light in irradiance and, from there, calculate the UV index.

4.2. Data Samples Acquisition

First, we started with the ECG Pulse Sensor of the invariant sample rate sensors. The sample rate was 1 kHz (Nyquist theorem) because the signal has main components until 100 Hz. Therefore, 1400 samples were obtained in 10 controlled experiments. Second, the variable sample rate sensors were exposed to their physical magnitude for 10 s, and then they returned to their initial condition (flexometer and force sensors). Consequently, this process was carried out ten times to store 1000 samples with a 100 Hz sample rate. A similar procedure was carried out with pulse train sensors, such as DTH11 and MQ135. Finally, communication protocol sensors (SCD30 and VMLE) were tested in 10 controlled experiments. As a result, we stored 500 samples with a 50 Hz sample rate because their response times are higher than those of the other sensors.

5. Results

The sensors were tested with statistical metrics according to the experiments performed for each one. Then, they were evaluated with functional metrics such as accuracy, reproducibility, repeatability, and stability. These metrics represent the sensors working in real conditions. Thus, for a better understanding of each metric result, four evaluation levels were established for the sensors: (i) excellent, (ii) good, (iii), normal, and (iv) poor.

Finally, Figure 3 shows all the sensors used in this research and their connections with a sample board, such as Arduino.

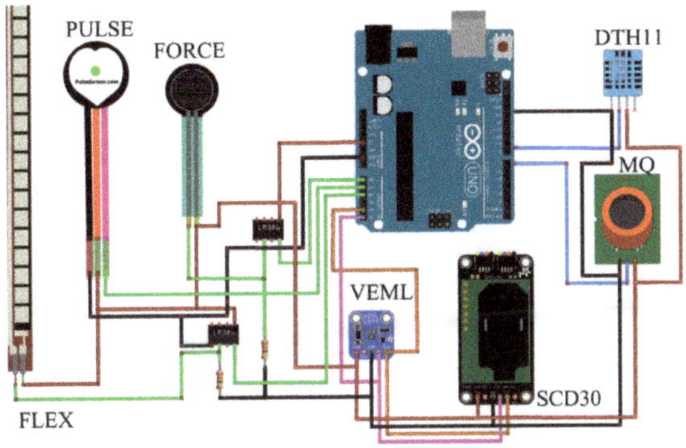

Figure 3. Sensors used in this work according to the new proposed taxonomy. (—) Analog connection, (—) Digital connection, (—) Prot. Communication connection (SDA), (—) Prot. Communication connection (SCL), (—) VCC, (—) GND.

5.1. Invariant Sampling Rate (ISR)

The signal needs to be converted to the frequency domain to detect the principal components. Therefore, a fast Fourier transform was implemented to define that the EMG components were between 5 and 40 Hz, which are presented in Figure 4. Then, IIR filters were the first approach with **Chebyshev**, **Butterworth**, and **Bessel** approximations with 3, 5, and 7 orders of band-pass filter design. We noticed that the filters in order 5 fit better than the rest. Table 2 summarizes the results of the statistical metrics mentioned before. The **Butterworth** filter demonstrated superior SNR, MAE, and R2 metrics. Additionally, it is visible that **Butterworth** reduced the noise with few signal alterations. The second approach was FIR filters. They focus on a time-domain analysis through windows. Reference [35] defines that using 10% as a window size of sample rate is recommended. Thus, we defined window sizes of 150, 250, and 300 components to compare with the ECG signal. Table 3 summarizes that windows size equal to 150 components produced a better SNR when **Nutall window** was applied. However, the differences between the windows were minimal when we tried to improve the signal. As a result, FIR filters are a better option than IIR. Finally, Figure 4 shows the components in the frequency domain and the graphical results of FIR and IIR filters.

Table 2. EMG signal statistical analysis and IIR filters.

Approximation	SNR (dB)	MSE	MAE	RMSE	R2
Butterworth	4.44	0.13	0.31	0.36	−6.83
Bessel	4.20	0.20	0.38	0.44	−10.66
Chebyshev	4.12	0.12	0.30	0.34	−6.26

Table 3. EMG signal statistical analysis and FIR filters.

Window	SNR (dB)	MSE	MAE	RMSE	R2
Nutall	4.48	0.04	0.18	0.20	−1.55
Hamming	3.77	0.13	0.33	0.36	−7.15
Taylor	4.21	0.80	0.81	0.9	−8.43
Blackman	4.09	0.06	0.22	0.25	−3.0

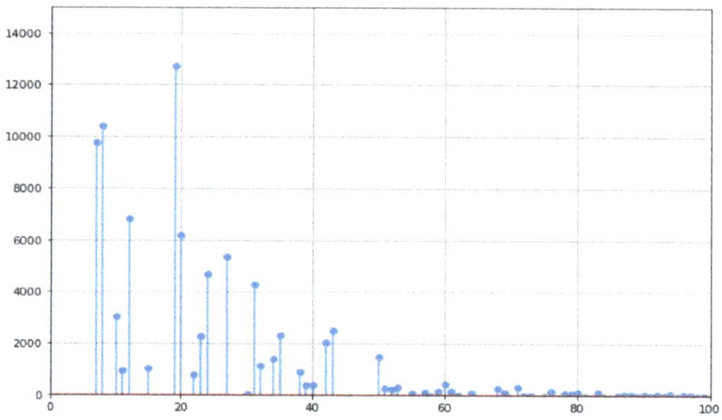

(**a**)

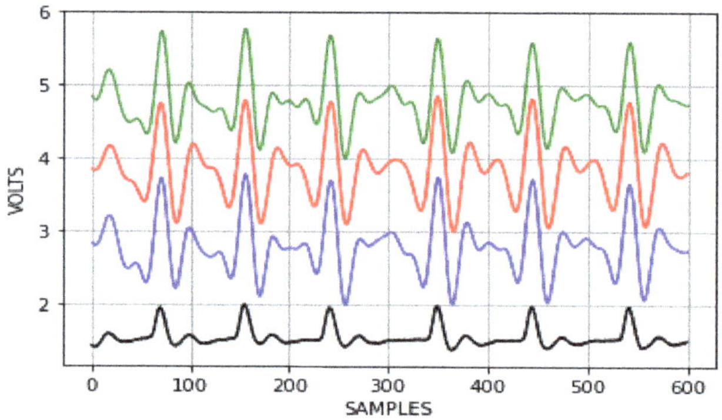

(**b**)

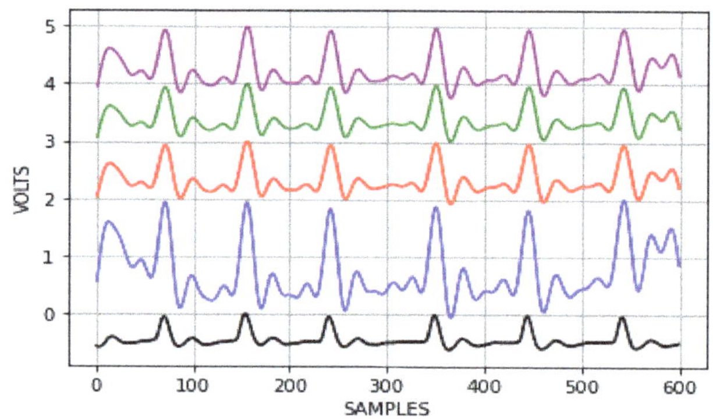

(**c**)

Figure 4. EMG signal analysis. (**a**) EMG signal in the frequency domain. (**b**) IIR filters: (—) Butterwoth, (—) Chebyshev, (—) Bessel, (—) original samples. (**c**) FIR filters: (—) Hamming, (—) Nutall, (—) Taylor, (—) Blackman, (—) original samples.

5.2. Variable Sample Rate (VSR)

For experimental purposes, the Force Sensitive Resistor sensor was tested with 40 lbs of pressure, and the Flex sensor bent it 45 degrees. Both were tested with the sample rate mentioned above (100 Hz). Additionally, their datasheet recommends using an analog amplifier in follow-up configuration to avoid DC voltage. Therefore, we applied smoothing filters. The average filter has a better SNR metric; however, the R2 score indicates that this filter affects the original signal. Additionally, the Gaussian filter tends to round off the maximum values obtained and modifies the output signal due to `sigma` parameter (Gaussian bell size). The **Savi–Golay** filter eliminates noise in VSR signals: see the strong results in R2 score and SNR metrics (Table 4). Figure 5 shows the graphical results of each smoothing filter.

Table 4. Statistical analysis of sensors with various sampling rates.

Sensor	Average k = 20	Medium k = 20	Gaussian Sigma = 7	Savi–Golay k = 9, Poly = 4	Statistical Metrics
Flex sensor	9.07	8.28	8.97	7.90	MSE
	1.49	1.60	0.65	1.27	MAE
	1.91	1.96	0.98	2.81	RMSE
	0.642	0.56	0.99	0.99	R2 score
	2.65	2.16	2.47	2.49	SNR
Force sensor	195.39	198.23	205.31	158.2	MSE
	5.25	5.29	3.20	4.96	MAE
	18.85	15.78	14.32	15.67	RMSE
	0.75	0.65	0.99	0.99	R2 score
	2.91	2.65	2.86	2.87	SNR

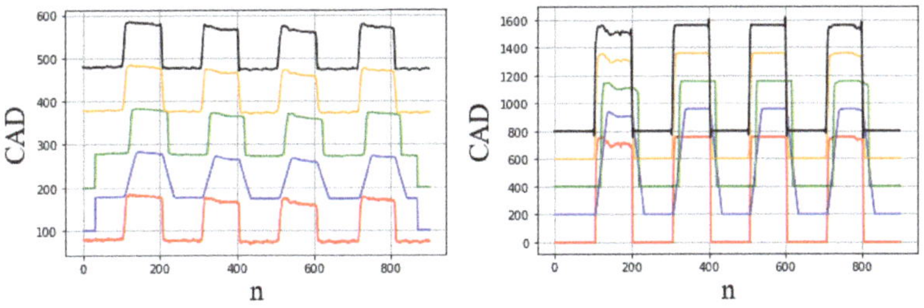

Figure 5. Smoothing graphical analysis in the proposed sensor taxonomy. (—) Original samples. (—) Average filter. (—) Medium filter. (—) Gaussian filter. (—) Savi–Golay filter. (**a**) FLEX sensor. (**b**) FORCE sensor.

Sensor performance metrics: These sensors have a variable resistor as their main component. Therefore, they are stable in operation, and similar data can be obtained in each data gathering process. However, their wear and tear is very high, subject to human activity. For this reason, they are dependent on their location and use, and their reproducibility tends to decrease over time.

5.3. Digital Pulse-Train

The data collection process was based on having a closed box with an incandescent bulb, a fan, and extra space for sensors. First, we used the `DTH11` to get measurements when the temperature inside the box increased due to the bulb and then decreased when the fan was powered. Then, for the gas sensor MQ135, we used a gas emitter (lighter) instead of the bulb and a fan, and a sensor inside to change the gas concentration inside quickly. These experiments demonstrated that Savi–Golay and **average** filters fit with these kinds of signals and have better SNR metrics. Consequently, we noticed that the average filter reduces the dc voltage (peaks), producing good R2 score, MAE, and MSE results (Table 5). Moreover, Figure 6 represents the smoothing signal applied in pulse train sensors, from which we can notice that medium and Savi–Golay filters do not modify the electric signal.

Sensor performance metrics: These sensors have standard accuracy and stability due to their calibrated modes. However, they have restrictions on repeatability and reproducibility metrics because they sense physical magnitudes that do not vary in short periods, such as temperature and humidity, among others.

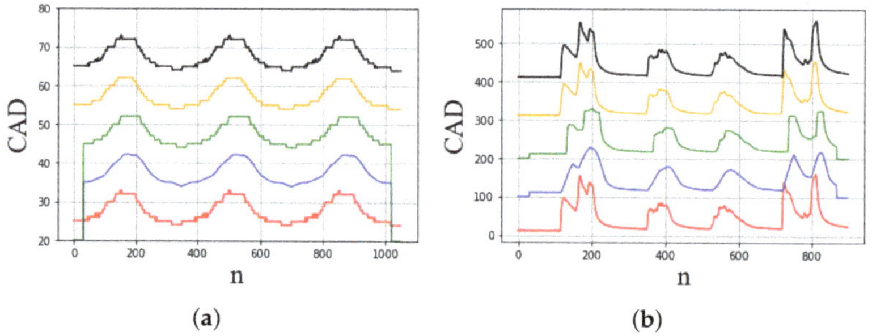

Figure 6. Smoothing graphical analysis in the proposed sensor taxonomy. (—) Original samples (—) Average filter. (—) Medium filter. (—) Gaussian filter. (—) Savi–Golay filter. (**a**) DHT-22 sensor. (**b**) MQ-135 sensor.

Table 5. Digital pulse-train sensors' statistical analysis.

Sensor	Average k = 30	Medium k = 30	Gaussian Sigma = 7	Savi–Golay k = 9, Poly = 4	Statistical Metrics
DHT-11	6.40	6.51	0.2	4.29	MSE
	2.15	2.14	0.07	0.29	MAE
	1.03	1.04	0.15	0.54	RMSE
	0.75	0.77	0.99	0.96	R2 score
	9.72	9.60	9.61	9.69	SNR
MQ-135	13.55	11.79	10.48	13.73	MSE
	1.35	2.29	0.29	1.62	MAE
	3.77	3.49	0.69	3.70	RMSE
	0.51	0.35	0.99	0.98	R2 score
	1.36	1.27	1.26	1.28	SNR

5.4. I2C Communication Protocol

These sensors were exposed to their corresponding physical features (UV rays and CO_2 gas). Gausian and Savi–Golay filters removed the noise better than the other algorithms. However, the Gausian modifies the signal output significantly. Additionally, the average does not fit with these types of electrical signals due to the sizes of their windows affecting the signal with few samples of data. Therefore, **medium** and **Savi–Golay** can be applied to these sensors. Table 6 represents the statistical analysis, and Figure 7 shows the graphical results.

Sensor performance metrics: They have poor repeatability and reproducibility because UV rays do not have considerable variability during the day. Moreover, CO_2 can increase exponentially in fires, smoking zones, etc., but it needs a few hours to normalize. As a result, the sensor has restrictions concerning returning to the initial state. Figure 7 shows the smoothing graphical results of both sensors.

Table 6. Communication protocol sensors' statistical analysis.

Sensor	Average k = 30	Medium k = 20	Gaussian Sigma = 7	Savi–Golay k = 9, Poly = 4	Statistical Metrics
SCD30	540.16	650.66	435.10	475.0	MSE
	47.89	69.5	62.14	105.78	MAE
	178.05	111.02	124.23	182.96	RMSE
	0.55	0.77	0.94	0.86	R2 score
	1.51	1.47	2.01	2.37	SNR
VEML6075	2.51	3.44	2.14	2.43	MSE
	0.97	0.98	0.39	0.9	MAE
	1.58	1.85	1.2	0.20	RMSE
	0.42	0.22	0.10	0.99	R2 score
	1.0	0.88	0.89	0.92	SNR

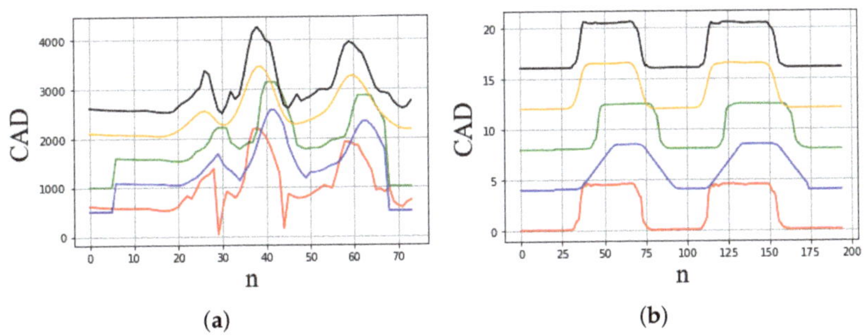

Figure 7. Smoothing graphical analysis in the proposed sensor taxonomy. (—) Original samples. (—) Average filter. (—) Medium filter. (—) Gaussian filter. (—) Savi–Golay filter. (**a**) SCD30 sensor. (**b**) VEML6075.

5.5. Real Tests

Sensors were evaluated under natural conditions to test each filter selected. In addition, we compare the voltage obtained through sensors using a multimeter KEYSIGHT DIGITAL MULTIMETER U1282A, which has a 0.025% voltage accuracy. Therefore, for a better understanding of each metric's result, four levels of evaluation were established for the sensors: (i) excellent, (ii) good, (iii), normal, and (iv) poor. Table 7 shows the results obtained.

Finally, we obtained the system response time for each sensor with the filter deployed on the device. For example, the Butterworth filter takes 2.5 ms to process an array with 300 samples, the Savi–Golay takes 1.2 ms to process the same number of samples, and the medium filter takes 0.68 ms. Therefore, this pre-processing technique is a suitable solution to run in real-time scenarios when the IoT system can define threads for each procedure to reduce the time response of each task. Additionally, filters have a small footprint in memory, leaving enough space to run the IoT application.

Table 7. Sensor performance metrics.

Performance Metrics	Sensor Taxonomy			
	Analog Sensors		Pulse	Comm.
	ISR	VSR	train	Protocol
Accuracy	Good	Good	Normal	Excellent
Reproducibility	Good	Poor	Poor	Excellent
Repeatability	Good	Normal	Excellent	Poor
Stability	Normal	Poor	Good	Normal
Noise	Poor	Normal	Good	Good

6. Conclusions and Future Works

This work introduced a new taxonomy of sensors focused on data pre-processing on-device to upload reliable data to the cloud. Furthermore, filter implementation criteria were established to prevent erroneous data from being part of the ML model. We now present the conclusions of this work:

- This taxonomy of sensors is appropriate for the new trend of executing some ML stages on-device. Therefore, this work prevents data that do not describe the phenomenon being studied from being part of the ML model. Thus, the sampling frequency used in the sensors is a fundamental part of implementing filters.
- The proposed methodology demonstrates which filter is adequate and does not deform the original signal.
- Performance metrics in real environments define the ability to reduce noise and provide new trends to improve this process for coming sensors.
- We declare the Butterworth filter suitable for analog sensors with invariant sampling rates. Savi–Golay fits analog sensors with variant sampling rates. The average filter is adequate for digital pulse train sensors. Regarding communication protocol sensors, Savi–Golay and medium filters remove noise and provide improved signal for the proposed data gathering.

Finally, we understand that the next step is to detect anomalies in sensor data due to manipulation or sensor failure.

Author Contributions: P.D.R.-M.: conceptualization, methodology, software, formal analysis, investigation, writing—original draft preparation, visualization, and resources; V.F.L.-B.: investigation, supervision, and project administration; D.H.P.-O: formal analysis, writing—review, visualization, project administration, and funding. All authors read and agreed to the published version of the manuscript.

Funding: This research was funded by Novo Nordisk Fonden, grant number NNF20OC0064411, with the project Privacy through Co-Design for Real-World Data Analytics in the cloud.

Institutional Review Board Statement: Not applicable.

Informed Consent Statement: Not applicable.

Data Availability Statement: Not applicable.

Acknowledgments: The authors are grateful for the support given by the SDAS Research Group (https://sdas-group.com/, accessed on 10 April 2022).

Conflicts of Interest: The authors declare no conflict of interest.

References

1. Debauche, O.; Mahmoudi, S.; Guttadauria, A. A New Edge Computing Architecture for IoT and Multimedia Data Management. *Information* **2022**, *13*, 89. [CrossRef]
2. Bose, T.; Bandyopadhyay, S.; Kumar, S.; Bhattacharyya, A.; Pal, A. Signal Characteristics on Sensor Data Compression in IoT—An Investigation. In Proceedings of the 2016 IEEE International Conference on Sensing, Communication and Networking (SECON Workshops), Sydney, Australia, 14–18 March 2016; pp. 1–6. [CrossRef]
3. Canziani, A.; Culurciello, E.; Paszke, A. Evaluation of neural network architectures for embedded systems. In Proceedings of the 2017 IEEE International Symposium on Circuits and Systems (ISCAS), Baltimore, MD, USA, 28–31 May 2017; pp. 1–4. [CrossRef]
4. Komatsu, N.; Nakano, M. Embedded Systems. In *Encyclopedia of Biometrics*; Springer: Boston, MA, USA, 2015; pp. 397–401. [CrossRef]
5. Dobrin, A.; Stamatescu, G.; Dragana, C.; Sgarciu, V. Cloud challenges for networked embedded systems: A review. In Proceedings of the 2016 20th International Conference on System Theory, Control and Computing (ICSTCC), Sinaia, Romania, 13–15 October 2016; pp. 866–871. [CrossRef]
6. Kalantar-zadeh, K. Sensors . In *Sensors*; Springer: Boston, MA, USA, 2013; pp. 11–28. [CrossRef]
7. Dasgupta, R.; Dey, S. A comprehensive sensor taxonomy and semantic knowledge representation: Energy meter use case. In Proceedings of the 2013 Seventh International Conference on Sensing Technology (ICST), Wellington, New Zealand, 3–5 December 2013; pp. 791–799. [CrossRef]
8. Moreau, T.; San Miguel, J.; Wyse, M.; Bornholt, J.; Alaghi, A.; Ceze, L.; Enright Jerger, N.; Sampson, A. A Taxonomy of General Purpose Approximate Computing Techniques. *IEEE Embed. Syst. Lett.* **2018**, *10*, 2–5. [CrossRef]
9. Lin, Y.L.; Kyung, C.M.; Yasuura, H.; Liu, Y. *Smart Sensors and Systems*; Springer International Publishing: Cham, Switzerland, 2015; Volume XII, p. 467. [CrossRef]
10. Raschka, S.; Patterson, J.; Nolet, C. Machine Learning in Python: Main Developments and Technology Trends in Data Science, Machine Learning, and Artificial Intelligence. *Information* **2020**, *11*, 193. [CrossRef]
11. Lin, L.; Liao, X.; Jin, H.; Li, P. Computation Offloading Toward Edge Computing. *Proc. IEEE* **2019**, *107*, 1584–1607. [CrossRef]
12. Huuhtanen, T.; Ambos, H.; Jung, A. Outlier Detection from Non-Smooth Sensor Data. In Proceedings of the 2019 27th European Signal Processing Conference (EUSIPCO), Coruña, Spain, 2–6 September 2019; pp. 1–5. [CrossRef]
13. França, C.M.; Couto, R.S.; Velloso, P.B. Missing Data Imputation in Internet of Things Gateways. *Information* **2021**, *12*, 425. [CrossRef]
14. Zhang, W.A.; Dong, H.; Guo, G.; Yu, L. Distributed Sampled-Data Filtering for Sensor Networks With Nonuniform Sampling Periods. *IEEE Trans. Ind. Inform.* **2014**, *10*, 871–881. [CrossRef]
15. Britton, R. *Digital Filter Designer's Handbook*; McGraw-Hill: New York, NY, USA, 2010. Available online: http://dsp-book.narod.ru/DFD/DFD0.pdf (accessed on 10 April 2022).
16. Williams, A. Analog Filter and Circuit Design Handbook. 2013. Available online: https://www.amazon.com/Analog-Filter-Circuit-Design-Handbook/dp/0071816712 (accessed on 10 April 2022).
17. Aslam, F.; Aimin, W.; Li, M.; Ur Rehman, K. Innovation in the Era of IoT and Industry 5.0: Absolute Innovation Management (AIM) Framework. *Information* **2020**, *11*, 124. [CrossRef]
18. MacRuairi, R.; Keane, M.T.; Coleman, G. A Wireless Sensor Network Application Requirements Taxonomy. In Proceedings of the 2008 Second International Conference on Sensor Technologies and Applications (Sensorcomm 2008), Washington, DC, USA, 25–31 August 2008; pp. 209–216. [CrossRef]
19. Fowler, K.R. The future of sensors and sensor networks survey results projecting the next 5 years. In Proceedings of the 2009 IEEE Sensors Applications Symposium, Atlanta, GA, USA, 12–14 February 2009; pp. 1–6. [CrossRef]
20. Tuukkanen, S.; Rajala, S. A survey of printable piezoelectric sensors. In Proceedings of the 2015 IEEE SENSORS, Busan, Korea, 1–4 November 2015; pp. 1–4. [CrossRef]
21. Noel, A.B.; Abdaoui, A.; Elfouly, T.; Ahmed, M.H.; Badawy, A.; Shehata, M.S. Structural Health Monitoring Using Wireless Sensor Networks: A Comprehensive Survey. *IEEE Commun. Surv. Tutor.* **2017**, *19*, 1403–1423. [CrossRef]
22. Cornacchia, M.; Ozcan, K.; Zheng, Y.; Velipasalar, S. A Survey on Activity Detection and Classification Using Wearable Sensors. *IEEE Sens. J.* **2017**, *17*, 386–403. [CrossRef]
23. Quy, V.K.; Hau, N.V.; Anh, D.V.; Quy, N.M.; Ban, N.T.; Lanza, S.; Randazzo, G.; Muzirafuti, A. IoT-Enabled Smart Agriculture: Architecture, Applications, and Challenges. *Appl. Sci.* **2022**, *12*, 3396. [CrossRef]
24. Latif, S.A.; Wen, F.B.X.; Iwendi, C.; li F. Wang, L.; Mohsin, S.M.; Han, Z.; Band, S.S. AI-empowered, blockchain and SDN integrated security architecture for IoT network of cyber physical systems. *Comput. Commun.* **2022**, *181*, 274–283. [CrossRef]
25. Anajemba, J.H.; Iwendi, C.; Razzak, M.; Ansere, J.A.; Okpalaoguchi, M.I. A Counter-Eavesdropping Technique for Optimized Privacy of Wireless Industrial IoT Communications. *IEEE Trans. Ind. Inform.* **2022**, 1. Available online: https://ieeexplore.ieee.org/document/9669024 (accessed on 10 April 2022). [CrossRef]
26. Morrison, W.; Guerdan, L.; Kanugo, J.; Trull, T.; Shang, Y. TigerAware: An Innovative Mobile Survey and Sensor Data Collection and Analytics System. In Proceedings of the 2018 IEEE Third International Conference on Data Science in Cyberspace (DSC), Guangzhou, China, 18–21 June 2018; pp. 115–122. [CrossRef]

27. Infanteena, S.D.; Anita, E.M. Survey on compressive data collection techniques for wireless sensor networks. In Proceedings of the 2017 International Conference on Information Communication and Embedded Systems (ICICES), Chennai, India, 23–24 February 2017; pp. 1–4. [CrossRef]
28. Tiboni, M.; Borboni, A.; Vérité, F.; Bregoli, C.; Amici, C. Sensors and Actuation Technologies in Exoskeletons: A Review. *Sensors* **2022**, *22*, 765. [CrossRef] [PubMed]
29. Bhat, D.; Kaur, A.; Singh, S. Wireless sensor network specific low power FIR filter design and implementation on FPGA. In Proceedings of the 2015 2nd International Conference on Computing for Sustainable Global Development (INDIACom), New Delhi, India, 11–13 March 2015; pp. 1534–1536.
30. Safaei, M.; Driss, M.; Boulila, W.; Sundararajan, E.A.; Safaei, M. Global Outliers Detection in Wireless Sensor Networks: A Novel Approach Integrating Time-Series Analysis, Entropy, and Random Forest-based Classification. *arXiv* **2021**, arXiv: 2107.10135.
31. Kowalski, P.; Smyk, R. Review and comparison of smoothing algorithms for one-dimensional data noise reduction. In Proceedings of the 2018 International Interdisciplinary PhD Workshop (IIPhDW), Swinoujscie, Poland, 9–12 May 2018; pp. 277–281. [CrossRef]
32. Saad, L.B.; Beferull-Lozano, B.; Isufi, E. Quantization Analysis and Robust Design for Distributed Graph Filters. *arXiv* **2020**, arXiv:2004.06692.
33. Zhang, M.; Li, X.; Wang, L. An Adaptive Outlier Detection and Processing Approach Towards Time Series Sensor Data. *IEEE Access* **2019**, *7*, 175192–175212. [CrossRef]
34. Iwendi, C.; Wang, G.G. Combined power generation and electricity storage device using deep learning and internet of things technologies. *Energy Rep.* **2022**, *8*, 5016–5025. [CrossRef]
35. Gizlenmistir, Y. Filter based analysis unit design for data acquisition systems. In Proceedings of the 2018 26th Signal Processing and Communications Applications Conference (SIU), Izmir, Turkey, 2–5 May 2018; pp. 1–4.. [CrossRef]

information

MDPI

Review

Pervasive Healthcare Internet of Things: A Survey

Kim Anh Phung [1,*], Cemil Kirbas [2], Leyla Dereci [3] and Tam V. Nguyen [2]

1 Independent Researcher, University of Dayton, Dayton, OH 45469, USA
2 Department of Computer Science, University of Dayton, Dayton, OH 45469, USA; ckirbas1@udayton.edu (C.K.); tamnguyen@udayton.edu (T.V.N.)
3 Heritage College of Osteopathic Medicine, Ohio University, Athens, OH 45701, USA; ld459916@ohio.edu
* Correspondence: kimaphung@gmail.com; Tel.: +1-937-303-1848

Abstract: Thanks to the proliferation of the Internet of Things (IoT), pervasive healthcare is gaining popularity day by day as it offers health support to patients irrespective of their location. In emergency medical situations, medical aid can be sent quickly. Though not yet standardized, this research direction, healthcare Internet of Things (H-IoT), attracts the attention of the research community, both academia and industry. In this article, we conduct a comprehensive survey of pervasive computing H-IoT. We would like to visit the wide range of applications. We provide a broad vision of key components, their roles, and connections in the big picture. We classify the vast amount of publications into different categories such as sensors, communication, artificial intelligence, infrastructure, and security. Intensively covering 118 research works, we survey (1) applications, (2) key components, their roles and connections, and (3) the challenges. Our survey also discusses the potential solutions to overcome the challenges in this research field.

Keywords: Internet of Things; healthcare; pervasive computing

Citation: Phung, K.A.; Kirbas, C.; Dereci, L.; Nguyen, T.V. Pervasive Healthcare Internet of Things: A Survey. *Information* **2022**, *13*, 360. https://doi.org/10.3390/info13080360

Academic Editors: Spyros Panagiotakis and Evangelos K. Markakis

Received: 30 May 2022
Accepted: 22 July 2022
Published: 28 July 2022

1. Introduction

Pervasive computing, or ubiquitous computing, is a computing paradigm that leverages the user interaction with microprocessors or gadgets in an "anywhere and anytime" manner. The users do not need to access a PC or laptop; instead, they can use their body-worn devices. Due to the rapid proliferation of handheld and wearable devices, the Internet of Things (IoT) enabled technology is evolving healthcare in the era of pervasive computing. The development of cloud technology empowers pervasive computing even more by providing communication across different objects for data sharing. Thanks to that, IoT offers a stage to associate heterogeneous devices from smart homes, and smart urban communities, to smart healthcare. These interconnected objects and sensors harvest the information for complicated tasks such as recognition, prediction, planning, and recommendation [1]. With the tremendous growth in recent years, smart IoT systems are expected to play a vital role in many pervasive healthcare applications. Indeed, IoT integrated with other advanced technologies could significantly transform the landscape of pervasive healthcare in an uninterrupted and ubiquitous monitoring manner. This new direction is regarded as the Healthcare Internet of Things (H-IoT), which is very important, especially during the pandemic era. For example, to prevent the spread of the virus, social distancing can be implemented by deploying IoT devices, including smart watches and monitoring devices. Here, we would like to highlight the differences between generic IoT and H-IoT. The generic IoT is usually deployed over a large-scale area such as smart cities or urban planning. On the other hand, H-IoT is usually deployed in a small-scale area such as the human body or a smart home or hospital. H-IoT nodes, miniaturized to be unobtrusive, are used to monitor human body vitals. These nodes can collect energy from a human via body heat or motion.

In this paper, our overarching goal is thus to provide a comprehensive survey of pervasive computing in H-IoT. We would like to provide a broad vision of its components and their connections. We classify the vast amount of publications into different categories such as applications, sensors, communication, storage infrastructure, security, and artificial intelligence. Intensively covering more than 100 publications, we survey (1) applications, (2) key components and their roles, and (3) the challenges.

The remainder of this paper is organized as follows. Section 2 compares this work to other existing surveys. Section 3 surveys the applications of pervasive computing in H-IoT. The key components of H-IoT are reviewed in Section 4. Section 5 reviews the existing challenges. Finally, Section 6 concludes this paper.

2. Comparison with Other Surveys

There have been many surveys on H-IoT in the literature. Qi et al. [2] discussed the applications, the data sensing and processing, and their challenges in H-IoT. However, this survey is outdated since it did not address the latest technology in cloud computing or security issues in personalized healthcare systems. In 2018, Alam et al. [3] surveyed the roles of communication technologies in H-IoT applications. They introduced four applications on infectious diseases, cardiovascular diseases, musculoskeletal disorders, and neuromuscular disorders. They also discussed the issues and challenges along with the emerging communication technologies. However, this survey disregards the impact of artificial intelligence, which is a key component in HIoT applications. Meanwhile, Shaikh et al. [4] reviewed smart healthcare systems using the Internet of Things, for example, e-health systems, telehealth and home monitoring systems, and RFID-based monitoring systems. They also discussed issues related to smart healthcare systems such as reliability, low-latency tolerance, and interoperability. However, Shaikh et al. [4] did not discuss the key components within HIoT systems. In another work, Ahmadi et al. [5] reviewed the applications of the Internet of Things in healthcare. In particular, they reviewed the main components and their functions along with the main issues and challenges. Still, they did not consider the security and privacy issues in H-IoT. In addition, artificial intelligence is not addressed in the survey. Similarly, Rajini [6] reviewed different applications and services in smart healthcare systems. However, the survey is lacking discussions regarding the key components of such systems.

Habibzadeh et al. [7] surveyed the H-IoT from a clinical perspective. They reviewed the key components such as sensing, communication, and data analytics. They also discussed the open issues and future trends in this field. This survey, however, did not discuss the importance of artificial intelligence and cloud computing in this research field. In a different survey, Usak et al. [8] reviewed the health care service delivery based on IoT. They also discussed the pros and cons of other surveys. However, the key components within the IoT healthcare system are not visited. This is not helpful for readers to gain an understanding of the existing IoT healthcare system. Dhanvijay and Patil [9] introduced a survey of technologies in H-IoT and their applications. Like other surveys, open issues and the main challenges are discussed. Similarly, artificial intelligence and cloud computing were not addressed in this survey. Recently, Ali Tunc et al. [10] compiled a survey on emerging technologies, applications, challenges, and future trends for IoT-smart healthcare. The key components such as sensors and privacy issues were not addressed in the survey.

Table 1 shows the comparison between our survey with the existing surveys in the literature. Our survey covers all the important content related to pervasive computing in H-IoT.

Table 1. Literature survey comparison. The checkmark (✓) denotes the availability of the mentioned content in the survey.

Surveys	Survey Comparison	Review of Applica-tions	Key Components					Challenges
			Devices/Sensors	Commu-nication	Artificial Intelligence	Cloud Computing	Security and Privacy	
Qi et al. [2]		✓	✓	✓	✓			✓
Alam et al. [3]		✓		✓		✓		✓
Shaikh et al. [4]		✓						✓
Ahmadi et al. [5]	✓			✓		✓		✓
Rajin [6]		✓						
Habibzadeh et al. [7]	✓	✓	✓	✓			✓	✓
Usak [8]	✓	✓						✓
Dhanvijay and Patil [9]	✓	✓		✓			✓	✓
Ali Tunc et al. [10]	✓	✓		✓	✓	✓		✓
Ours	✓	✓	✓	✓	✓	✓	✓	✓

3. Applications of Pervasive Healthcare Internet of Things

Pervasive computing in H-IoT can be found in various application domains as shown in Figure 1. We summarize some notable applications below.

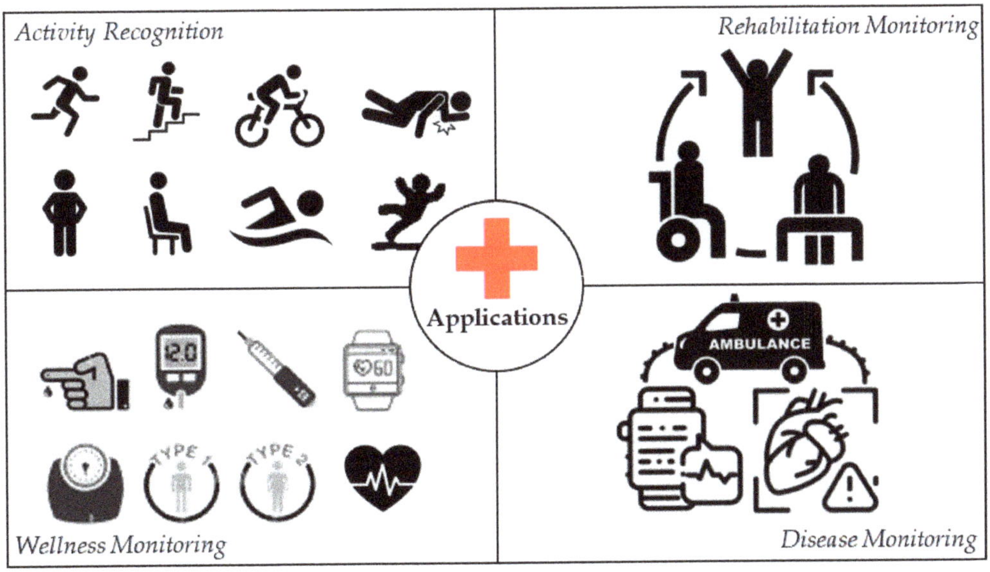

Figure 1. Some applications of pervasive computing in the healthcare Internet of Things. Section 3 reviews these components in detail.

3.1. Activity Recognition

Activity recognition is crucial in the sense that it provides a context of what is happening so that the IoT system can respond appropriately. Note that the activity can be recognized via various sensors. Nguyen et al. [11] used a spatial-temporal attention-aware pooling for action recognition in video. First, the visual saliency is predicted from the input video. Saliency-aware matching kernels are thus derived as the similarity measurement of these channels. The kernels are then fed into support vector machines for activity classification. Falls are one of the major health threats to independent living, especially for elderly people. Note that the elderly often live alone and receive only irregular visits. Thus there is a legitimate need to detect a fall or abnormal activities. Ni et al. [12] used Kinect, an RGB-D camera for fall detection in hospitals. From both color and depth video frames, the motion and shape features are extracted. Then, the extracted features are fused via a multiple kernel learning model to detect the anomalous events. Once the system detects a patient getting up from the bed, nursing staff are informed to provide immediate assistance. Wang et al. [13] detect falls by using a WiFi signal. In particular, they take advantage of the wireless physical information-Channel State Information (CSI) in widely deployed commercial wireless infrastructure. They found that the static human body does not affect CSI in the time domain. Instead, human activities, such as walking, sitting, standing up, and falling will change the variance of CSI. Therefore, the variance change of CSI can be used to detect the anomaly of human activities. Ruan et al. [14] used Radio-frequency identification (RFID) tags to determine the fall in an unobstructed manner. The RFID tags are used to sense regular actions and fall events simultaneously. When a person falls from standing, the Received Signal Strength Indicators (RSSI) show different fluctuation patterns, indicating the potential for detecting a fall. The detected fall event is sent to caregivers along with the contexts of fall orientations. Uddin and Soylu [15] proposed a body sensor-based activity modeling and recognition system using a time-sequential information-based deep learning algorithm. First, data are obtained from multiple wearable sensors while the subjects perform their daily routine activities. The collected time-sequential information then goes through the feature extraction component. The extracted time-sequential features are later fed to the deep learning model, i.e., long short-term memory (LSTM) for activity recognition. Wang et al. [16] recognized multiuser activities by using wireless body sensor networks. An RFID reader is located on each hand to detect the presence of a tagged object within a few centimeters. There is an ultra-high frequency (UHF) RFID reader located in each room to sense the proximity of a person wearing an UHF tag. The sensor data consist of the 3-axis acceleration data for both hands, object use for both hands, temperature, humidity, light, and user location. The sensor data is then used to recognize the activity via a pattern recognition model. Zhu et al. [17] proposed ambient radar for indoor human activity recognition. In particular, they used a 7.8 GHz radar to emit 16 pulse signals per second and sample the reflected signals at 128 kHz to recognize the fine dynamics of human activities.

3.2. Rehabilitation Monitoring

The assessment after stroke is very important for patients, especially for outpatients who need to be evaluated often like inpatients but can still come back to their normal lives as soon as possible [18]. Furthermore, in America, the risk of a second stroke in the first year is 23% in 2022 [19]. Standen et al. [20] introduced a virtual reality system for home-based arm rehabilitation for a post-stroke patient. The proposed system uses Kinect and smart glass to monitor patients after stroke to predict and assess the recovery process. In [21], Hoda et al. designed and developed a prototype to simulate real post-stroke rehabilitation exercises. To find the correlation between the kinematics of the upper limb and the muscle strength, they use least-square regression method. Like [20], a Kinect depth sensor and a force sensing resistors glove are used to track the subject's data such as limb joints and muscle strength. The data are collected while the subjects are performing their exercises. The evaluation on 13 subjects demonstrates the usefulness of the system in recognizing the muscle strength of stroke patients without wearing any devices. Meanwhile, Bobin et al. [22] introduced a system to monitor and guide stroke patients. The system consists of a smart mug that tracks the patient's drinking activities. For example, the information such as drinking frequency, liquid level, drinking orientation, and liquid type, i.e., water, coffee. This solution allows therapists to monitor the patient and assign the suitable exercises for rehabilitation sessions.

3.3. Wellness Monitoring

According to the American Diabetes Association, about 37.3 million people (11.3% US population) suffer from diabetes [23]. Blood sugar monitors play an important role in managing the blood glucose level. Al-Taee et al. [24] suggested a method to assist in improving diabetes in young people by using smart robots. Fioravanti et al. [25] used a texting system for helping patients who have problems with abnormally high blood sugar levels. Kaiya et al. [26] used wireless devices to set up a diabetic meal plan by gathering the figures from IoT tag systems. About 116 million (nearly 47%) adults in the United States have problems with blood pressure, especially hypertension (high blood pressure) [27]. Janjua et al. [28] introduced a Bluetooth chest wearable device to monitor vital signs. Since then, we can detect and prevent the chance of hypertension. This early intervention can significantly reduce the risk of mortality caused by abnormally high blood pressure. Meanwhile, Iakovakis and Hadjileontiadis [29] monitored orthostatic hypotension (low blood pressure caused by postural changes) by using smartwatch sensors.

Cardiovascular disease is one of the most common causes of mortality globally, occurring in both men and women, and it is also the number one killer in the United States [30]. There are 659,000 people in America who die from heart disease every year, comprising 25% of total deaths [31]. Kiranyaz et al. [32] designated an individual monitoring device and advanced alert notification system for patients with an irregular heartbeat (cardiac arrhythmias). Schmier et al. [33] developed a small sensor (the size of a paper clip) placed in the pulmonary artery to check the heart rate with the cardioMEMS HF system like a remote home monitoring unit. Xia et al. [34] presented an automatic wearable electrocardiogram (ECG) to classify and monitor patients with the diagnosis of cardiac arrhythmia. Hijazi et al. [35] proposed the effectiveness of electronic monitoring by using machines for supporting victims of a cardiac-related disease. Arppana et al. [36] analyzed cardiac rate and rhythm from real-time face images to extract the activity of the heart in a cycle by a non-contact-based method.

Respiration Rate Monitoring: respiration is one of five vital signs (heart rate, temperature, blood pressure, oxygen saturation, and blood glucose level) that reflect patient breathing problems. The respiration rate also plays an important role which is useful to be admitted to ICU (Intensive Care Unit). Tan et al. [37] presented a real-time vision-based respiration activity monitoring platform. Ferreira et al. [38] presented a smart system dubbed Baby Night Watch to protect children from SIDS (Sudden Infant Death Syndrome) by wearing a chest belt. In addition, Raji et al. [39] introduced a system for doctors to regularly monitor asthmatic patients by using multiple remote sensors through an Android application.

Sleep is an essential state of rest that recharges the body and refreshes the mind. Appropriate sleep can help one stay healthy and fight off diseases [40]. Therefore, pervasive H-IoT systems can be used to monitor patients during sleep. Phan et al. [41] proposed SeqSleepNet to automatically recognize the sleep stages. Nguyen et al. [42] developed a system called LIBS (Light-weight and Inexpensive In-ear Bioelectrical Sensing System) to monitor patients' whole-night sleep and then classify four stages in sleep cycles through the activity of the brain, eye, and muscle. Meanwhile, Yang et al. [43] used millimeter waves to monitor vital signs, particularly with regard to detecting posture and irregular breathing rhythms during sleep.

With the fast rhythms of developing society, work-related stress becomes more and more common. People living with high stress suffer the risk of cardiovascular diseases, mental health problems, eating disorders, and menstrual problems. Many pervasive H-IoT systems have been designed to detect stress using wearable sensors. Clarke et al. [44] presented a recommendation mobile application for just-in-time adaptive interventions to recognize and reduce stress by detecting heart rate and suggesting the adapted treatment model. McWhorter et al. [45] innovated a remote wearable sensor for PTSD (Post Traumatic Stress Disorder) patients. Vidal et al. [46] developed a smartwatch-based platform to support autistic people's self-control of their emotions. Oti et al. [47] presented a real-time stress level estimation approach for pregnant women. They adopted an unlabeled response method to estimate the stress level from the heart rate during pregnancy.

3.4. Disease Monitoring

In 2020, it is estimated that 5.8 million Americans are diagnosed with Alzheimer's Disease [48]. One new dementia case occurs every three seconds. Alzheimer's patients generally require daily assistance during their lives given their existing condition. IoT-based systems can provide day-to-day support in many areas. Ishii et al. [49] introduced a system for early dementia recognition using the machine-to-machine/IoT platform. Tamamizu et al. [50] proposed a device to detect anomalous activities for home dementia care. The users can define the anomaly and the corresponding cares. Once the anomaly is detected, the corresponding care will be provided. In another work, Szeles and Kubota [51] used smartphones for location monitoring for elderly people. The mobile application gives them reminders based on their current location.

Parkinson's and Huntington's diseases are neurodegenerative and are characterized by movement disorders. Huntington's disease is an inherited condition caused by a mutated gene from a parent. However, Parkinson's disease may happen as a result of many genetic or non-genetic factors. About one million people live with Parkinson's disease (PD) in the US [52]. This number is projected to rise to 1.2 million in the next eight years. Nearly 60,000 Americans are living with PD every year and more than 10 million people are diagnosed with PD worldwide [52]. Meanwhile, about 30,000 Americans suffer from Huntington's disease (HD) and another 200,000 have a chance of developing the condition [53]. The IoT systems provide a significant platform for screening the movement disorders associated with PD/HD and managing disease status, advancement, and treatment effectiveness. Dinesh et al. [54] and Adams et al. [55] proposed methods to affix multiple wearable sensors detecting unusual movement symptoms in people with PD and HD. These sensors can distinguish between individuals with motor symptoms and those that do not have them. The combination of medication and the sensor-based system is needed to apply in treatments effectively. Flagg et al. [56] introduced real-time virtual monitoring of posture and gait valuation for PD.

From the summarized applications, we found that sensors, communication, artificial intelligence, storage and computing infrastructure, security, and privacy are the key components. Figure 2 shows the key components inside an H-IoT system. The following section will review the key components in detail.

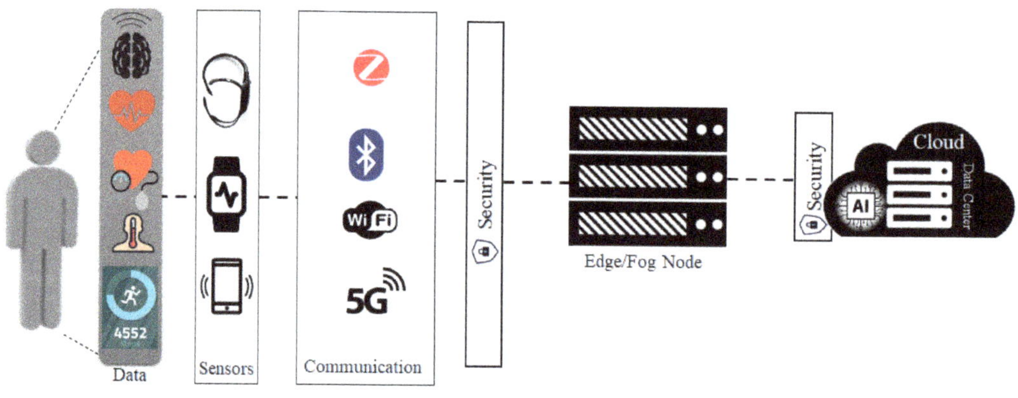

Figure 2. The key components in an H-IoT system, namely sensors, communication, AI, fog/edge/cloud computing, and security. Section 4 reviews these components in detail.

4. Key Components in the Pervasive Healthcare Internet of Things

4.1. Sensors

Sensors are essential in pervasive H-IoT systems. They are used to collect user data such as heart rate and body temperature. In addition, sensors are used for sensing environmental information such as humidity, temperature, light, noise, and air quality. In this subsection, we consider two types of sensors, namely, wearable sensors and environmental sensors [57–82].

Regarding wearable sensors, the most common inertial sensors yield valuable information such as accelerometers [70] (position change) and gyroscopes [71] (rotational change) are suitable for assessing human physical movement. They are worn on different body parts to detect human motions such as bending the knees and walking up stairs. Moncada-Torres et al. [57] categorized activity based on inertial and barometric pressure sensors at separate positions of body parts. In addition, identifying the location of humans is very important. There are many location sensors such as the Global Positioning System (GPS) [72], the Death Reckoning Module (DRM) [73], and RFID [74,83]. Next, the sensors reading vital signs are very crucial in any pervasive H-IoT. These sensors read the heart rate, blood oxygen, and pressure [75]. The advancement of hardware integrates these sensors into wearable devices for convenience, and the data can be transmitted via the Internet. Bulling et al. [61] fused different data modalities from body-worn sensors to recognize human activity. Recently, ego-centric cameras have become more popular. Therefore, the images/videos captured from the head-mounted camera can be used in H-IoT for activity analysis [76]. Journal et al. [63] used a wearable sensor for limbic encephalitis patients to improve their biographical memory by using a camera as an image diary. After viewing a visual diary, the user is able to recall approximately 80% of recent, personally experienced events. Meanwhile, 49% of an event can be remembered by reading a written diary.

Regarding environmental sensors, modern thermostats [77] can be connected to WiFi and provide information such as temperature and humidity. In addition, there are position trackers such as Beacon [78], AirTags [79], and RFID tags [80] which are used to localize certain physical objects. For example, Yang et al. [68] efficiently found objects by using sparsely distributed passive RFID tags. Lastly, other sensors such as the doorbell, bulb, and motion sensors [80–82] are important within an H-IoT system. Table 2 shows the categories of sensors in pervasive H-IoT.

Table 2. Categories of sensors and devices used in pervasive healthcare Internet of Things, namely wearable sensors and environmental sensors.

Sensors						
Wearable Sensors				Environmental Sensors		
Inertial	**Location**	**Vital Signs**	**Imaging**	**Thermostat**	**Position Tracker**	**Others**
Accelerometer, gyroscopes, altitude	GPS, Death Reckoning Module, RFID	Blood oxygen, pressure, heart rate	Live photos, videos	Temperature, humidity	Beacon AirTag RFID	Smart Doorbell, bulb, motion sensors

4.2. Communication

The data retrieved from the aforementioned sensors will be transmitted for further processes such as activity recognition, anomaly detection, or recommendation. In this subsection, we review the popular communication standards used in pervasive H-IoT.

4.2.1. Wireless Sensor Networks and Smart Body Area Networks

Wireless sensor networks (WSNs) consist of spatially dispersed sensors that collect environmental data such as temperature, sound, pollution levels, humidity, and wind. The collected data will then be forwarded to a central station [84]. With the ubiquitous setting, WSN suits pervasive IoT systems with many devices or sensors [65–67]. The sensors can communicate with each other via WiFi, Bluetooth, or Zigbee connections. However, there exist several well-known limitations in WSNs such as power consumption, communication range, and body-to-body communications. Therefore, Smart Body Area Network (SmartBAN) technology [83,85–87] was proposed to support a range of medical, health improvement, personal safety, and wellbeing via a network of small, low-power devices. In particular, SmartBAN is designed for supporting body-to-body communications. SmartBAN is based on a multi-radio approach to connect devices via radio standards.

Furthermore, SmartBAN has separate control channels, namely, a data channel and a control channel for data transmission and control message transmission. Takabayashi et al. [87] evaluated SmartBAN in H-IoT applications. Javaid et al. [83] studied the feasibility of using SmartBAN in monitoring COVID-19 [88] cases.

4.2.2. Cellular Technologies

The aforementioned WSNs and SmartBAN can be deployed for a small area such as a home or hospital. To increase the deployment scale, cellular communication technology needs to be involved [89]. Cellular technology has evolved from 1G to 5G. At the moment, 5G supports faster and more secure connectivity that is advancing everything from self-driving vehicles to AI-enabled robots. This benefits a massive-scale IoT ecosystem of billions of connected devices. There have been many studies on 5G and IoT architecture [90,91].

4.3. *Artificial Intelligence*

Artificial intelligence (AI) and machine learning (ML) is the brain of any health IoT system. Figure 3 illustrates the application of AI/ML in H-IoT. We categorized the AI/ML applications into the following groups.

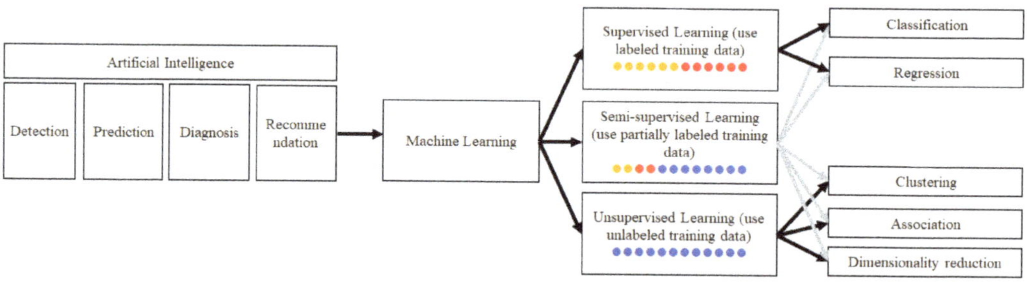

Figure 3. The application of artificial intelligence in a pervasive healthcare Internet of Things system. **Orange** and **red** dots denote the annotated training data, whereas the **blue** dots represent the unlabeled training data.

4.3.1. Activity Detection

Machine learning techniques such as modern deep learning can be used to understand and analyze the users' heterogeneous data. In [92], Zhou et al. introduced deep learning to improve activity recognition in IoT. They proposed a semi-supervised deep learning framework to recognize motion from sensor data. Note that semi-supervised learning only uses partially labeled data, so it saves labor costs for data annotation. Nguyen et al. [11] used a supervised method, namely, kernel Support Vector Machine (SVM) to classify the input features such as visual and motion cues into different action labels. Meanwhile, Ni et al. [12] extracted 3D block features from RGB-D videos to efficiently classify the fall activity in hospitals.

4.3.2. Disease Diagnosis

Machine learning is used in diagnosing some of the most popular eye problem disorders such as angle-closure glaucoma and age-related macular degeneration [93]. Some of the ML techniques used in various disease diagnosis applications include support vector machines, deep learning systems, convolutional neural calculates, and backpropagation calculates [94]. The input information varies based on the diagnoses being established. Machine learning is normally used in media image processing to give the diagnosis. In parallel, chronic diseases are diagnosed based on time progress data, including other factors such as patient history, demographics, genes, and symptoms [48].

The most-reported information includes vital signs, age, X-ray features, and blood cell count [60]. Machine learning algorithms are significantly effective, as they can access many pulmonary images of patients during the COVID-19 pandemic and can distinguish between those infected by COVID-19 or not. Therefore, images, tabular, text, and time series are the input data types for these prediction models. For example, X-rays can be supported with ML classifiers to recognize COVID-19 [90]. Wynants et al. [95] developed prediction models to diagnose the COVID-19 cases. Another study demonstrated significant accuracy in the diagnosis of COVID-19 by using eight binary features: sexagenarian, gender, exposure to COVID-19 virus, and five other vital signs [96].

4.3.3. Disease Prediction

AI can help physicians by suggesting more options for diagnosis or prediction. Yan et al. [97] studied the application of AI and ML in predicting heart disease. The physicians can give the diagnosis through an AI model for another round of screening. This may lower the risk of misdiagnosis. In [98], Almustafa predicted cardiovascular disease by assessing different ML methods such as KNN, SVM, Adaboost, and Decision Tree. The user enters the features such as age, gender, cholesterol level, blood pressure, and fasting glucose. The experiments show that the KNN classification is the best since it is a data-driven method. Another patient with similar health conditions tends to have the same diagnosis.

4.3.4. Medical Decision Recommendation

Artificial intelligence can analyze the activity of users and recommend changes; for example, doing exercise, changing one's diet, or visiting the doctor. Michie et al. [99] analyzed the behavior of individuals to determine the necessary changes. In particular, they modeled the features and the series of activities in the form of ontologies and they executed ontology reasoning for the final recommendation outcome. Asthana et al. [100] proposed a recommendation system for personalized advised wearables. Given the user's medical history, the system identifies the diseases that this person is at risk for. The system then provides recommendations via a computational model. Almeida et al. [101] proposed a recommendation system to automatically discover cohorts of interest. The cohort here is a group of users sharing common information. The system uses context-aware retrieval and collaborative filtering to localize relevant cohorts regarding Alzheimer's disease. Erdeniz et al. [102] introduced virtual nurses to help chronic patients (i.e., diabetes, asthma) reach their goals. The system reads the patient's data via IoT sensors such as a wristband or smartwatch. It then calculates the distance from the current health condition to the predefined target and provides suitable recommendations.

4.4. Cloud Computing Infrastructure

Cloud computing (cloud storage) [103] is an available system model to deliver different services for enabling ubiquitous, convenient, on-demand internet access to a shared pool of customizable computing resources: networks (horizontal, physical, or virtual), servers, storage, software, databases, analytics, and intelligence. This method can minimize the management and interaction from the provider rapidly and effectively [104]. Large clouds usually have functions diffused over various regions/locations, each region being a data center (traditional cloud). According to the National Institute of Standards and Technology (NIST), fog computing is also known as an architecture located between traditional cloud and smart end-devices. This paradigm delivers vertically isolated, latency-sensitive services for ubiquitous, scalable, federated, and distributed computing. The cloud now becomes a hierarchical structure since the edge is usually confined to some peripheral layers. In practical terms, edge computing can be described as the system layer including the peripheral devices and their users. This network encourages the edge to support local computing proficiencies for mIoT devices. These edge and fog infrastructures show a tendency to bypass the gap between the data and the end user. Recently, Laroui et al. [105] conducted research on current activities and future directions of cloud and edge computing for many different fields. Figure 4 shows the relationship between edge, fog, and cloud computing in the context of IoT-based smart healthcare.

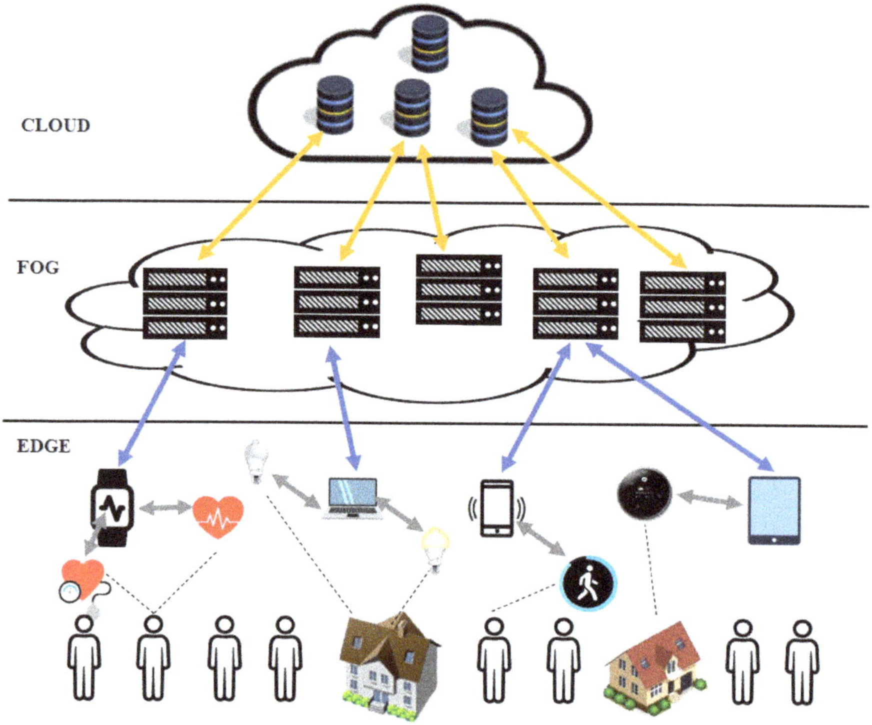

Figure 4. The relationship between edge (sensors), fog (nodes), and cloud (data centers/cloud services) computing in the context of pervasive healthcare Internet of Things.

4.5. Security

Security issues include security, authentication, privacy, identity management, and ethical challenges.

Security is always considered in H-IoT communication, whenever the communications happen: in body, on body, or around body. Synchronizing those of any other type of communication device protecting vulnerable data is not simple. Security problems such as accuracy and data privacy must be addressed to ameliorate confidentiality. Diminution of a variety of threats and attacks (some types of attack are active, authentication, access control, and availability) becomes more and more essential in cybersecurity [106]. Moreover, evil twin access points, eavesdropping, and man-in-the-middle attacks are common strategies that can threaten the security of a system, while replay attacks, denial-of-service (DoS), or frame injection attacks confront the system's integrity. Beacon flood, radio frequency jamming, and association/authentication flood can damage the availability and constitute a menace to the individuality of patients. In some cases, vindictive hackers may abuse other ways of extreme and occasionally undignifying types to obtain private data. As discussed by Yu et al. [107], audio signal processing and machine learning are secretly used to eavesdrop on handwriting and can be executed using nearby smart devices. The authors emphasize that they can obtain up to 50–60% of word recognition with certain support conditions. This approach has been customized to other areas, such as a hand motion tracking technique to upgrade the work of the eavesdropping, thus enhancing the performance to 70–80% [108]. The variety of attacks prove that cybersecurity and individuality for H-IoT are seriously important for the viability of smart healthcare.

Authentication is the act of proving an assertion, such as the identity of a system user. Recently, biometric information such as face photos and fingerprints has been used for authentication. However, deepfakes, synthesized images, and videos generated by generative adversarial networks (GAN) [109] pose a cybersecurity threat. A deepfake superimposes one person's face and voice onto another to create fake videos that appear authentic. Thus deepfakes may fool electronic devices/sensors for authentication.

Privacy in the context of IoT-based healthcare may be considered as the right of each individual to decide how much of their private information is shared. Therefore, patient privacy may be in danger when security is violated. Privacy issues have become the biggest challenge for analysts and also for patients using the smart healthcare service [110], where they share their s-Health records (SHRs) [111] and trust that only authorized professional healthcare staff can access them. As discussed in [111], in the situation where traditional access conduct techniques are applied, either data security is violated or only coarse-grained access policies are approved. To attenuate this matter, Zhang et al. [111] suggested a privacy-aware s-health access control system (PASH). In PASH, privacy information related to access permissions is invisible, and generic attribute names are available. A competitive decoding test is attached before full decoding to boost efficiency. Hackers may access any wearable or in-body, off-body sensor devices that can penetrate the privacy data of patients (e.g., loss or change of data) or loss of device managing rights. This can lead to the unexpected operation of the medical device system [112]. It is therefore compulsory to improve the protection of data security and privacy.

Identity management is very important since patients expect their personal data, such as social security numbers, their medical records, and even credit card numbers, to remain confidential. However, such valuable information can be illegally retrieved through ransomware and phishing attempts, unrestricted access to computers, and even patients' lack of adequate knowledge. Abdullah et al. [113] proposed using digital signatures for identity verification. The signing operation utilizes a hash function and a private key to encrypt the data. Meanwhile, the verification operation utilizes the hash function and the public key to decrypt the data. If the output of the hash function and the data decryption match, the signature is valid.

Ethics challenges and legal issues are worthy of being addressed in H-IoT. Most of the ethical challenges are about accessibility rights and the private use of information. In the IoT attacks, the losses will reach the point that they will affect people's lives. For example, if an attacker can log in to a medical application and make a small change in a patient's file. The unauthorized change may result in the wrong medication, which could affect the patient's life. Meanwhile, there are many legal issues related to questions raised in H-IoT. For example, who will be responsible when the Internet is down in medical applications? What will happen if a medical service provider goes out of business? How does this affect patients? And how will the patient's data be used? AboBakr and Azer [114] introduced new policies to address the ethical challenges and legal issues. New laws and standards should be introduced to maintain complete security and privacy and cover all legal issues.

5. Existing Challenges

From the extensive review of previous surveys, applications, and key components, we observe the following challenges. The first challenge is computational intensity. We have a loss of information; for example, the interruption of data collection from wearable devices. The second challenge comes from the restricted storage in mobile devices. For example, thanks to wearable cameras, people can easily do lifelogging. However, how to design an architecture to store and update huge volumes of image data is a big challenge. Such time-series data requires a huge volume of storage space. One potential solution is to apply machine learning to retrieve the intrinsic information of the data by reducing the number of feature dimensions.

The training process of artificial intelligence models also requires large amounts of data and high computational costs. Furthermore, we have difficulty in filtering an authentic and reliable correlation among various health information. Due to heterogeneous data, the network overhead may continuously change patient data. In the future, we expect a real-time and interactive decision-making IoMT system that provides accurate monitoring and diagnosis in medical fields. Such systems can support data from various sources and multimodal deep learning frameworks for decision making.

Regarding the recommendation of healthcare systems, the suggestions should be characterized and related to the guideline of every user's specific data. The suggester could play an important role in gathering data from patients and characterizing a plan that accommodates available objectives. In addition, a physician (semi-supervised recommendation) should be the key in the filtering stage and in collecting the information that is more related to the health issues of the patient.

Security and privacy become more severe when there are more and more users in pervasive H-IoT. There may be an abuse of a patient's personal information, false medical data, and unauthorized access to data. Lately, 5G massive machine-type communications was introduced to link millions of different medical kinds of sensors together to the service network. Patient lives might be in danger if such sensor security is vulnerable. 5G networks are estimated to be attacked more and more in comparison to the previous 4G networks. In the future, with the powerful processing capabilities of quantum computing, public-key cryptography will be vulnerable. Therefore, this challenge must be addressed with more advanced cryptography [115] in the near future. For other communication standards for both short-range and long-range distance communication technologies, please refer to the extensive reviews [116,117].

Lastly, power may pose a big issue. On the one hand, for smart devices to execute multiple tasks, a significant amount of power must be used. In particular, devices and nodes must be charged with sufficient energy. On the other hand, the high-power consumption of smart devices and device shutdown may result in vital consequences. Therefore, the research on batteries for prolonged periods will attract more attention in the near future. In addition, the introduction of compact models [118] is appreciated since they can greatly reduce storage space and computational cost resulting in less power consumption.

6. Conclusions

In this paper, we conducted an extensive review of 118 papers on pervasive computing in H-IoT. In particular, we compared our survey with others with a strong emphasis on applications, key components, and existing challenges in this field. We observed the wide range of applications of H-IoT. We summarized the key components with cutting-edge technologies. We discussed many applications such as fall detection, rehab monitoring, and early medical disease detection which will greatly benefit future healthcare systems. Such systems will minimize the need for dedicated medical personnel for patient monitoring and provide the patients with high-quality medical services.

We believe this survey will significantly contribute to the existing body of research. There are a few research challenges in H-IoT systems, including privacy, security, and trust issues. In addition, there exist energy optimization issues and challenges of dealing with big data. We can see a lot of stakeholders affected by this research direction such as AI researchers, policy makers, governments, municipalities, and healthcare organizations. We foresee the future of medical IoT systems with more innovative and modern kinds of sensors such as nanotechnology. We also expect the future utilization of advanced communication technologies as well as advanced artificial intelligence in terms of accuracy, speed, and human-centric personalization.

Author Contributions: Conceptualization, K.A.P. and T.V.N.; methodology, K.A.P.; validation, K.A.P. and T.V.N.; investigation, K.A.P., C.K., L.D. and T.V.N.; resources, K.A.P.; writing—original draft preparation, K.A.P., C.K., L.D. and T.V.N.; writing—review and editing, K.A.P., C.K., L.D. and T.V.N.; visualization, T.V.N.; supervision, C.K. and T.V.N.; project administration, C.K. and T.V.N.; funding acquisition, T.V.N. All authors have read and agreed to the published version of the manuscript.

Funding: This research received no external funding.

Institutional Review Board Statement: Not applicable.

Informed Consent Statement: Not applicable.

Data Availability Statement: Not applicable.

Conflicts of Interest: The authors declare that they have no conflict of interest.

References

1. Isravel, D.P.; Silas, S. A comprehensive review on the emerging IoT-cloud based technologies for smart healthcare. In Proceedings of the 2020 6th International Conference on Advanced Computing and Communication Systems (ICACCS), Coimbatore, India, 6–7 March 2020; pp. 606–611.
2. Qi, J.; Yang, P.; Min, G.; Amft, O.; Dong, F.; Xu, L. Advanced internet of things for personalised healthcare systems: A survey. *Pervasive Mob. Comput.* **2017**, *41*, 132–149. [CrossRef]
3. Alam, M.M.; Malik, H.; Khan, M.I.; Pardy, T.; Kuusik, A.; le Moullec, Y. A survey on the roles of communication technologies in IoT-based personalized healthcare applications. *IEEE Access* **2018**, *6*, 36611–36631. [CrossRef]
4. Shaikh, Y.; Parvati, V.K.; Biradar, S.R. Survey of smart healthcare systems using internet of things (IoT). In Proceedings of the 2018 International Conference on Communication, Computing and Internet of Things (IC3IoT), Chennai, India, 15–17 February 2018; pp. 508–513.
5. Ahmadi, H.; Arji, G.; Shahmoradi, L.; Safdari, R.; Nilashi, M.; Alizadeh, M. The application of internet of things in healthcare: A systematic literature review and classification. *Univers. Access Inf. Soc.* **2019**, *18*, 837–869. [CrossRef]
6. Rajini, N.H. A comprehensive survey on internet of things based healthcare services and its applications. In Proceedings of the 2019 3rd International Conference on Computing Methodologies and Communication (ICCMC), Erode, India, 27–29 March 2019; pp. 483–488.
7. Habibzadeh, H.; Dinesh, K.; Shishvan, O.R.; Boggio-Dandry, A.; Sharma, G.; Soyata, T. A survey of healthcare internet of things (HIoT): A clinical perspective. *IEEE Internet Things J.* **2020**, *7*, 53–71. [CrossRef] [PubMed]
8. Usak, M.; Kubiatko, M.; Shabbir, M.S.; Viktorovna Dudnik, O.; Jermsittiparsert, K.; Rajabion, L. Health care service delivery based on the internet of things: A systematic and comprehensive study. *Int. J. Commun. Syst.* **2020**, *33*, e4179. [CrossRef]
9. Dhanvijay, M.M.; Patil, S.C. Internet of things: A survey of enabling technologies in healthcare and its applications. *Comput. Netw.* **2019**, *153*, 113–131. [CrossRef]
10. Tunc, M.A.; Gures, E.; Shayea, I. A Survey on IoT Smart Healthcare: Emerging Technologies, Applications, Challenges, and Future Trends. *arXiv* **2021**, arXiv:CoRR/2109.02042.
11. Nguyen, T.V.; Song, Z.; Yan, S. STAP: Spatial-Temporal Attention-Aware Pooling for Action Recognition. *IEEE Trans. Circuits Syst. Video Technol.* **2015**, *25*, 77–86. [CrossRef]
12. Ni, B.; Nguyen, C.D.; Moulin, P. RGBD-camera based get-up event detection for hospital fall prevention. In Proceedings of the 2012 IEEE International Conference on Acoustics, Speech and Signal Processing (ICASSP), Kyoto, Japan, 25–30 March 2012; pp. 1405–1408.
13. Wang, Y.; Wu, K.; Ni, L.M. WiFall: Device-free fall detection by wireless networks. *IEEE Trans. Mobile Comput.* **2017**, *16*, 581–594. [CrossRef]
14. Ruan, W.; Yao, L.; Sheng, Q.Z.; Falkner, N.; Li, X.; Gu, T. Tagfall: Towards unobstructive fine-grained fall detection based on UHF passive RFID tags. In Proceedings of the 12th EAI nternational Conference Mobile and Ubiquitous Systems: Computing, Networking, and Services, Coimbra, Portugal, 22–24 July 2015; pp. 140–149.
15. Uddin, M.Z.; Soylu, A. Human activity recognition using wearable sensors, discriminant analysis, and long short-term memory-based neural structured learning. *Sci. Rep.* **2021**, *11*, 16455. [CrossRef] [PubMed]
16. Gu, T.; Wang, L.; Chen, H.; Tao, X.; Lu, J. Recognizing multiuser activities using wireless body sensor networks. *IEEE Trans. Mobile Comput.* **2011**, *10*, 1618–1631. [CrossRef]
17. Zhu, S.; Xu, J.; Guo, H.; Liu, Q.; Wu, S.; Wang, H. Indoor human activity recognition based on ambient radar with signal processing and machine learning. In Proceedings of the 2018 IEEE International Conference on Communications (ICC), Kansas City, MO, USA, 20–24 May 2018; pp. 1–6.
18. Teasell, R.; Meyer, M.J.; McClure, A.; Pan, C.; Murie-Fernandez, M.; Foley, N.; Salter, K. Stroke rehabilitation: An international perspective. *Top Stroke Rehabil.* **2009**, *16*, 44–56. [CrossRef]
19. 3 Ways to Avoid a Second Stroke. 2022. Available online: https://www.hopkinsmedicine.org/health/conditions-and-diseases/stroke/3-ways-to-avoid-a-second-stroke (accessed on 29 May 2022).
20. Standen, P.; Threapleton, K.; Richardson, A.; Connell, L.; Brown, D.; Battersby, S.; Platts, F.; Burton, A. A low cost virtual reality system for home based rehabilitation of the arm following stroke: A randomised controlled feasibility trial. *Clin. Rehabil.* **2017**, *31*, 340–350. [CrossRef]
21. Hoda, M.; Hoda, Y.; Hafidh, B.; El Saddik, A. Predicting muscle forces measurements from kinematics data using Kinect in stroke rehabilitation. *Multimed. Tools Appl.* **2018**, *77*, 1885–1903. [CrossRef]
22. Bobin, M.; Anastassova, M.; Boukallel, M.; Ammi, M. SyMPATHy: Smart glass for monitoring and guiding stroke patients in a home-based context. In Proceedings of the 8th ACM SIGCHI Symposium on Engineering Interactive Computing Systems, Brussels, Belgium, 21–24 June 2016; ACM: New York, NY, USA, 2016; pp. 281–286.

23. National Diabetes Statistics Report. 2022. Available online: https://www.cdc.gov/diabetes/data/statistics-report/index.html (accessed on 29 May 2022).
24. Al-Taee, M.A.; Al-Nuaimy, W.; Muhsin, Z.J.; Al-Ataby, A. Robot assistant in management of diabetes in children based on the internet of things. *IEEE Internet Things J.* **2017**, *4*, 437–445. [CrossRef]
25. Fioravanti, A.; Fico, G.; Salvi, D.; García-Betances, R.I.; Arredondo, M.T. Automatic messaging for improving patients engagement in diabetes management: An exploratory study. *Med. Biol. Eng. Comput.* **2015**, *53*, 1285–1294. [CrossRef]
26. Kaiya, K.; Koyama, A. Design and implementation of meal information collection system using IoT wireless tags. In Proceedings of the 2016 10th International Conference on Complex, Intelligent, and Software Intensive Systems (CISIS), Fukuoka, Japan, 6–8 July 2016; pp. 503–508.
27. Centers for Disease Control and Prevention. *Hypertension Cascade: Hypertension Prevalence, Treatment and Control Estimates among U.S. Adults Aged 18 Years and Older Applying the Criteria from the American College of Cardiology and American Heart Association's 2017 Hypertension Guideline—NHANES 2015–2018external Icon*; U.S. Department of Health and Human Services: Atlanta, GA, USA, 2021.
28. Janjua, G.; Guldenring, D.; Finlay, D.; McLaughlin, J. Wireless chest wearable vital sign monitoring platform for hypertension. In Proceedings of the 2017 39th Annual International Conference IEEE Engineering in Medicine and Biology Society (EMBC), Jeju Island, Korea, 11–15 July 2017; pp. 821–824.
29. Iakovakis, D.; Hadjileontiadis, L. Standing hypotension prediction based on smartwatch heart rate variability data: A novel approach. In Proceedings of the 18th International Conference Human-Computer Interaction with Mobile Devices and Services, Florence, Italy, 6–9 September 2016; pp. 1109–1112.
30. Centers for Disease Control and Prevention. *Underlying Cause of Death, 1999–2018*; CDC WONDER Online Database; Centers for Disease Control and Prevention: Atlanta, GA, USA, 2018.
31. Virani, S.S.; Alonso, A.; Aparicio, H.J.; Benjamin, E.J.; Bittencourt, M.S.; Callaway, C.W.; Carson, A.P.; Chamberlain, A.M.; Cheng, S.; Delling, F.N.; et al. Heart disease and stroke statistics—2021 update: A report from the American Heart Association. *Circulation* **2021**, *143*, e254–e743. [CrossRef] [PubMed]
32. Kiranyaz, S.; Ince, T.; Gabbouj, M. Personalized monitoring and advance warning system for cardiac arrhythmias. *Sci. Rep.* **2017**, *7*, 9270. [CrossRef]
33. Schmier, J.K.; Ong, K.L.; Fonarow, G.C. Cost-effectiveness of remote cardiac monitoring with the cardioMEMS heart failure system. *Clin. Cardiol.* **2017**, *40*, 430–436. [CrossRef] [PubMed]
34. Xia, Y.; Zhang, H.; Xu, L.; Gao, Z.; Zhang, H.; Liu, H.; Li, S. An automatic cardiac arrhythmia classification system with wearable electrocardiogram. *IEEE Access* **2018**, *6*, 16529–16538. [CrossRef]
35. Hijazi, S.; Page, A.; Kantarci, B.; Soyata, T. Machine learning in cardiac health monitoring and decision support. *IEEE Comput. Mag.* **2016**, *49*, 38–48. [CrossRef]
36. Arppana, A.R.; Reshmma, N.K.; Raghu, G.; Mathew, N.; Nair, H.R.; Aneesh, R.P. Real Time Heart Beat Monitoring Using Computer Vision. In Proceedings of the 2021 Seventh International Conference on Bio Signals, Images, and Instrumentation (ICBSII), Chennai, India, 25–27 March 2021; pp. 1–6.
37. Tan, K.S.; Saatchi, R.; Elphick, H.; Burke, D. Real-time vision based respiration monitoring system. In Proceedings of the 2010 7th International Symposium Communication Systems Networks and Digital Signal Processing (CSNDSP), Porto, Portugal, 20–22 July 2010; pp. 770–774.
38. Ferreira, A.G.; Fernandes, D.; Branco, S.; Monteiro, J.L.; Cabral, J.; Catarino, A.P.; Rocha, A.M. A smart wearable system for sudden infant death syndrome monitoring. In Proceedings of the 2016 IEEE International Conference Industrial Technology (ICIT), Taipei, Taiwan, 14–17 March 2016; pp. 1920–1925.
39. Raji, A.; Devi, P.K.; Jeyaseeli, P.G.; Balaganesh, N. Respiratory monitoring system for asthma patients based on IoT. In Proceedings of the 2016 Online Int. Conf. Green Engineering and Technologies (IC-GET), Coimbatore, India, 19 November 2016; pp. 1–6.
40. Why Do We Need Sleep? 2022. Available online: https://www.sleepfoundation.org/how-sleep-works/why-do-we-need-sleep (accessed on 29 May 2022).
41. Phan, H.; Andreotti, F.; Cooray, N.; Chén, O.Y.; de Vos, M. SeqSleepNet: End-to-End Hierarchical Recurrent Neural Network for Sequence-to-Sequence Automatic Sleep Staging. *IEEE Trans. Neural Syst. Rehabil. Eng.* **2019**, *27*, 400–410. [CrossRef] [PubMed]
42. Nguyen, A.; Alqurashi, R.; Raghebi, Z.; Banaei-Kashani, F.; Halbower, A.C.; Vu, T. A lightweight and inexpensive in-ear sensing system for automatic whole-night sleep stage monitoring. In Proceedings of the 14th ACM Conference on Embedded Network Sensor Systems CD-ROM, Stanford, CA, USA, 14–16 November 2016; ACM: New York, NY, USA; pp. 230–244.
43. Yang, Z.; Pathak, P.H.; Zeng, Y.; Liran, X.; Mohapatra, P. Vital sign and sleep monitoring using millimeter wave. *ACM Trans. Sens. Netw. (TOSN)* **2017**, *13*, 14. [CrossRef]
44. Clarke, S.; Jaimes, L.G.; Labrador, M.A. mStress: A mobile recommender system for just-in-time interventions for stress. In Proceedings of the 2017 14th IEEE Annu. Consumer Communications & Networking Conference (CCNC), Las Vegas, NV, USA, 8–11 January 2017; pp. 1–5.
45. McWhorter, J.; Brown, L.; Khansa, L. A wearable health monitoring system for posttraumatic stress disorder. *Biol. Inspired Cogn. Archit.* **2017**, *22*, 44–50. [CrossRef]

46. Vidal, J.C.T.; Montoro, G.; Gomez, J. The potential of smartwatches for emotional self-regulation of people with autism spectrum disorder, BIOSTEC 2016. In Proceedings of the 9th Internationl Joint Conference Biomedical Engineering Systems and Technologies: Health Infornation, Rome, Italy, 21–23 February 2016.
47. Oti, O.; Azimi, I.; Anzanpour, A.; Rahmani, A.M.; Axelin, A.; Liljeberg, P. IoT-Based Healthcare System for Real-Time Maternal Stress Monitoring. In Proceedings of the 2018 IEEE/ACM International Conference on Connected Health: Applications, Systems and Engineering Technologies (CHASE), Washington, DC, USA, 26–28 September 2018; pp. 57–62.
48. Alzheimer's Disease and Related Dementias. 2022. Available online: https://www.cdc.gov/aging/aginginfo/alzheimers.htm (accessed on 29 May 2022).
49. Ishii, H.; Kimino, K.; Aljehani, M.; Ohe, N.; Inoue, M. An early detection system for dementia using the M2M/IoT platform. *Procedia Comput. Sci.* **2016**, *96*, 1332–1340. [CrossRef]
50. Tamamizu, K.; Tokunaga, S.; Saiki, S.; Matsumoto, S.; Nakamura, M.; Yasuda, K. Towards person-centered anomaly detection and support system for home dementia care. In Proceedings of the International Conference on Human-Computer Interaction (HCI International), Toronto, ON, Canada, 17–22 July 2016; pp. 274–285.
51. Szeles, J.; Kubota, N. Location monitoring support application in smart phones for elderly people, using suitable interface design. In Proceedings of the International Conference Intelligent Robotics and Applications, Yantai, China, 22–25 October 2016; pp. 3–14.
52. Parkinson's Statistics. 2022. Available online: https://www.parkinson.org/Understanding-Parkinsons/Statistics (accessed on 29 May 2022).
53. Huntington's Disease. 2022. Available online: https://rarediseases.org/rare-diseases/huntingtons-disease (accessed on 29 May 2022).
54. Dinesh, K.; Xiong, M.; Adams, J.; Dorsey, R.; Sharma, G. Signal analysis for detecting motor symptoms in Parkinson's and Huntington's disease using multiple body-affixed sensors: A pilot study. In Proceedings of the 2016 IEEE Western New York Image and Signal Processing Workshop (WNYISPW), Rochester, NY, USA, 18 November 2016; pp. 1–5.
55. Adams, J.; Dinesh, K.; Xiong, M.; Tarolli, C.; Sharma, S.; Sheth, N.; Aranyosi, A.J.; Zhu, W.; Goldenthal, S.; Biglan, K.; et al. Multiple wearable sensors in Parkinson and Huntington disease individuals: A pilot study in clinic and at home. *Digit. Biomark.* **2017**, *1*, 52–63. [CrossRef] [PubMed]
56. Cristopher, F.; Ophir, F.; Sean, M.; Gholam, M. Real-time Streaming of Gait Assessment for Parkinson's Disease. In Proceedings of the 14th ACM International Conference on Web Search and Data Mining, Jerusalem, Israel, 8–12 March 2021; pp. 1081–1084.
57. Moncada-Torres, A.; Leuenberger, K.; Gonzenbach, R.; Luft, A.; Gassert, R. Activity classification based on inertial and barometric pressure sensors at different anatomical locations. *Physiol. Meas.* **2014**, *35*, 1245–1263. [CrossRef] [PubMed]
58. Shoaib, M.; Bosch, S.; Incel, O.; Scholten, H.; Havinga, P. A Survey of Online Activity Recognition Using Mobile Phones. *Sensors* **2015**, *15*, 2059–2085. [CrossRef] [PubMed]
59. Pawar, T.; Anantakrishnan, N.S.; Chaudhuri, S.; Duttagupta, S.P. Impact of Ambulation in Wearable-ECG. *Ann. Biomed. Eng.* **2008**, *36*, 1547–1557. [CrossRef] [PubMed]
60. Wensley, D.; Silverman, M. Peak Flow Monitoring for Guided Self-management A Randomized Controlled Trial. *Am. J. Respir. Crit. Care Med.* **2004**, *170*, 606–612. [CrossRef]
61. Bulling, A.; Ward, J.A.; Gellersen, H. Multimodal recognition of reading activity in transit using body-worn sensors. *ACM Trans. Appl. Percept.* **2012**, *9*, 1–21. [CrossRef]
62. Sun, F.; Kuo, C.; Cheng, H.; Buthpitiya, S.; Collins, P.; Griss, M. Activity-aware Mental Stress Detection Using Physiological Sensors. In Proceedings of the International Conference on Mobile Computing, Applications, and Services, Seattle, WA, USA, 11–12 October 2012; Volume 76, pp. 1–20.
63. Berry, E.; Kapur, N.; Williams, L.; Hodges, S.; Watson, P.; Smyth, G.; Srinivasan, J.; Smith, R.; Wilson, B.; Wood, K. The use of a wearable camera, SenseCam, as a pictorial diary to improve autobiographical memory in a patient with limbic encephalitis: A preliminary report. *Eur. Rehabil.* **2016**, *2011*, 582–601. [CrossRef]
64. Doherty, A.R.; Caprani, N.; Conaire, C.Ó.; Kalnikaite, V.; Gurrin, C.; Smeaton, A.F.; Connor, N.E.O. Computers in Human Behavior Passively recognising human activities through lifelogging. *Comput. Hum. Behav.* **2011**, *27*, 1948–1958. [CrossRef]
65. Sugimoto, C.; Kohno, R. Wireless Sensing System for Healthcare Monitoring Thermal Physiological State and Recognizing Behavior. In Proceedings of the 2011 International Conference on Broadband and Wireless Computing, Communication and Applications, Barcelona, Spain, 26–28 October 2011; pp. 285–291.
66. Sixsmith, A.; Johnson, N. A smart sensor to detect the falls of the elderly. *IEEE Pervasive Comput.* **2004**, *3*, 42–47. [CrossRef]
67. Jong, H.; Hee, S.; Ha, K.; Chul, H.; Chung, W.; Young, J.; Chang, Y.; Hyun, D. Ubiquitous healthcare service using Zigbee and mobile phone for elderly patients. *Int. J. Med. Inform.* **2008**, *8*, 193–198.
68. Yang, P.; Wu, W.; Moniri, M.; Chibelushi, C.C. Efficient object localization using sparsely distributed passive RFID tags. *IEEE Trans. Ind. Electron.* **2013**, *60*, 5914–5924. [CrossRef]
69. Rafferty, J.; Member, C.N.-I.; Member, L.C.-I.; Qi, J.; Dutton, R.; Zirk, A.; Boye, L.T.; Kohn, M.; Hellman, R. NFC based provisioning of instructional videos to assist with instrumental activities of daily living. In Proceedings of the 2014 36th Annual International Conference of the IEEE Engineering in Medicine and Biology Society, Chicago, IL, USA, 26–30 August 2014; pp. 4131–4134.
70. Bouten, C.V.; Koekkoek, K.T.; Verduin, M.; Kodde, R.; Janssen, J.D. A triaxial accelerometer and portable data processing unit for the assessment of daily physical activity. *IEEE Trans. Biomed. Eng.* **1997**, *44*, 136–147. [CrossRef]

71. Dejnabadi, H.; Jolles, B.M.; Aminian, K. A new approach to accurate measurement of uniaxial joint angles based on a combination of accelerometers and gyroscopes. *IEEE Trans. Biomed. Eng.* **2005**, *52*, 1478–1484. [CrossRef]
72. Liao, L.; Fox, D.; Kautz, H. Hierarchical Conditional Random Fields for GPS-Based Activity Recognition. *Star* **2007**, *28*, 487–506.
73. Honeywell Introduces DRM 4000 Dead Reckoning Module. 2022. Available online: https://www.fierceelectronics.com/components/honeywell-introduces-drm-4000-dead-reckoning-module (accessed on 29 May 2022).
74. Sangwan, R.S.; Qiu, R.G.; Member, S.K.; Jessen, D. Using RFID Tags for Tracking Patients, Charts and Medical Equipment within an Integrated Health Delivery Network. In Proceedings of the 2005 IEEE Networking, Sensing and Control, Tucson, AZ, USA, 19–22 March 2005; pp. 70–74.
75. Davies, R.; Galway, L.; Nugent, C.; Jamison, C.; Gawley, R.; McCullagh, P.; Zhang, H.; Black, N. A Platform for Self-Management Supported by Assistive, Rehabilitation and Telecare Technologies. In Proceedings of the 2011 5th International Conference on Pervasive Computing Technologies for Healthcare (PervasiveHealth) and Workshops, Dublin, Ireland, 23–26 May 2011; pp. 458–460.
76. Núñez-Marcos, A.; Azkune, G.; Arganda-Carreras, I. Egocentric Vision-based Action Recognition: A survey. *Neurocomputing* **2022**, *472*, 175–197. [CrossRef]
77. Nest Thermostat. 2022. Available online: https://nest.com/thermostats/ (accessed on 29 May 2022).
78. Estimote Beacon. 2022. Available online: https://estimote.com/ (accessed on 29 May 2022).
79. AirTag. 2022. Available online: https://www.apple.com/airtag (accessed on 29 May 2022).
80. Smart Doorbell. 2022. Available online: https://ring.com/doorbell-cameras (accessed on 29 May 2022).
81. Smart Bulbs. 2022. Available online: https://www.amazon.com/smart-bulbs (accessed on 29 May 2022).
82. de Toledo, P.; Sanchis, A. Activity Recognition Using Hybrid Generative/Discriminative Models on Home Environments Using Binary Sensors. *Sensors* **2013**, *13*, 5460–5477.
83. Javaid, M.; Khan, I.H. Internet of Things (IoT) enabled healthcare helps to take the challenges of COVID-19 Pandemic. *J. Oral Biol. Craniofacial Res.* **2021**, *11*, 209–214. [CrossRef] [PubMed]
84. Ullo, S.L.; Sinha, G.R. Advances in Smart Environment Monitoring Systems Using IoT and Sensors. *Sensors* **2020**, *20*, 3113. [CrossRef] [PubMed]
85. Smart Body Area Network (SmartBAN). Enhanced Ultra-Low Power Physical Layer. document ETSI TS 103 326 V1.1.1. *ETSI TC* **2015**, *13*, V1.
86. Hämäläinen, M.; Mucchi, L.; Girod-Genet, M.; Paso, T.; Farserotu, J.; Tanaka, H.; Anzai, D.; Pierucci, L.; Khan, R.; Alam, M.M.; et al. ETSI SmartBAN Architecture: The Global Vision for Smart Body Area Networks. *IEEE Access* **2020**, *8*, 150611–150625. [CrossRef]
87. Takabayashi, K.; Tanaka, H.; Sakakibara, K. Integrated Performance Evaluation of the Smart Body Area Networks Physical Layer for Future Medical and Healthcare IoT. *Sensors* **2018**, *19*, 30. [CrossRef]
88. Coronavirus Disease (COVID-19) Pandemic. Available online: https://www.who.int/emergencies/diseases/novel-coronavirus-2019 (accessed on 29 May 2022).
89. Xiong, X.; Zheng, K.; Xu, R.; Xiang, W.; Chatzimisios, P. Low power wide area machine-to-machine networks: Key techniques and prototype. *IEEE Commun. Mag.* **2015**, *53*, 64–71. [CrossRef]
90. Kar, S.; Mishra, P.; Wang, K.-C. 5G-IoT Architecture for Next Generation Smart Systems. In Proceedings of the 2021 IEEE 4th 5G World Forum (5GWF), Montreal, QC, Canada, 13–15 October 2021; pp. 241–246. [CrossRef]
91. Gupta, N.; Juneja, P.K.; Sharma, S.; Garg, U. Future Aspect of 5G-IoT Architecture in Smart Healthcare System. In Proceedings of the 2021 5th International Conference on Intelligent Computing and Control Systems (ICICCS), Madurai, India, 6–8 May 2021; pp. 406–411.
92. Zhou, X.; Liang, W.; Wang, K.I.; Wang, H.; Yang, L.T.; Jin, Q. Deep learning-enhanced human activity recognition for internet of healthcare things. *IEEE Internet Things J.* **2020**, *7*, 6429–6438. [CrossRef]
93. Xu, J.; Xue, K.; Zhang, K. Current status and future trends of clinical diagnoses via image-based deep learning. *Theranostics* **2019**, *9*, 7556–7565. [CrossRef]
94. Battineni, G.; Sagaro, G.G.; Chinatalapudi, N.; Amenta, F. Applications of Machine Learning Predictive Models in the Chronic Disease Diagnosis. *J. Pers. Med.* **2020**, *10*, 21. [CrossRef]
95. Wynants, L.; Van Calster, B.; Collins, G.S.; Riley, R.D.; Heinze, G.; Schuit, E.; Bonten, M.M.J.; Dahly, D.L.; Damen, J.A.; Debray, T.P.A.; et al. Prediction models for diagnosis and prognosis of COVID-19: Systematic review and critical appraisal. *BMJ* **2020**, *369*, m1328. [CrossRef] [PubMed]
96. Zoabi, Y.; Deri-Rozov, S.; Shomron, N. Machine learning-based prediction of COVID-19 diagnosis based on symptoms. *NPJ Digit. Med.* **2021**, *4*, 1–5. [CrossRef] [PubMed]
97. Yan, Y.; Zhang, J.-W.; Zang, G.-Y.; Pu, J. The primary use of artificial intelligence in cardiovascular diseases: What kind of potential role does artificial intelligence play in future medicine? *J. Geriatr. Cardiol. JGC* **2019**, *16*, 585–591.
98. Almustafa, K.M. Prediction of heart disease and classifiers' sensitivity analysis. *BMC Bioinform.* **2020**, *21*, 278. [CrossRef]
99. Michie, S.; Thomas, J.; John, S.-T.; Mac Aonghusa, P.; Shawe-Taylor, J.; Kelly, M.P.; Deleris, L.A.; Finnerty, A.N.; Marques, M.M.; Norris, E.; et al. The Human Behavior-Change Project: Harnessing the power of artificial intelligence and machine learning for evidence synthesis and interpretation. *Implement. Sci.* **2017**, *12*, 1–12. [CrossRef] [PubMed]

100. Asthana, S.; Megahed, A.; Strong, R. A Recommendation System for Proactive Health Monitoring Using IoT and Wearable Technologies. In Proceedings of the 2017 IEEE International Conference on AI & Mobile Services (AIMS), Honolulu, HI, USA, 25–30 June 2017; pp. 14–21. [CrossRef]
101. Almeida, J.R.; Monteiro, E.; Silva, L.B.; Sierra, A.P.; Oliveira, J.L. A Recommender System to Help Discovering Cohorts in Rare Diseases. In Proceedings of the 2020 IEEE 33rd International Symposium on Computer-Based Medical Systems (CBMS), Rochester, MN, USA, 28–30 July 2020; pp. 25–28. [CrossRef]
102. Seda, P.E.; Ilias, M.; Andreas, M.; Alexander, F.; Tran, T.N.T. Recommender Systems for IoT Enabled m-Health Applications. In Proceedings of the IFIP International Conference on Artificial Intelligence Applications and Innovations, Rhodes, Greece, 25–27 May 2018; Springer: Cham, Switzerland, 2018; pp. 227–237.
103. Ray, P.P. An Introduction to Dew Computing: Definition, Concept and Implications. *IEEE Access* **2018**, *6*, 723–737. [CrossRef]
104. Montazerolghaem, A.; Yaghmaee, M.H.; Leon-Garcia, A. Green Cloud Multimedia Networking: NFV/SDN Based Energy-Efficient Resource Allocation. *IEEE Trans. Green Commun. Netw.* **2020**, *4*, 873–889. [CrossRef]
105. Laroui, M.; Nour, B.; Moungla, H.; Cherif, M.A.; Afifi, H. Mohsen Guizanie. Edge and fog computing for IoT: A survey on current research activities & future directions. *Comput. Commun.* **2021**, *180*, 210–231.
106. Zou, Y.; Zhu, J.; Wang, X.; Hanzo, L. A Survey on Wireless Security: Technical Challenges, Recent Advances, and Future Trends. *Proc. IEEE* **2016**, *104*, 1727–1765. [CrossRef]
107. Yu, T.; Jin, H.; Nahrstedt, K. WritingHacker: Audio based eavesdropping of handwriting via mobile devices. In Proceedings of the UbiComp 2016—Proceedings of the 2016 ACM International Joint Conference on Pervasive and Ubiquitous Computing, Heidelberg, Germany, 12 September 2016.
108. Yu, T.; Jin, H.; Nahrstedt, K. Mobile Devices based Eavesdropping of Handwriting. *IEEE Trans. Mob. Comput.* **2019**, *19*, 1649–1663. [CrossRef]
109. Goodfellow, I.J.; Pouget-Abadie, J.; Mirza, M.; Xu, B.; Warde-Farley, D.; Ozair, S.; Courville, A.C. Yoshua Bengio: Generative adversarial networks. *Commun. ACM* **2020**, *63*, 139–144. [CrossRef]
110. Solanas, A.; Patsakis, C.; Conti, M.; Vlachos, I.S.; Ramos, V.; Falcone, F.; Postolache, O.; Perez-martinez, P.A.; Pietro, R.D.; Perrea, D.N.; et al. Smart health: A context-aware health paradigm within smart cities. *IEEE Commun. Mag.* **2014**, *52*, 74–81. [CrossRef]
111. Zhang, Y.; Zheng, D.; Deng, R.H. Security and Privacy in Smart Health: Efficient Policy-Hiding Attribute-Based Access Control. *IEEE Internet Things J.* **2018**, *5*, 2130–2145. [CrossRef]
112. Katzis, K.; Jones, R.; Despotou, G. The challenges of balancing safety and security in implantable medical devices. In *Unifying the Applications and Foundations of Biomedical and Health Informatics*; IOS Press: Amsterdam, The Netherlands, 2016; pp. 25–28.
113. Abdullah, G.M.; Mehmood, Q.; Khan, C.B.A. Adoption of lamport signature scheme to implement digital signatures in IoT. In Proceedings of the 2018 International Conference on Computing, Mathematics and Engineering Technologies (iCoMET), Sukkur, Pakistan, 3–4 March 2018; pp. 1–4.
114. AboBakr, A.; Azer, M.A. IoT ethics challenges and legal issues. In Proceedings of the 2017 12th International Conference on Computer Engineering and Systems (ICCES), Cairo, Egypt, 19–20 December 2017; pp. 233–237.
115. Lewis, A.M.; Travagnin, M. *A Secure Quantum Communications Infrastructure for Europe*; JRC Technical Papers; JRC: Ispra, Italy, 2019.
116. Vidakis, K.; Mavrogiorgou, A.; Kiourtis, A.; Kyriazis, D. A comparative study of short-range wireless communication technologies for health information exchange. In Proceedings of the 2020 International Conference on Electrical, Communication, and Computer Engineering (ICECCE), Istanbul, Turkey, 12–13 June 2020; pp. 1–6.
117. Koufos, K.; EI Haloui, K.; Dianati, M.; Higgins, M.; Elmirghani, J.; Imran, M.A.; Tafazolli, R. Trends in Intelligent Communication Systems: Review of Standards, Major Research Projects, and Identification of Research Gaps. *J. Sens. Actuator Netw.* **2021**, *10*, 60. [CrossRef]
118. Do, T.T.; Le, K.; Hoang, T.; Le, H.; Nguyen, T.V.; Cheung, N.M. Simultaneous feature aggregating and hashing for compact binary code learning. *IEEE Trans. Image Process.* **2019**, *28*, 4954–4969. [CrossRef] [PubMed]

MDPI AG
Grosspeteranlage 5
4052 Basel
Switzerland
Tel.: +41 61 683 77 34

Information Editorial Office
E-mail: information@mdpi.com
www.mdpi.com/journal/information

www.ingramcontent.com/pod-product-compliance
Lightning Source LLC
Chambersburg PA
CBHW061828210926
51850CB00006B/834
9783725814879